DIPEPTIDYL AMINOPEPTIDASES IN HEALTH AND DISEASE

ADVANCES IN EXPERIMENTAL MEDICINE AND BIOLOGY

Recent Volumes in this Series

Volume 515
NEUROPILIN: From Nervous System to Vascular and Tumor Biology
Edited by Dominique Bagnard

Volume 516
TRIPLE REPEAT DISEASES OF THE NERVOUS SYSTEM
Edited by Lubov T. Timchenko

Volume 517
DOPAMINERGIC NEURON TRANSPLANTATION IN THE WEAVER MOUSE MODEL OF PARKINSON'S DISEASE
Edited by Lazaros C. Triarhou

Volume 518
ADVANCES IN MALE MEDIATED DEVELOPMENTAL TOXICITY
Edited by Bernard Robaire and Barbara F. Hales

Volume 519
POLYMER DRUGS IN THE CLINICAL STAGE: Advantages and Prospects
Edited by Hiroshi Maeda, Alexander Kabanov, Kazurori Kataoka, and Teruo Okano

Volume 520
CYTOKINES AND AUTOIMMUNE DISEASE
Edited by Pere Santamaria

Volume 521
IMMUNE MECHANISMS IN PAIN AND ANALGESIA
Edited by Halina Machelska and Christoph Stein

Volume 522
NOVEL ANGIOGENIC MECHANISMS: Role of Circulating Progenitor Endothelial Cells
Edited by Nicanor I. Moldovan

Volume 523
ADVANCES IN MODELLING AND CLINICAL APPLICATION OF INTRAVENOUS ANAESTHESIA
Edited by Jaap Vuyk and Stefan Schraag

Volume 524
DIPEPTIDYL AMINOPEPTIDASES IN HEALTH AND DISEASE
Edited by Martin Hildebrandt, Burghard F. Klapp, Torsten Hoffmann, and Hans-Ulrich Demuth

---

DIPEPTIDYL AMINOPEPTIDASES IN HEALTH AND DISEASE

Edited by

Martin Hildebrandt, Burghard F. Klapp
Medical Faculty Charité, Humboldt University Berlin
Berlin, Germany

Torsten Hoffman and Hans-Ulrich Demuth
Probiodrug SG
Halle (Saale), Germany

Springer Science+Business Media, LLC

Proceedings of the International Conference on Dipeptidyl Aminopeptidases, held September 26–28, 2002, in Berlin, Germany

ISSN 0065 2598

DOI 10.1007/978-0-306-47920-5

Originally published by Kluwer Academic / Plenum Publishers in 2003
MyCopy version of the original edition 2003

http://www.wkap.nl/

10 9 8 7 6 5 4 3 2 1

A C.I.P. record for this book is available from the Library of Congress

Contributors

Catherine A. Abbott
School of Biological Sciences, Flinders University, GPO Box 2100,
Adelaide SA 5001, Australia

Katerina Ajami
Centenary Institute, Royal Prince Alfred Hospital, Locked Bag No 6,
Newtown, NSW, 2042, Sydney, Australia

Siegfried Ansorge
IMTM, Magdeburg, Germany

Dorit Anthonsen
Biochemistry Laboratory C, IMBG, The Panum Institute, Blegdamsvej 3,
DK-2200 Copenhagen N, Denmark

Petra C. Arck
Department of Internal Medicine/Psychosomatics, Charité Campus Mitte,
Luisenstr. 13A, Berlin, Germany

Marco Arndt
Institute of Experimental Internal Medicine, Otto-von-Guericke University
Magdeburg, Leipziger Strasse 44, D-39120 Magdeburg, Germany

Koen Augustyns
Laboratory of Medical Biochemistry, Dept. of Pharmaceutical Sciences,
University of Antwerp, Universiteitsplein 1 S-6, 2610 Wilrijk, Belgium

William W. Bachovchin
Department of Biochemistry, Tufts University School of Medicine, 136 Harrison Avenue, Boston, MA 02111, USA

Joachim Bär
Probiodrug AG, Weinbergweg 22, 06120 Halle, Germany

Oliver Baum
Institut für Molekularbiologie und Biochemie, Freie Universität Berlin, Arnimallee 22, 14195 Berlin, Germany

A. Belyaev
Dept of Pharmaceutical Sciences, University of Antwerp, Universiteitsplein 1, B-2610 Wilryk, Belgium

Sergei A. Belyakov
R & D, Guilford Pharmaceuticals Inc., 6611 Tributary St., 21224 Baltimore, MD, USA

Felix Bermpohl
Institut für Molekularbiologie und Biochemie, Freie Universität Berlin, Arnimallee 22, 14195 Berlin, Germany

Maria G. Bernengo
Department of Medical and Surgical Specialities, 1st Dermatologic Clinic, University of Turin, Via Cherasco 23, Turin 10126, Italy

Gregory A. Bohach
Department of Microbiology, Molecular Biology and Biochemistry, University of Idaho, S. Line St./P.O. Box 443052, 83843 Moscow, ID, USA

Ilona Born
Institut für Biochemie, Martin-Luther-Universität Halle-Wittenberg, Kurt-Mothes-Str. 3, 06120 Halle, Germany

Ernst Brandt
Research Center Borstel, Parkalle 22, 23845 Borstel, Germany

Wolfgang Brandt
Institute of Plant Biochemistry, Leibniz-Institute Halle, Weinberg 3, D-06120 Halle, Germany

Alicja Bukowska
Institute of Experimental Internal Medicine, Otto-von-Guericke University Magdeburg, Leipziger Strasse 44, D-39120 Magdeburg, Germany

Petr Bušek
1st Faculty of Medicine of Charles University, U Nemocnice 5, 12853 Prague 2, Czech Republic

Richard D. Carr
Pharmacology Research 1, Novo Nordisk A/S, Novo Allé, DK-2880 Bagsværd, Denmark

Tong Chen
School of Biological Sciences, Flinders University of South Australia, Adelaide, Australia

Wen-Tien Chen
Dept of Medicine/Medical Oncology, SUNY, HSC T-15, Rm 053, 11794-8154 Stony Brook, N.Y., USA

Blas Cigic
Institute Jozef Stefan, Department of Biochemistry and Molecular Biology Jamova 39, 1000 Ljubljana, Slovenia

Alessandra Comessatti
Department of Medical and Surgical Specialities, 1st Dermatologic Clinic, University of Turin, Via Cherasco 23, Turin 10126, Italy

Oscar J. Cordero
Departamento de Bioquímica e Bioloxía Mol., Facultade de Bioloxía, Universidade de Santiago de Compostela, Campus Sur, 15782 Santiago de Compostela, Spain

Guy Cox
Electron Microscope Unit, University of Sydney, Sydney, Australia

William C. Davis
Washington State University, Pullman, WA 99163, USA

Ingrid De Meester
Dept of Pharmaceutical Sciences, University of Antwerp, Universiteitsplein 1, B-2610 Wilryk, Belgium

Carolyn F. Deacon
Department of Medical Physiology, Panum Institute, Blegdamsvej 3, DK-2200 Copenhagen N, Denmark

Hans-Ulrich Demuth
Probiodrug AG, Biozentrum, Weinbergweg 22, 6120 Halle (Saale), Germany

H. Dong
Department of Medicine/Medical Oncology, State University of New York, Stony Brook, New York 11794-8154, USA

Timothy Doty
Department of Physiology, University of British Columbia, 2329 West Mall, V6T 1Z4 Vancouver, Canada

Jan A. Ehses
Department of Physiology, University of British Columbia, 2329 West Mall, V6T 1Z4 Vancouver, Canada

Lori Estes
Department of Physiology and Biophysics, Georgetown University Medical Center, 3900 Reservoir Rd., NW, Washington, DC, USA

Hua Fan
Department of Molecular Biology and Biochemistry, Freie Universität Berlin, Arnimallee 22, 14195 Berlin, Germany

Sabiha Fatima
Department of Biochemistry, Aligarh Muslim University, 202002 Aligarh U.P., India

Jürgen Faust
Institut für Biochemie, Martin-Luther-Universität Halle-Wittenberg, Kurt-Mothes-Str. 3, 06120 Halle, Germany

Maria T. Fierro
Department of Medical and Surgical Specialities, 1st Dermatologic Clinic, University of Turin, Via Cherasco 23, Turin 10126, Italy

Axel Fischer
Dept. of Internal Medicine, Psychosomatics/Psychotherapie, Humboldt-University, Charite Campus Mitte, Liusenstraße13A, 10117 Berlin, Germany

Herbert Fliege
Dept of Internal Medicine/ Psychosomatics, Charité, Humboldt University Berlin, Luisenstraße 13A, 10117 Berlin, Germany

Lawrence K. Fox
Washington State University, Pullman, WA 99163, USA

Daniel Friedrich
Probiodrug AG, Weinbergweg 22, 06120 Halle, Germany

Petra Fuchs
Institute of Biochemistry, Department of Biochemistry/Biotechnology, Martin-Luther-University, Halle-Wittenberg, Germany

Vilmos Fülöp
Department of Biological Sciences, University of Warwick, Gibbet Hill Road, Coventry, UK

Bernd Gerhartz
Novartis Pharma AG, Klybeckstr. 141, CH-4057 Basel, Switzerland

Giulio Ghersi
Department of Cellular and Developmental Biology, Università di Palermo, Viale delle Scienze, 90138 Palermo, Italy

L.A. Goldstein
Department of Medicine/Medical Oncology, State University of New York, Stony Brook, New York 11794-8154, USA

Mark D. Gorrell
Dept of Gastroenterology, Royal Prince Alfred Hospital, Locked Bag No 6, Sydney Newtown, NSW, 2042, Australia

David A. Groneberg
Dept. of Internal Medicine, Psychosomatics/Psychotherapie, Humboldt-University, Charite Campus Mitte, Liusenstraße13A, 10117 Berlin, Germany

Vanessa Gysbers
Dept of Gastroenterology, Royal Prince Alfred Hospital, Locked Bag No 6, Sydney Newtown, NSW, 2042, Australia

Achiel Haemers
Laboratory of Medical Biochemistry, Dept. of Pharmaceutical Sciences, University of Antwerp, Universiteitsplein 1 S-6, 2610 Wilrijk, Belgium

L. Hakkinen
University of British Columbia, Division of Periodontics, 2199 Westbrook Mall, Vancouver, BC V6T 1Z3, CANADA

Ross Haller
VA Medical Center, University of Minnesota, Minneapolis, MN, USA

Gregory S. Hamilton
Guilford Pharmaceuticals, Inc., Tributary St., Baltimore, MD, USA

Ulrich Heiser
Probiodrug AG, Weinbergweg 22, 06120 Halle, Germany

Martin Hildebrandt
Dept of Internal Medicine/ Psychosomatics, Charité, Humboldt University Berlin, Luisenstraße 13A, 10117 Berlin, Germany

Simon A. Hinke
Department of Physiology, University of British Columbia, 2329 West Mall, V6T 1Z4 Vancouver, Canada

Maw Hliang
Department of Biochemistry, Tufts University School of Medicine, 136 Harrison Avenue, Boston, MA 02111, USA

Torsten Hoffmann
Probiodrug AG, Weinbergweg 22, 06120 Halle, Germany

Jens J. Holst
Department of Endocrinology, The Panum Institute, Blegdamsvej 3, 2200 Copenhagen, Denmark

Kazuhiko Ino
Department of Obstetrics and Gynecology, Nagoya University Graduate School of Medicine, Tsurumai-cho 65, Showa-ku, Nagoya 466-8550, Japan

Annelore Ittenson
Institute of Immunology, Otto-von-Guericke University Magdeburg, Leipziger Strasse 44, D-39120 Magdeburg, Germany

Paul Jackson
Guilford Pharmaceuticals, Inc., Tributary St., Baltimore, MD, USA

Eleanor Kable
Electron Microscope Unit, University of Sydney, Sydney, Australia

Thilo Kähne
Research Center of Immunology, Institute of Experimental Internal Medicine, University of Magdeburg, Magdeburg, Germany

Hiroaki Kajiyama
Department of Obstetrics and Gynecology, Nagoya University Graduate School of Medicine, Tsurumai-cho 65, Showa-ku, 466-8550 Nagoya, Japan

Fumitaka Kikkawa
Department of Obstetrics and Gynecology, Nagoya University Graduate School of Medicine, Tsurumai-cho 65, Showa-ku, 466-8550 Nagoya, Japan

Joanna Kitlinska
Department of Physiology and Biophysics, Georgetown University Medical Center, 3900 Reservoir Rd., NW, Washington, DC, USA

Burghard F. Klapp
Department of Internal Medicine/Psychosomatics, Charité Campus Mitte, Luisenstr. 13A, Berlin, Germany

Zdeněk Kleibl
1st Faculty of Medicine of Charles University, U Nemocnice 5, 12853 Prague 2, Czech Republic

Sina Koch
Biochemistry Laboratory C, IMBG, The Panum Institute, Blegdamsvej 3, DK-2200 Copenhagen N, Denmark

Stephan Korom
Abteilung f. Thoraxchirurgie, Universitätsspital Zürich, Rämistrasse 100, 8091 Zürich, Switzerland

Kerstin Kühn-Wache
Probiodrug AG, Weinbergweg 22, 06120 Halle, Germany

Yumi Kumagai
Department of Microbiology, Nippon Dental University, Tokyo, Japan

Jack Lai
Department of Biochemistry, Tufts University School of Medicine, 136 Harrison Avenue, Boston, MA 02111, USA

Anne-Marie Lambeir
Laboratory of Medical Biochemistry, Dept. of Pharmaceutical Sciences, University of Antwerp, Universiteitsplein 1 S-6, 2610 Wilrijk, Belgium

H.S. Larjava
University of British Columbia, Division of Periodontics, 2199 Westbrook Mall, Vancouver, BC V6T 1Z3, CANADA

Edward W. Lee
Department of Physiology and Biophysics, Georgetown University Medical Center, 3900 Reservoir Rd., NW, Washington, DC, USA

Sang-Un Lee
University of Idaho, Moscow, ID 83844, USA

Uwe Lendeckel
Institut für Experimentelle Innere Medizin, Universitätsklinikum, Otto-von-Guericke-Universität, Leipziger Str. 44, 39120 Magdeburg, Germany

Miriam T. Levy
Dept of Gastroenterology, Royal Prince Alfred Hospital, Locked Bag No 6, Sydney Newtown, NSW, 2042, Australia

Lijun Li
Department of Physiology and Biophysics, Georgetown University Medical Center, 3900 Reservoir Rd., NW, Washington, DC, USA

David C. Limburg
Guilford Pharmaceuticals, Inc., Tributary St., Baltimore, MD, USA

Juan Lojo
University of Santiago de Compostela, Department of Biochemistry and Molecular Biology, 15782 Santiago de Compostela, Galicia, Spain

X.M. Lu
Guilford Pharmaceuticals Inc., 6611 Tributary Street, Baltimore, Maryland 21224, USA

Andreas Ludwig
Research Center Borstel, Parkalle 22, 23845 Borstel, Germany

Francis Lynn
Department of Physiology, University of British Columbia, 2329 West Mall, V6T 1Z4 Vancouver, Canada

P. Majer
Guilford Pharmaceuticals Inc., 6611 Tributary Street, Baltimore, Maryland 21224, USA

Radek Malík
1st Faculty of Medicine of Charles University, U Nemocnice 5, 12853 Prague 2, Czech Republic

Susanne Manhart
Probiodrug AG, Weinbergweg 22, 06120 Halle, Germany

Vladislav Mareš
1st Faculty of Medicine of Charles University, U Nemocnice 5, 12853 Prague 2, Czech Republic

Didier Marguet
Centre d'Immunologie INSERM-CNRS de Marseille-Luminy, Marseille, France

George Marinos
Gastroenterology Department, Prince of Wales Hospital Sydney, Sydney, Australia

Geoffrey W. McCaughan
Dept of Gastroenterology, Royal Prince Alfred Hospital, Locked Bag No 6, Sydney Newtown, NSW, 2042, Australia

Christopher H.S. McIntosh
Department of Physiology, University of British Columbia, 2329 West Mall, V6T 1Z4 Vancouver, Canada

Rolf Mentlein
Anatomisches Institut, Universität Kiel, Olshausenstr. 40, 24098 Kiel, Germany

Shigehiko Mizutami
Department of Obstetrics and Gynecology, Nagoya University Graduate School of Medicine, Tsurumai-cho 65, Showa-ku, 466-8550 Nagoya, Japan

Hubert Mönnikes
Department of Internal Medicine, Division of Gastroenterology, Charité, Humboldt University, Luisenstrasse 13a, Berlin, Germany

Carmen Mrestani-Klaus
Institute of Biochemistry, Martin-Luther-University Halle-Wittenberg, Kurt-Mothes-Str. 3, 06120 Halle, Germany

Klaus Neubert
Institut für Biochemie, Martin-Luther-Universität Halle-Wittenberg, Kurt-Mothes-Str. 3, 06120 Halle, Germany

Jan H. Niess
Department of Internal Medicine, Division of Psychosomatics and Psychotherapy, Charité, Humboldt University, Luisenstrasse 13a, Berlin, Germany

André Niestroj
Probiodrug AG, Weinbergweg 22, 06120 Halle, Germany

Montserrat Nogueira
University of Santiago de Compostela, Department of Biochemistry and Molecular Biology, 15782 Santiago de Compostela, Galicia, Spain

Mauro Novelli
Department of Medical and Surgical Specialities, 1st Dermatologic Clinic, University of Turin, Via Cherasco 23, Turin 10126, Italy

Nathalie Pamir
Department of Physiology, University of British Columbia, 2329 West Mall, V6T 1Z4 Vancouver, Canada

Yong-Ho Park
Seoul National University, Seoul, Korea

Raymond A. Pederson
Department of Physiology, University of British Columbia, 2329 West Mall, V6T 1Z4 Vancouver, Canada

Christian Peiser
Dept. of Internal Medicine, Psychosomatics/Psychotherapie, Humboldt-University, Charite Campus Mitte, Liusenstraße13A, 10117 Berlin, Germany

Astrid Plamboeck
Department of Medical Physiology, Panum Institute, Blegdamsvej 3, DK-2200 Copenhagen N, Denmark

Jennifer Pons
Department of Physiology and Biophysics, Georgetown University Medical Center, 3900 Reservoir Rd., NW, Washington, DC, USA

Sarah Poplawski
Department of Biochemistry, Tufts University School of Medicine, 136 Harrison Avenue, Boston, MA 02111, USA

J. Andrew Pospisilik
Department of Physiology, University of British Columbia, 2329 West Mall, V6T 1Z4 Vancouver, Canada

Paul Proost
Rega Institute of Medical Science, Catholic University of Leuven, Leuven, Belgium

Pietro Quaglino
Department of Medical and Surgical Specialities, 1st Dermatologic Clinic, University of Turin, Via Cherasco 23, Turin 10126, Italy

Dean Rea
Department of Biological Sciences, University of Warwick, Gibbet Hill Road, Coventry, UK

Dirk Reinhold
Institute of Immunology, Otto-von-Guericke-University Magdeburg, Leipziger Strasse 44, Magdeburg, Germany

Werner Reutter
Institut für Molekularbiologie und Biochemie, Freie Universität Berlin Arnimallee 22, 14195 Berlin, Germany

Camilo J. Rojas
R & D, Guildford Pharmaceuticals Inc., 6611 Tributary St., 21224 Baltimore, MD, USA

Fred Rosche
Probiodrug AG, Weinberweg 22, D-06120 Halle, Germany

Matthias Rose
Dept of Internal Medicine/ Psychosomatics, Charité, Humboldt University Berlin, Luisenstraße 13A, 10117 Berlin, Germany

Jens Rüter
Dept of Internal Medicine/ Psychosomatics, Charité, Humboldt University Berlin, Luisenstraße 13A, 10117 Berlin, Germany

Francisco J. Salgado
University of Santiago de Compostela, Department of Biochemistry and Molecular Biology, 15782 Santiago de Compostela, Galicia, Spain

David G. Sanford
Department of Biochemistry, Tufts University School of Medicine, 136 Harrison Avenue, Boston, MA 02111, USA

Paola Savoia
Department of Medical and Surgical Specialities, 1st Dermatologic Clinic, University of Turin, Via Cherasco 23, Turin 10126, Italy

Simon Scharpé
Laboratory of Medical Biochemistry, Dept. of Pharmaceutical Sciences, University of Antwerp, Universiteitsplein 1 S-6, 2610 Wilrijk, Belgium

Florian Schiemann
Research Center Borstel, Parkalle 22, 23845 Borstel, Germany

Dagmar Schlenzig
Probiodrug AG, Weinbergweg 22, 06120 Halle, Germany

Georg Schmidbauer
Department of General and Thoracic Surgery, Justus Liebig University, Giessen, Germany

Detlef Schuppan
Klinik für Innere Medizin I, Abt. Hepatologie, Universität Erlangen-Nürnberg, Ulmenweg 18, D-91054 Erlangen, Germany

Konrad Schwemmle
Department of General and Thoracic Surgery, Justus Liebig University, Giessen, Germany

Aleksi Sedo
Institute of Biochemistry and Experimental Oncology, 1st Faculty of Medicine of Charles University, U Nemocnice 5, 12853 Prague 2, Czech Republic

Jan Ševčík
1st Faculty of Medicine of Charles University, U Nemocnice 5, 12853 Prague 2, Czech Republic

Kiyosumi Shibata
Department of Obstetrics and Gynecology, Nagoya University Graduate School of Medicine, Tsurumai-cho 65, Showa-ku, Nagoya 466-8550, Japan

Akhouri A. Sinha
University of Minnesota Cancer Center, University of Minnesota, Minneapolis, MN, USA

Hans Sjöström
Biochemistry Laboratory C, IMBG, The Panum Institute, Blegdamsvej 3, DK-2200 Copenhagen N, Denmark

Hanne Skovbjerg
Biochemistry Laboratory C, IMBG, The Panum Institute, Blegdamsvej 3, DK-2200 Copenhagen N, Denmark

Joel W. Slaton
University of Minnesota Cancer Center, University of Minnesota, Minneapolis, MN, USA

B.S. Slusher
Guilford Pharmaceuticals Inc., 6611 Tributary Street, Baltimore, Maryland 21224, USA

Sabine Stehling
Institut für Molekularbiologie und Biochemie, UKBF, Freie Universität Berlin, Arnimallee 22, D-14195 Berlin, Germany

Joseph P. Steiner
Guilford Pharmaceuticals, Inc., Tributary St., Baltimore, MD, USA

Beate Stiebitz
Institut für Biochemie, Martin-Luther-Universität Halle-Wittenberg, Kurt-Mothes-Str. 3, 06120 Halle, Germany

Andrea Stöckel-Maschek
Department of Biochemistry and Biotechnology, Martin-Luther-University Halle-Wittenberg, Postfach 8, 06099 Halle, Germany

Jörg Stork
Probiodrug AG, Weinbergweg 22, 06120 Halle, Germany

James L. Sudmeier
Department of Biochemistry, Tufts University School of Medicine, 136 Harrison Avenue, Boston, MA 02111, USA

Takahiro Suzuki
Department of Obstetrics and Gynecology, Nagoya University Graduate School of Medicine, Tsurumai-cho 65, Showa-ku, Nagoya 466-8550, Japan

Janine Tadje
Institute of Experimental Internal Medicine, Otto-von-Guericke University Magdeburg, Leipziger Strasse 44, D-39120 Magdeburg, Germany

Q. Thai Dinh
Dept. of Internal Medicine, Psychosomatics/Psychotherapie, Humboldt-University, Charite Campus Mitte, Liusenstraße13A, 10117 Berlin, Germany

A.G. Thomas
Guilford Pharmaceuticals Inc., 6611 Tributary Street, Baltimore, Maryland 21224, USA

T. Tsukamoto
Guilford Pharmaceuticals Inc., 6611 Tributary Street, Baltimore, Maryland 21224, USA

J.J. Vornov
Guilford Pharmaceuticals Inc., 6611 Tributary Street, Baltimore, Maryland 21224, USA

Otto B. Walter
Department of Internal Medicine, Division of Psychosomatics and Psychotherapie, Charité, Humboldt University, Luisenstrasse 13a, Berlin, Germany

Xin M. Wang
Dept of Gastroenterology, Royal Prince Alfred Hospital, Locked Bag No 6, Sydney Newtown, NSW, 2042, Australia

Neil F. Wasserman
University of Minnesota Cancer Center, University of Minnesota, Minneapolis, MN, USA

Michael Werman
Probiodrug AG, Weinbergweg 22, 06120 Halle, Germany

Douglas E. Wilkinson
Guilford Pharmaceuticals, Inc., Tributary St., Baltimore, MD, USA

Michael J. Wilson
Research Service, University of Minnesota, VA Medical Center, One Veterans Drive, 55417 Minneapolis, MN, USA

Carmen Wolke
Institute of Experimental Internal Medicine, Otto-von-Guericke University Magdeburg, Leipziger Strasse 44, D-39120 Magdeburg, Germany

K.M. Wozniak
Guilford Pharmaceuticals Inc., 6611 Tributary Street, Baltimore, Maryland 21224, USA

Sabine Wrenger
Institute of Immunology, Otto-von-Guericke-University Magdeburg, Leipziger Strasse 44, Magdeburg, Germany

Wengen Wu
Department of Biochemistry, Tufts University School of Medicine, 136 Harrison Avenue, Boston, MA 02111, USA

Yong-Qian Wu
Guilford Pharmaceuticals, Inc., Tributary St., Baltimore, MD, USA

Shuling Yan
Institut für Molekularbiologie und Biochemie, UKBF, Freie Universität Berlin, Arnimallee 22, D-14195 Berlin, Germany

Y. Yeh
Department of Medicine/Medical Oncology, State University of New York, Stony Brook, New York 11794-8154, USA

Yuhong Zhou
Department of Biochemistry, Tufts University School of Medicine, 136 Harrison Avenue, Boston, MA 02111, USA

Zofia Zukowska
Department of Physiology and Biophysics, Georgetown University Medical Center, 3900 Reservoir Rd., NW, Washington, DC, USA

Preface

In recent years, the research on Dipeptidyl Aminopeptidases has led from basic science to first concepts of pharmaceutical applications. The complex role of these enzymes in physiological processes as well as in the pathophysiology of diseases such as diabetes mellitus, rheumatoid arthritis, autoimmune disease, transplant rejections and cytokine-mediated abortions has become evident and provides a rationale for therapeutic interventions by a modulation of their respective enzymatic activity.

The *International Conference on Dipeptidyl Aminopeptidases,* held in Berlin, Germany, in September 2002, was initiated in an attempt to bridge basic science, results of clinical studies and strategies for therapeutic applications. Experts in the field presented a state-of-the-art view on the various aspects of Dipeptidyl Aminopeptidases, thus contributing to a sound basis for decisions on research and therapeutic strategies. The conference, fueled by the presentation of the crystallographic structure of DPP IV by Dr. D. Webb, was concluded by a panel discussion on perspectives and limitations of DPP IV inhibitors.

Selected authors have shared their opinion on basic science, clinical applications and therapeutic interventions with the audience, and they extend their view to the readers of this book. By doing so, they have generously supported us in our attempt to provide a comprehensive overview over Dipeptidyl Aminopeptidases at a time when the therapeutic implications of enzyme inhibitors are imminent, demanding a critical risk assessment based on a detailed understanding of the complex function that these enzymes appear to exert.

A conference on Dipeptidyl Aminopeptidases hosted by a Department of Internal Medicine/Psychosomatics with an interest in these enzymes may be surprising, but reflects the current state of research on these enzymes, with a growing understanding of specific clinical implications and therapeutic options. A glance at a preliminary list of substrates of DPP IV (Table 1) may help to understand such interest: DPP IV-mediated turnover of peptide hormones in various areas of relevance for the body-mind relationship must thrill every one interested in functional thinking. Especially a psychosomaticist in psychoanalytical tradition is reminded of the expectation of Freud, the founder of 20th century's mainstream psychosomatics, that one day every psychological event will be understood from the underlying physiological and biochemical processes. The contributions to our conference and to this book nurture the idea that research on DPP IV is an example for a step towards a biochemical, or even molecular founded psychosomatics.

Table 1. Examples of substrates for DPP IV activity

Substrate	Conceivable implications of DPP IV activity
Growth hormone releasing factor	***Nutritional Control***
Glucagon-like peptide 1 (GLP-1)	▪ hunger
Glucagon-like peptide 2 (GLP-2)	▪ satiety
Gastric inhibitory peptide (GIP)	
Gastrin-releasing peptide (GRP)	
Procolipase	
Enterostatin	
Pancreatic Polypeptide	
Trypsinogen	
Neuropeptide Y (NPY)	
SDF-1α (stromal-cell-derived factor-1α)	***Immunomodulation:***
Lymphotoxin (aa 1-5)	Mechanisms of defence
Eotaxin	
TNF-α	
IP-10	
Monocyte chemotactic protein (MCP-2)	
RANTES	
CLIP (ACTH fragment 18-39)	
Endomorphin-1	***Mood:***
ß-Casomorphin	▪ elation
Kentsin	▪ dysphoria
Morphiceptin	▪ anxiety
Substance P	▪ anger
Tyr-MIF-1	▪ pain

For reviews cf. [1,2]

Although hypothetical at this stage, an integrative concept as presented in Figure 1 may help to appreciate consequences of changes in DPP IV activity or of DPP IV inhibitors. The potential of such inhibitors for novel treatment approaches of type II diabetes has served as a spearhead in the search for substances with inhibitory activity, and continues to fuel research on structure and function of DPP IV as well as on the growing number of DPP IV-like enzymes. The targeted modulation of these enzymes in distinct sites of the organism represents a pharmacodynamic and pharmakokinetic challenge for the development of novel drugs and prodrugs. Given the complexity of the biological processes affected by inhibitors of Dipeptidyl aminopeptidases, their clinical use cannot be fully appreciated yet.

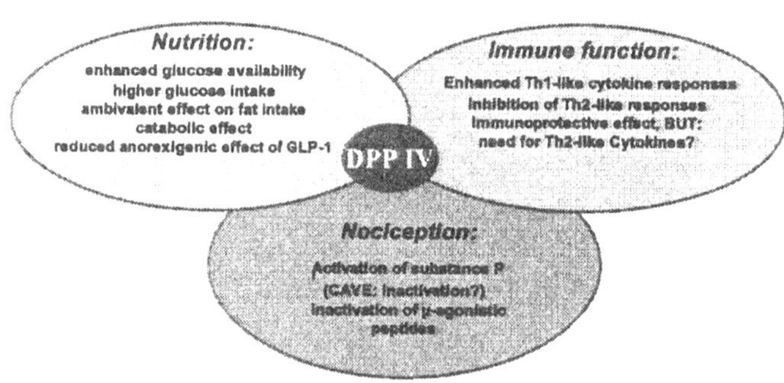

Figure 1. Pleiotropic involvement of DPP IV activity (modified after [1])

At this stage, it may be helpful to label these novel and potent substances *Peptide modulators*, denominating a process rather than a clinical indication and thus avoiding a premature narrowing on a single disease entity. Furthermore, the risk assessment in the exploitation of inhibitory substances should take potential side effects such as an impact on hunger and satiety, pain perception or immunomodulation into account.

The Berlin Conference has continued a series of similar conferences on this topic; a further continuation in 2004 was announced by Uwe Lendeckel and Dirk Reinhold from Magdeburg, members of a team that pioneered in the research on Dipeptidyl Aminopeptidases. We are convinced that major advancements justifying further conferences can be expected in the next future.

Burghard F. Klapp and Martin Hildebrandt
Department of Internal Medicine/ Psychosomatics
Charité, Humboldt University
Berlin, Germany

1. Hildebrandt, M., Reutter, W., Arck, *et al.*, 2000, *Clin Sci (Colch).*, **99(2):** 93-104.
2. De Meester, I., Durinx, C., Bal, G., *et al.*, 2000, *Adv.Exp.Med.Biol.* **477:** 67-87.

Acknowledgments

We are indebted to Joanna Lawrence and Joanne Duggan from Kluwer Academic/Plenum Publishers, for their continuing support in the editing process.

Furthermore, we thank Dr. Jens Rüter, Medical Faculty Charité, Berlin, and Florian König, Kingsize Event, Berlin, as members of the Organizing Committee of the *International Conference on Dipeptidyl Aminopeptidases*, Berlin, September 2002. The conference formed the basis from which the decision was made to compile and edit this volume. The success of this conference is clearly attributable to their commitment and highly professional work.

The *International Conference on Dipeptidyl Aminopeptidases* received support from the following organisations:

Abbott Laboratories, Abbott Park, IL, USA
Essex Pharma GmbH, München, Germany
Fonds der Chemischen Industrie im Verband der Chemischen Industrie e.V., Frankfurt, Germany
Merck Research Laboratories, Rahway, NJ, USA
Probiodrug AG, Halle/S., Germany

Berlin, October 2002
The Editors

Contents

Structure and Function of Dipeptidyl Aminopeptidases

Angiogenesis and Cancer

Diabetes and Metabolism

1

STRUCTURE AND FUNCTION OF DIPEPTIDYL AMINOPEPTIDASES

Dipeptidyl Peptidase IV Substrates

An update on in vitro peptide hydrolysis by human DPPIV

INGRID DE MEESTER, ANNE-MARIE LAMBEIR, PAUL PROOST[#] and SIMON SCHARPÉ

[*]*Department of Pharmaceutical Sciences, University of Antwerp, Universiteitsplein 1, Antwerp, Belgium;* [#]*Rega Institute of Medical Science, Catholic University of Leuven, Leuven, Belgium*

1. INTRODUCTION

Dipeptidyl-peptidase IV/CD26 (DPPIV) is a cell-surface serine protease belonging to the prolyloligopeptidase family. It selectively removes the N-terminal dipeptide from peptides with proline or alanine on the second position. Apart from its catalytic activity it contains several protein binding sites, for instance for adenosine deaminase, the HIV gp120 protein, fibronectin, collagen, the chemokine receptor CXCR4, and the tyrosine phosphatase CD45. DPPIV is expressed on a specific set of T lymphocytes, where it is up-regulated upon activation. It is also expressed in a variety of tissues, primarily on endothelial and epithelial cells. A soluble form is present in plasma and other body fluids[1-4].

DPPIV truncates many bio-active peptides of medical importance and this subject has been reviewed recently[5-7]. It plays a role in glucose homeostasis through proteolytic inactivation of the incretins[8]. DPPIV inhibitors improve glucose tolerance, and pancreatic islet function in general, in animal models of type 2 diabetes and in patients[9-10]. The role of DPPIV/CD26 within the immune system most probably results from a combination of its exopeptidase activity and its capacity to interact with different molecules. Here we give an update on the *in vitro* truncation of peptides by DPPIV.

Dipeptidyl Aminopeptidases in Health and Disease, Edited by Hildebrandt et al.
Kluwer Academic/Plenum Publishers, New York, 2003

2. SUBSTRATE SPECIFICITY OF DPPIV

DPPIV preferentially cleaves off X-Pro and X-Ala dipeptides from substrates consisting of three or more amino acids or from dipeptides linked to C-terminal chromogenic or fluorogenic compounds such as the *p*-nitroanilides (*p*-NA). The active site Ser^{630} of human DPPIV is located in the sequence Gly-Trp-Ser-Tyr-Gly, which corresponds to the motif Gly-X-Ser-X-Gly, common to serine proteases. The other residues forming the catalytic triad are Asp^{708} and His^{740}. The linear order of the catalytic triad in the α/β hydrolase fold is inversed, compared to the trypsin or subtilisin-like enzymes. The similar catalytic behaviour of DPPIV and PO suggests resemblance in the architecture of their active sites[11]. However as long as there is no crystal structure available, every model remains an approximation as the sequence homology between DPPIV and PO is far too low to allow a good prediction of the loops and turns that constitute the major sites of interaction.

The active site of an enzyme performs the twofold function of binding the substrate and catalyzing the reaction. The efficiency of these actions determines the overall activity of the enzyme towards a particular substrate, *i.e.* determines the specificity of the enzyme. Steady-state kinetic analysis of substrates classically produces three parameters: the K_m (M) or the *Michaelis-Menten constant*, k_{cat} (s^{-1}) or the *catalytic rate constant*, k_{cat}/K_m ($M^{-1}s^{-1}$) or the *specificity constant*.

The substrate specificity of DPPIV was determined by experiments with synthetic peptides, natural substrates and *p*-nitroanilide analogs.

Originally, DPPIV was considered to cleave specifically after a proline or an alanine at the *penultimate position* (P_1). Meanwhile, the substrate spectrum has been enlarged. The P_1 residue can also be hydroxyproline, dehydroproline, serine, glycine, valine, threonine or leucine[12-16]. The k_{cat}/K_m values for X-Pro and X-Ala-*p*-NA differ with a factor 10 to 100. The discrimination between Pro^2 and Ala^2 is generally much greater for the dipeptide chromogenic and fluorogenic substrates than for natural substrates. Even the difference between Ala^2 and Ser^2 is not always discernible in larger peptide substrates (see further).

At the *amino-terminal position,* DPPIV accepts all residues, on the condition that they have a protonated amine group. In general, hydrophobic aliphatic residues are favored at the amino-terminal position[1]. The influence on the substrate specificity is rather small. Substitution of the hydroxyl group of Ser^1 or Thr^1 with a phosphate group prevents truncation of the substrate.

DPPIV is unable to hydrolyse substrates with proline, hydroxyproline or N-methyl glycine on the *third position* (P_1')[17,18]. Since tripeptides with Pro^3 are DPPIV inhibitors, the substrate-binding pocket can accommodate proline on the P_1' position but the X-Pro bond is not hydrolyzed[19].

DPPIV has *S_1', S_2' ... S_n' binding subsites* probably as far as S_4'. The shortening of a peptide substrate can therefore both increase and decrease the proteolysis rate. Modifications further away from the scissile bond can also affect the kinetic parameters[15,20].

Catalysis by DPPIV is strongly *stereospecific*. The scissile and P_2-P_1 bonds must be in *trans* configuration[21].

DPPIV was reported to display weak *endopeptidase* activity: it would be able to digest certain N-blocked peptides and denatured collagen at intramolecular sites[22-24].

3. KINETIC STUDY OF THE *IN VITRO* TRUNCATION OF BIOACTIVE PEPTIDES BY DIPEPTIDYL-PEPTIDASE IV

The *in vitro* kinetic study of the truncation of bioactive peptides by DPPIV/CD26 resulted in the identification of several excellent substrates with high specificity constants (see figure).

Several conclusions concerning the substrate specificity of DPPIV can be drawn from *in vitro* kinetic studies. The amino acid sequence surrounding the scissile bond is not the only determinant for selectivity. Specific structural features of the substrate influence the catalytic parameters. This is illustrated by the improved cleavage of the long form of pituitary adenylate cyclase-activating peptide (PACAP38) compared to PACAP27 and the rapid truncation after a penultimate glycine in the truncated form of Macrophage Derived Chemokine, MDC(3-69)[14,15]. The interaction with Glucagon-like peptide–1 (GLP-1) also contains a component that is independent of the catalytic activity of DPPIV[25]. There is no clear link between the peptide length (up to about 100 amino acids) and the kinetic constants. A free and flexible N-terminus is required. The formation of dimers and higher aggregates at high (μM-mM) concentrations interferes with the *in vitro* truncation of the chemokines RANTES, LD78β and presumably also MIP-1β.

In vivo, bioactive peptides act in the pico- or nanomolar range, orders of magnitudes below their K_m values. At this low substrate concentration [S], the reaction rate (v) of the enzyme (E) is given by $v = [E][S]*k_{cat}/K_m$. At a given enzyme concentration, the specificity constant is the sole determinant of the half-life. The substrate concentration has no influence. The upper limit of k_{cat}/K_m is the diffusion-controlled association of the substrate (around 10^7 $M^{-1}s^{-1}$ for large molecules). The experimental values are all smaller. From the various publications, it is obvious that results are not easily reproduced between research groups and/or experimental techniques. This may be partially due to the fact that steady-state kinetics do not allow to determine absolute rate constants. The exact concentration of active sites in the reaction mixture needs to be known. As the specific activity and the used molecular weight of DPPIV differ considerably, this is presumably the most important cause of variability[8,15,26-28]. To circumvent the problem, the amount of DPPIV used is sometimes related to the activity in human serum. At least this reference is commonly available and does not depend on the methods used to determine the specific activity. In figure 1 we show the K_m, k_{cat} and specificity constants of a selection of substrates that were all measured in identical experimental conditions and with the same enzyme preparation (human natural DPPIV purified from semen). The figure includes the recent data on GLP-1 and GLP-2 obtained under conditions identical to the ones used for the truncation of all other peptides mentioned[28]. The in vitro kinetic data correspond with the *in vivo* findings that intact GLP-1 disappears faster than GLP-2. The K_m is very similar for both peptides and the major difference resides in the k_{cat} value.

Table 1. Kinetic data on peptide hydrolysis by DPP IV

Peptide	N-terminus	#aa	K_m *10^6 (M)	k_{cat} (s^{-1})	k_{cat}/K_m *10^{-6} ($M^{-1}s^{-1}$)	Ref.
PANCREATIC POLYPEPTIDE FAMILY						
Neuropeptide Y	YP↓SKP-	36	52	40	0.76	14, 15
PACAP/GLUCAGON FAMILY						
GLP-1	HA↓EGT-	30	36	7.1	0.20	28
GLP-2	HA↓DGS-	33	36	0.87	0.02	28
VIP	HS↓DA-	28	59	0.74	0.012	15
	DA↓VF-	26			0.0017	15
PACAP27	HS↓DG-	27	120	0.49	0.0042	15
	DG↓IF-	25			0.00010	15
PACAP38	HS↓DG-	38	35	0.85	0.024	15
CHEMOKINES						
Mig (CXCL9)	TP↓VVR-	103	<1	0.34	>0.4	14
IP-10 (CXCL10)	VP↓LSR-	77	4.0	1.8	0.50	14
I-TAC (CXCL11)	FP↓MFK-	73	9.0	10	1.2	14
SDF-1α (CXCL12)	KP↓VSL-	68	2.0	12	5	14
LD78β (CCL3L1)	AP↓LA-	70	1.0	0.0026	0.003	14
RANTES (CCL5)	SP↓YS-	68	1.0	0.04	0.040	14
Eotaxin (CCL11)	GP↓AS-	74	25	1.9	0.080	14
MDC (CCL22)	GP↓YG-	69	2.5	9.6	4	14
	YG↓AN-	67	6.0	3.2	0.5	
GRP	VP↓LP-	27	6.0	11	1.8	15
	LP↓AG-	25	20	41	2.1	15

All data were obtained using the same batch of human natural DPP IV, purified from semen. Values of k_{cat} were calculated using a molecular mass of 85 kDa per subunit of DPPIV.

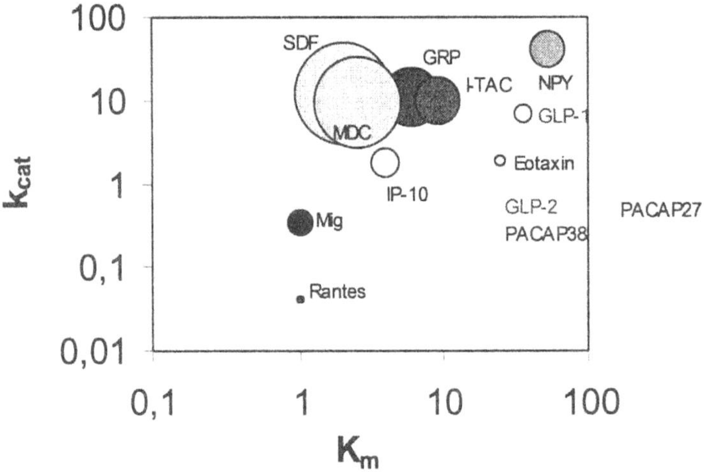

Figure 1. In vitro peptide truncation by DPPIV. The position of the bubble is determined by the K_m (x value) and the k_{cat} (y value) while the dimensions of the circles are proportional to the specificity constant (k_{cat} / K_m) for the peptides

4. PHYSIOLOGICAL RELEVANCE OF PEPTIDE TRUNCATION BY DPPIV

To decide upon the physiological relevance of DPPIV for the activity of a particular peptide substrate, several questions need to be answered: (1) have truncated forms been observed in biological samples, (2) is DPPIV present at same the sites, (3) does truncation affect receptor binding or biological properties of the peptide, (4) how does the *in vitro* truncation rate compare with the *in vivo* findings. In the following paragraphs we attempt to highlight recent results for a selection of peptides.

4.1 The PACAP/glucagon family of peptides

The PACAP/glucagon family includes glucagon, the glucagon-like peptides GLP-1 and GLP-2, secretin, vasoactive intestinal peptide (VIP), pituitary adenylate cyclase-activating peptide (PACAP), glucose-dependent insulinotropic peptide (GIP, also referred to as gastric inhibitory peptide), growth hormone-releasing factor (GRF), and peptide histidine methionine (PHM). These peptides are related in the structure of their N-terminal region, distribution, function and receptors[29].

In this family, the intact N-terminus (Tyr-Ala, His-Ala or His-Ser) is necessary for biological activity and truncation by DPPIV causes inactivation[30-33]. Since the PACAP/glucagon family members with a penultimate serine were considered to be 'DPPIV-resistant'[8], the truncation of glucagon by DPPIV was unexpected[13,34]. However, cleavage by DPPIV after a serine had been observed before for analogs of growth hormone-releasing factor (GRF)[12,17].

The main biological effect of GLP-1 is the action as an incretin: it amplifies meal-induced insulin release and synthesis in a glucose-dependent manner[33].

GIP (gastric inhibitory peptide, renamed glucose-dependent insulinotropic polypeptide, 42-amino acids) is secreted by the endocrine K cells of the proximal intestine in response to nutrients, especially fats[32]. GIP acts through a G protein-coupled receptor in a large array of tissues. The truncation of GIP by DPPIV, resulting in GIP(3-42), has been shown *in vitro* and *in vivo*[35, 36].

VIP is a 28-amino acid neuropeptide and PACAP38 is a 38-amino acid peptide with a C-terminally truncated 27-residue splicing variant PACAP27. VIP and PACAP show 70% identity and they share G protein-coupled receptors. We showed that DPPIV sequentially cleaves off two dipeptides from VIP, PACAP27 and PACAP38[15, 37]. Since residues 2 and 4 are important for receptor activation, the sequential truncation by DPPIV is likely to affect the biological activity of VIP and PACAP[38,39]. VIP and PACAP are rather poor substrates with turnover numbers comparable to those reported earlier for glucagon, but with higher K_m values. PACAP was shown to be a very strong insulinotropic peptide[40]. This means that small differences in active peptide levels can have important biological consequences. It is not excluded that in microenvironments with high DPPIV activity, the truncation is of physiological importance.

GLP-2 plays an essential role in intestinal mucosa regeneration and permeability. The highly localized expression of the GLP-2 receptor in the gastrointestinal tract, may add to the high specificity of GLP-2 for induction of intestinal growth, without affecting other peripheral tissues[41].

4.2 Gastrin releasing peptide

Gastrin-releasing peptide (GRP) is a member of the bombesin family of peptide hormones. GRP is produced in the brain, the intrinsic neurons of the gut and the parasympathic neurons of the pancreas[42]. The GRP receptor belongs to the G protein-coupled receptor family and is present on the epithelial cells lining the gastric antrum and in the pancreas[43,44]. GRP is considered as a general 'on-switch' in the gastrointestinal tract with actions that include stimulation of the secretion of gastrointestinal hormones and of the endocrine/exocrine pancreas[45, 46]. The insulinotropic action of GRP is due to direct stimulation of the β cells, to activation of postganglionic parasympathic nerves at the ganglionic level, and to stimulation of GLP-1 release [47, 48]. In the pancreas, GRP is released with VIP and PACAP upon stimulation of the parasympathic nerves, causing an increase in insulin secretion from the β cells[49]. Intact GRP receptors are required for an adequate insulin response and normal glucose tolerance after enteral glucose intake[50]. The neuropeptide has vasodilator effects on the (intestinal) circulation[51].

Human GRP is an excellent substrate for DPPIV, which sequentially removes two dipeptides with almost equal efficiency[15]. The specificity constant for GRP is significantly higher than for GLP-1 and GIP, whose active intact peptide levels are regulated by DPPIV. The *in vitro* half-life for truncation in circumstances where the DPPIV activity equals the normal serum activity is 2 minutes. The truncated GRP(5-27) form has been isolated from dog intestine and brain but the smallest active degradation product is GRP(18-27). Minimal ligand analysis showed that the eight C-terminal amino acids of GRP are sufficient for full receptor agonist activity[52]. The penultimate proline of GRP has been conserved through evolution but no function has been attributed to the N-terminus.

4.3 Chemokine processing by DPPIV

Chemokines or 'chemotactic cytokines' are produced by leukocytes, epithelium, endothelium and tissue cells, either constitutively or after induction. Chemokines exert their effects locally in a paracrine or autocrine fashion. The main subfamilies, CXC, CC, C and CX_3C, are distinguished based on the number of cysteines and the sequence surrounding the first cysteines. Chemokines act via heptahelical G protein-coupled receptors, designated as CXCR, CCR, CR and CX_3CR, followed by a number [53]. Recent nomenclature refers to the chemokines as ligands belonging to a specific class, for example CCL5 (RANTES) and CXCL10 (IP-10).

The region N-terminal from the first cysteine (usually less than 11 amino acids) has almost maximal flexibility and contains critical residues for receptor triggering. Alterations to the N-terminus of a chemokine do not necessarily affect receptor binding, but strongly influence the ability to induce signaling and can therefore result in antagonists.

Chemokines play fundamental roles in the development, homeostasis and function of the immune system. They act as regulatory molecules in leukocyte maturation and traffic and in homing of lymphocytes and the development of lymphoid tissues. Soluble chemokines bind to proteoglycans on endothelial cell surfaces and of the extracellular matrix. Bound chemokines retain their full chemotactic activity and thus form an immobilized concentration gradient along which the leukocytes migrate.

The importance of the N-terminus for chemokine signalling, the occurrence of a penultimate proline or alanine in a great number of chemokines and the natural occurrence of the truncated forms, led to the study of the cleavage by DPPIV/CD26. The truncated forms of the following chemokines have been isolated: MDC, eotaxin, RANTES, LD78β and GCP-2 [54-58]. The processing by DPPIV/CD26 has an important impact on the biological activity of several chemokines. DPPIV/CD26 plays a role in the post-translational regulation of chemokine activity by reducing the inflammatory properties of most chemokines and enhancing those of LD78β[58]. DPPIV/CD26-mediated truncation of chemokines reduces the redundancy in their target cell specificity and influences the antiviral response[56, 59].

The processing of the interferon-γ-induced chemokines (IP-10, Mig and I-TAC) decreases CXCR3 binding and abolishes chemotaxis of their main target cells, activated T_{H1} lymphocytes. The high specificity constants obtained for I-TAC and IP-10, together with the elevated levels of DPPIV/CD26 on their target cells, support a role for the enzyme in the fine-tuning of these chemokines[60,61]. The anti-angiogenic properties of IP-10, Mig and I-TAC were not significantly altered[60].

Several of these chemokines were *in vitro* shown to be excellent substrates that are rapidly cleaved by DPPIV at levels likely to be encountered *in vivo*. Not all chemokines were cleaved at the same rate; some of them are clearly preferred by the enzyme, in particular SDF-1, MDC, I-TAC and IP-10. The specificity constants of most other chemokine substrates (see table 1 and figure 1) are in the same order of magnitude as the incretins (GLP-1 and GIP) [14].

In vivo, the net outcome of DPPIV-mediated truncation depends on the local availability of DPPIV/CD26, the concentration of the chemokine, competition between substrates, their specificity constants and receptor density. However, due to the redundancy in the chemokine-chemokine receptor network, it is possible that an alternative pathway will correct for modulating one type of interaction. Contrary to the inflammatory chemokines, SDF-1 is constitutively expressed in many tissues. *In vitro*, SDF-1α is one of the best DPPIV substrates. Whether the enzyme contributes significantly to the metabolism of SDF-1 needs to be further evaluated. N-terminal inactivation by other peptidases was recently reported [62-64].

Since chemokine receptors are exploited by HIV-1 for cell entry, several chemokines provide a certain degree of protection against HIV-1 infection. CCL5/RANTES inhibits R5 HIV-1 infection by binding to the CCR5 receptor. Truncation by DPPIV/CD26 increases the affinity of RANTES for the CCR5 receptor and thereby improves its anti-HIV-1 activity [56,65]. The relative inefficiency of RANTES in inhibiting the HIV-1 infection of monocytes[66] can be partially explained by absence of DPPIV/CD26 on these cells, leading to a reduced activation of the anti-HIV-1 properties of RANTES.

CCL3L1/LD78β prevents the infection of mononuclear cells by R5 HIV-1 strains very effectively[57,69]. Truncation by DPPIV/CD26 strongly increases the binding to the CCR5 receptor. LD78β(3-70) is currently the chemokine that most powerfully blocks the infection with R5 HIV-1 strains[69].

CXCL12/SDF-1 is a ligand for CXCR4 and it blocks the X4 HIV-1 entry. After DPPIV-mediated cleavage, SDF-1 looses its ability to bind to the CXCR4 and also its anti-HIV-1 properties[67, 68].

Antiviral activity against X4 and R5 HIV-1 strains was also reported for CCL22/MDC. The sequential removal by DPPIV/CD26 of two dipeptides from the amino-terminus of MDC does abolish its antiviral activity[70].

5. CONCLUSIONS

Many bioactive peptides qualify to be DPPIV substrates. Considering the ubiquitous presence of the enzyme – on cells, on vesicles, in fluids – it is hardly surprising that many of them are indeed found truncated *in vivo*. However, not all substrates are cleaved with the same efficiency. The molecular properties of DPPIV involved in substrate recognition are still poorly understood.

ACKNOWLEDGEMENTS

We thank Christine Durinx for her contributions to DPP IV-substrate related research, and Nicole Lamoen for skillful technical assistance. The work is supported by grants from the Fund for Scientific Research Flanders and by a 'BOF'-grant from the University of Antwerp.

REFERENCES

1. Yaron, A., Naider, F., 1993, Proline-dependent structural and biological properties of peptides and proteins. *Crit.Rev.Biochem.Mol.Biol.* **28:** 31-81.
2. De Meester, I., Korom, S., Van Damme, J., *et al.*, 1999, CD26, let it cut or cut it down. *Immunol.Today,* **20:** 367-75.
3. Augustyns, K., Bal, G., Thonus, G., *et al.*, 1999,The unique properties of dipeptidyl-peptidase IV (DPP IV/CD26) and the therapeutic potential of DPP IV inhibitors. *Curr.Med Chem.* **6:** 311-27.
4. Kähne, T., Lendeckel, U., Wrenger, S., *et al.*,1999, Dipeptidyl peptidase IV: a cell surface peptidase involved in regulating T cell growth (review). *Int.J.Mol.Med.* **4:** 3-15.
5. Mentlein, R., 1999, Dipeptidyl-peptidase IV (CD26) - role in the inactivation of regulatory peptides. *Regul.Pept.* **85:** 9-24.
6. De Meester, I., Durinx, C., Bal, G., *et al.*, 2000, Natural substrates of dipeptidyl peptidase IV. *Adv.Exp.Med.Biol.* **477:** 67-87.
7. De Meester, I., Durinx, C., Proost, P., *et al.*, 2002, DPIV - Natural Substrates of Medical Importance. In: Langner, J. and Ansorge, S., eds. Ectopeptidases. CD13/Aminopeptidase N and CD26/Dipeptidylpeptidase IV in Medicine and Biology. Pp 223-257. New York: Kluwer Academic/Plenum Publishers.
8. Mentlein, R., Gallwitz, B., Schmidt, W.E., 1993, Dipeptidyl-peptidase IV hydrolyses gastric inhibitory polypeptide, glucagon-like peptide-1(7-36)amide, peptide histidine methionine and is responsible for their degradation in human serum. *Eur.J.Biochem.* **214:** 829-35.
9. Deacon, C.F., Holst, J.J.,2002, Dipeptidyl peptidase IV inhibition as an approach to the treatment and prevention of type 2 diabetes: a historical perspective. *Biochem Biophys Res Commun.* **294:**1-4.
10. Demuth, H., Hinke, S., Pederson, R., McIntosh, C., 2002, Rebuttal to Deacon and Holst: "Metformin effects on dipeptidyl peptidase IV degradation of glucagon-like peptide-1" versus "Dipeptidyl peptidase inhibition as an approach to the treatment and prevention of type 2 diabetes: a historical perspective". Biochem Biophys Res Commun. **296:** 229.

11. Rawlings, N.D., Polgar, L., Barrett, A.J., 1991, A new family of serine-type peptidases related to prolyl oligopeptidase. *Biochem.J.* **279:** 907-8.
12. Martin, R.A., Cleary, D.L., Guido, D.M., *et al.*, 1993, Dipeptidyl peptidase IV (DPP-IV) from pig kidney cleaves analogs of bovine growth hormone-releasing factor (bGRF) modified at position 2 with Ser, Thr or Val. Extended DPP-IV substrate specificity? *Biochim.Biophys.Acta* **1164:** 252-60
13. Pospisilik, J.A., Hinke, S.A., Pederson, R.A., *et al.* 2001, Metabolism of glucagon by dipeptidyl peptidase IV (CD26). *Regul.Pept.* **96:** 133-41.
14. Lambeir, A.M., Proost, P., Durinx, C., *et al.*, 2001, Kinetic investigation of chemokine truncation by CD26/dipeptidyl peptidase IV reveals a striking selectivity within the chemokine family. *J.Biol.Chem.* **276:** 29839-45.
15. Lambeir, A.M., Durinx, C., Proost, P., *et al.*, 2001, Kinetic study of the processing by dipeptidyl-peptidase IV/CD26 of neuropeptides involved in pancreatic insulin secretion. *FEBS Lett.* **507:** 327-30.
16. Rahfeld, J., Schutkowski, M., Faust, J., *et al.*, 1991, Extended investigation of the substrate specificity of dipeptidyl peptidase IV from pig kidney. *Biol.Chem.Hoppe-Seyler* **372:** 313-18.
17. Bongers, J., Lambros, T., Ahmad, M., *et al.*, 1992, Kinetics of dipeptidyl peptidase IV proteolysis of growth hormone-releasing factor and analogs. *Biochim.Biophys.Acta,* **1122:** 147-53.
18. Püschel, G., Mentlein, R., Heymann, E., 1982, Isolation and characterization of dipeptidyl peptidase IV from human placenta. *Eur.J.Biochem.* **126:** 359-65.
19. Hoffmann, T., Reinhold, D., Kähne, T., *et al.*, 1995, Inhibition of dipeptidyl peptidase IV (DPPIV) by anti-DP IV antibodies and non-substrate X-X-Pro-oligopeptides ascertained by capillary electrophoresis. *J.Chromatogr.A* **716:** 355-62.
20. Kikuchi, M., Fukuyama, K., Epstein, W.L., 1988, Soluble dipeptidyl peptidase IV from terminal differentiated rat epidermal cells: purification and its activity on synthetic and natural peptides. *Arch.Biochem.Biophys.* **266:** 369-76.
21. Fischer, G., Heins, J., Barth, A., 1983, The conformation around the peptide bond between the P1- and P2-positions is important for catalytic activity of some proline-specific proteases. *Biochim.Biophys.Acta* **742:** 452-62.
22. Reutter, W., Baum, O., Löster, K., *et al.*, 1995, Functional aspects of the three extracellular domains of dipeptidyl peptidase IV: characterization of glycolysation events, of the collagen-binding site and of endopeptidase activity. In: Fleischer, B., eds. Dipeptidyl peptidase IV (CD26) in metabolism and the immune response. Pp 55-78. Heidelberg: Springer-Verlag.
23. Bauvois, B., 1995, Modulation and functional diversity of dipeptidyl peptidase IV in murine and human systems. In: Fleischer, B., eds. Dipeptidyl peptidase IV (CD26) in metabolism and the immune response. Pp 99-110. Heidelberg: Springer-Verlag.
24. Bermpohl, F., Löster, K., Reutter, W., *et al.*, 1998, Rat dipeptidyl peptidase IV (DPP IV) exhibits endopeptidase activity with specificity for denatured fibrillar collagens. *FEBS Lett.* **428:** 152-6.
25. Hinke, S.A., Kühn-Wache, K., Hoffmann, T., *et al.*, 2002, Metformin effects on dipeptidylpeptidase IV degradation of glucagon- like peptide-1. *Biochem.Biophys.Res.Commun.* **291:** 1302-8.
26. Mentlein, R., Dahms, P., Grandt, D., *et al.*, 1993, Proteolytic processing of neuropeptide Y and peptide YY by dipeptidyl peptidase IV. *Regul.Pept.* **49:** 133-44.
27. Pauly, R.P., Rosche, F., Wermann, M., *et al.*, 1996, Investigation of glucose-dependent insulinotropic polypeptide-(1-42) and glucagon-like peptide-1-(7-36) degradation in

vitro by dipeptidyl peptidase IV using matrix-assisted laser desorption/ionization-time of flight mass spectrometry. A novel kinetic approach. *J.Biol.Chem.* **271:** 23222-9.
28. Lambeir, A.M., Proost, P., Scharpé, S. and De Meester, I., 2002, A kinetic study of Glucagon-like peptide 1 and Glucagon-like peptide 2 truncation by DPP IV, in vitro. *Biochem. Pharmacol.*, in press
29. Sherwood, N.M., Krueckl, S.L., McRory, J.E., 2000, The origin and function of the pituitary adenylate cyclase-activating polypeptide (PACAP)/glucagon superfamily. *Endocr.Rev.* **21:** 619-70.
30. Drucker, D.J., Shi, Q., Crivici, A., *et al.*, 1997, Regulation of the biological activity of glucagon-like peptide 2 in vivo by dipeptidyl peptidase IV. *Nat.Biotechnol.* **15:** 673-7.
31. Frohman, L.A., Jansson, J.O., 1986, Growth hormone-releasing hormone. *Endocr.Rev.* **7:** 223-53.
32. Brown, J.C., Dahl, M., Kwauk, S., *et al.*, 1981, Actions of GIP. *Peptides* 1981; 2: 241-5.
33. Drucker, D.J., 2002, Biological actions and therapeutic potential of the glucagon-like peptides. *Gastroenterology* **122:** 531-44.
34. Hinke, S.A., Pospisilik, J.A., Demuth, H.U., *et al.*, 2000, Dipeptidyl peptidase IV (DPIV/CD26) degradation of glucagon. Characterization of glucagon degradation products and DPIV-resistant analogs. *J.Biol.Chem.* **275:** 3827-34.
35. Kieffer, T.J., McIntosh, C.H., Pederson, R.A., 1995, Degradation of glucose-dependent insulinotropic polypeptide and truncated glucagon-like peptide 1 in vitro and in vivo by dipeptidyl peptidase IV. *Endocrinology* **136:** 3585-96.
36. Deacon, C.F., Nauck, M.A., Meier, J., *et al.*, 2000, Degradation of endogenous and exogenous gastric inhibitory polypeptide in healthy and in type 2 diabetic subjects as revealed using a new assay for the intact peptide. *J.Clin.Endocrinol.Metab.* **85:** 3575-81.
37. Kühn-Wache, K., Manhardt, S., Rosche, F. *et al.*, 1999, 2nd symposium on Cellular Peptidases in Immune Functions and Diseases, Magdeburg.
38. Nicole, P., Lins, L., Rouyer-Fessard, C., *et al.*, 2000, Identification of key residues for interaction of vasoactive intestinal peptide with human VPAC1 and VPAC2 receptors and development of a highly selective VPAC1 receptor agonist. Alanine scanning and molecular modeling of the peptide. *J.Biol.Chem.* **275:** 24003-12.
39. Robberecht, P., Gourlet, P., de Neef, P., *et al.*, 1992, Structural requirements for the occupancy of pituitary adenylate-cyclase-activating-peptide (PACAP) receptors and adenylate cyclase activation in human neuroblastoma NB-OK-1 cell membranes. Discovery of PACAP(6-38) as a potent antagonist. *Eur.J.Biochem.* **207:** 239-46.
40. Yada, T., Sakurada, M., Ihida, K., *et al.*, 1994, Pituitary adenylate cyclase activating polypeptide is an extraordinarily potent intra-pancreatic regulator of insulin secretion from islet beta-cells. *J.Biol.Chem.* **269:** 1290-3.
41. L'Heureux, M.C., Brubaker, P.L., 2001, Therapeutic potential of the intestinotropic hormone, glucagon-like peptide-2. *Ann.Med.* **33:** 229-35.
42. McDonald, T.J., Jornvall, H., Nilsson, G., *et al.*, 1979, Characterization of a gastrin releasing peptide from porcine non-antral gastric tissue. *Biochem.Biophys.Res.Commun.* **90:** 227-33.
43. Ferris, H.A., Carroll, R.E., Lorimer, D.L., *et al.*, 1997, Location and characterization of the human GRP receptor expressed by gastrointestinal epithelial cells. *Peptides,* **18:** 663-72.
44. Xiao, D., Wang, J., Hampton, L.L., *et al.*, 2001, The human gastrin-releasing peptide receptor gene structure, its tissue expression and promoter. *Gene* **264:** 95-103.
45. Horstmann, O., Nustede, R., Schmidt, W., *et al.*, 1999, On the role of gastrin-releasing peptide in meal-stimulated exocrine pancreatic secretion. *Pancreas* **19:** 126-32.

46. Jensen, R.T., Coy, D.H., Saeed, Z.A., *et al.*, 1988, Interaction of bombesin and related peptides with receptors on pancreatic acinar cells. *Ann.NY Acad.Sci.* **547**: 138-49
47. Karlsson, S., Sundler, F., Ahrén, B., 1998, Insulin secretion by gastrin-releasing peptide in mice: ganglionic versus direct islet effect. *Am.J.Physiol.* **274**: E124-9.
48. Roberge, J.N., Gronau, K.A., Brubaker, P.L., 1996, Gastrin-releasing peptide is a novel mediator of proximal nutrient-induced proglucagon-derived peptide secretion from the distal gut. *Endocrinology* **137**: 2383-8.
49. Karlsson, S., Sundler, F., Ahrén, B., 2001, Direct cytoplasmic CA(2+) responses to gastrin-releasing peptide in single beta cells. *Biochem.Biophys.Res.Commun.* **280**: 610-4.
50. Persson, K., Gingerich, R.L., Nayak, S., *et al.*, 2000, Reduced GLP-1 and insulin responses and glucose intolerance after gastric glucose in GRP receptor-deleted mice. *Am.J.Physiol.Endocrinol.Metab.* **279**: E956-62.
51. Clive, S., Jodrell, D., Webb, D., 2001, Gastrin-releasing peptide is a potent vasodilator in humans. *Clin.Pharmacol.Ther.* **69**: 252-9.
52. Heimbrook, D.C., Boyer, M.E., Garsky, V.M., *et al.*, 1988, Minimal ligand analysis of gastrin releasing peptide. Receptor binding and mitogenesis. *J.Biol.Chem.* **263**: 7016-9.
53. Zlotnik, A., Yoshie, O., 2000, Chemokines: a new classification system and their role in immunity. *Immunity* **12**: 121-7.
54. Noso, N., Sticherling, M., Bartels, J., *et al.*, 1996, Identification of an N-terminally truncated form of the chemokine RANTES and granulocyte-macrophage colony-stimulating factor as major eosinophil attractants released by cytokine-stimulated dermal fibroblasts. *J.Immunol.* **156**: 1946-53.
55. Struyf, S., De Meester, I., Scharpé, S, *et al.*, 1998, Natural truncation of RANTES abolishes signaling through the CC chemokine receptors CCR1 and CCR3, impairs its chemotactic potency and generates a CC chemokine inhibitor. *Eur.J.Immunol.* **28**: 1262-71.
56. Proost, P., De Meester, I., Schols, D., *et al.*, 1998, Amino-terminal truncation of chemokines by CD26/dipeptidyl-peptidase IV. Conversion of RANTES into a potent inhibitor of monocyte chemotaxis and HIV-1-infection. *J.Biol.Chem.* **273**: 7222-7.
57. Menten, P., Struyf, S., Schutyser, E., *et al.*, 1999, The LD78beta isoform of MIP-1alpha is the most potent CCR5 agonist and HIV-1-inhibiting chemokine. *J.Clin.Invest.* **104**: R1-5.
58. Proost, P., Menten, P., Struyf, S., *et al.*, 2000, Cleavage by CD26/dipeptidyl peptidase IV converts the chemokine LD78beta into a most efficient monocyte attractant and CCR1 agonist. *Blood* **96**: 1674-80.
59. Oravecz, T., Pall, M., Roderiquez, G., *et al.*, 1997, Regulation of the receptor specificity and function of the chemokine RANTES (regulated on activation, normal T cell expressed and secreted) by dipeptidyl peptidase IV (CD26)-mediated cleavage. *J.Exp.Med.* **186**: 1865-72.
60. Proost, P., Schutyser, E., Menten, P., *et al.*, 2001, Aminoterminal truncation of CXCR3 agonists impairs receptor signaling and lymphocyte chemotaxis, whilst preserving anti-angiogenic properties. *Blood* **98**: 3554-61.
61. Ludwig, A., Schiemann, F., Mentlein, R., *et al.*, 2002, Dipeptidyl peptidase IV (CD26) on T cells cleaves the CXC chemokine CXCL11 (I-TAC) and abolishes the stimulating but not the desensitizing potential of the chemokine. *J.Leukoc.Biol.* **72**: 183-91.
62. Delgado, M.B., Clark-Lewis, I., Loetscher, P., *et al.*, 2001, Rapid inactivation of stromal cell-derived factor-1 by cathepsin G associated with lymphocytes. *Eur.J.Immunol.* **31**: 699-707.

63. McQuibban, G.A., Butler, G.S., Gong, J.H., *et al.*, 2001, Matrix metalloproteinase activity inactivates the CXC chemokine stromal cell-derived factor-1. *J.Biol.Chem.* **276:** 43503-8.
64. Valenzuela-Fernandez, A., Planchenault, T., Baleux, F., *et al.*, 2002, Leukocyte elastase negatively regulates Stromal cell-derived factor-1 (SDF-1)/CXCR4 binding and functions by amino-terminal processing of SDF- 1 and CXCR4. *J.Biol.Chem.* **277:** 15677-89.
65. Schols, D., Proost, P., Struyf, S., *et al.*, 1998, CD26-processed RANTES(3-68), but not intact RANTES, has potent anti-HIV-1 activity. *Antiviral Res.* **39:** 175-87.
66. Simmons, G., Clapham, P.R., Picard, L., *et al.*, 1997, Potent inhibition of HIV-1 infectivity in macrophages and lymphocytes by a novel CCR5 antagonist. *Science* **276:** 276-9.
67. Shioda, T., Kato, H., Ohnishi, Y., *et al.*, 1998, Anti-HIV-1 and chemotactic activities of human stromal cell-derived factor 1alpha (SDF-1alpha) and SDF-1beta are abolished by CD26/dipeptidyl peptidase IV-mediated cleavage. *Proc.Natl.Acad.Sci.USA* **95:** 6331-6.
68. Proost, P., Struyf, S., Schols, D., *et al.*, 1998, Processing by CD26/dipeptidyl-peptidase IV reduces the chemotactic and anti-HIV-1 activity of stromal-cell-derived factor-1alpha. *FEBS Lett.* **432:** 73-6.
69. Struyf, S., Menten, P., Lenaerts, J.P., *et al.*, 2001, Diverging binding capacities of natural LD78beta isoforms of macrophage inflammatory protein-1alpha to the CC chemokine receptors 1, 3 and 5 affect their anti-HIV-1 activity and chemotactic potencies for neutrophils and eosinophils. *Eur.J.Immunol.* **31:** 2170-8.
70. Proost, P., Struyf, S., Schols, D., *et al.*, 1999, Truncation of macrophage-derived chemokine by CD26/dipeptidyl-peptidase IV beyond its predicted cleavage site affects chemotactic activity and CC chemokine receptor 4 interaction. *J.Biol.Chem.* **274:** 3988-93.

Structure-Function Relationship of DPP IV: Insights into its Dimerisation and Gelatinase Activity

OLIVER BAUM, WERNER REUTTER, and FELIX BERMPOHL
Institut für Molekularbiologie und Biochemie, Universitätsklinikum Benjamin Franklin, Freie Universität Berlin, Arnimallee 22, 14195 Berlin (Dahlem), Germany

1. INTRODUCTION

Dipeptidyl peptidase IV (DPP IV) which is also known as CD 26 is a serine peptidase with specificity for many bioactive hormones, chemokines and neuropeptides as glucagon-like peptide-1 (GLP), RANTES, and substance P[1]. As an integral ectoenzyme, DPP IV possesses five structural domains[2]. These comprise the only six amino acids long intracellular domain, the transmembrane domain and three extracellular domains, namely a carbohydrate-rich domain, a cysteine-rich domain and the C-terminal domain containing the catalytic triade. Furthermore, DPP IV is able to form homodimers.

For investigations of structure/function relationships, mutant DPP IV molecules derived from the rat strain Fischer 344 (supplied by a German distributor) as well as recombinant DPP IV proteins generated by site-directed mutagenesis of the gene were used. The molecular analyses have shown that even small alterations in each extracellular part of the primary structure of DPP IV profoundly affect the biological stability, enzyme activity and the ability to dimerise[3-7]. However, defined parts of the primary structure of DPP IV involved in the formation of homodimers were not identified so far.

DPP IV cleaves dipeptides from the N-terminus of peptides in which proline is the penultimate amino acid. However, peptides in which alanine or hydroxyproline are the penultimate amino acids are also degraded by DPP

IV but to a lower rate whereas peptides with proline as the antepenultimate amino acid represent no substrates for DPP IV.

When Hopsu-Havu and Glenner[8] discovered DPP IV, it already was considered that collagen might be digested by DPP IV since both proteins are widely co-distributed in tissues of mammals. Binding assays and cell adhesion assays also suggest that collagen is a substrate for DPP IV[9]. Furthermore, collagens contain a large number of prolyl residues which form preferred cleavage sites for DPP IV exopeptidase activity. However, since prolyl residues occur in collagens regularly at each third position, DPP IV would have to exhibit an endopeptidase activity apart from its exopeptidase activity to degrade collagen chains. Such a combination of exo- and endopeptidase activity has been demonstrated for the fibroblast activation protein-alpha (FAP) which reveals remarkable structural similarities to DPP IV[10]. However, no direct evidence has been obtained so far to support the hypothesis that collagen or gelatin (the denatured form of collagen) are substrates of DPP IV.

Therefore, we investigated whether DPP IV exhibits a gelatinase activity in addition to its well-characterised exopeptidase activity. Immunopurified, active DPP IV was subjected to both gelatin zymography and soluble proteolytic assays to demonstrate and, should the occasion arise, characterise such an endopeptidase activity. Furthermore, we performed histochemical studies on DPP IV-deficient Fischer rats in comparison to Wistar rats to search for differences in the collagen expression probably caused by the altered levels of DPP IV present in the tissues.

2. ELECTROPHORETIC PROPERTIES OF DPP IV ISOLATED FROM RAT KIDNEY

For the detection of its gelatinase activity, DPP IV isolated from rat kidney was used[11]. Therefore, rat kidneys were minced and homogenised. By centrifugation, a crude membrane fraction was obtained which was solubilised with Triton X-100 and, subsequently, subjected to both Concanavalin A (Con A)-lectin and immunoaffinity chromatography on mab 13.4-immoblized to protein A-sepharose. The final fraction was obtained by a pH-shift with diethylamine to pH 11.5. Eluted fractions were rapidly neutralised.

Each fraction of the purification procedure was analysed by SDS-PAGE under denaturing and reducing conditions and consecutive silver staining. In the eluate fraction of the immunoaffinity chromatography step, a 105 kDa band (denatured DPP IV) as well as a 60 kDa fragment were detected. N-terminal amino acid sequencing revealed that the 60 kDa protein represents a fragment of DPP IV starting at amino acid position 281 of the primary structure (Ile-Pro-Met-Gln-Ile). Therefore, the 60 kDa fragment starts immediately before the cysteine-rich domain begins and contains all presumed disulphide bridges. Interestingly, Iwaki-Egawa *et al.*[12] reported on the co-purification of an identical fragment during the isolation of DPP IV from rat kidneys. It is not clear if the 60 kDa DPP IV-fragment is produced by autocleavage or by the limited proteolytic activity of another peptidase. However, the generation of this fragment might have a (so far unknown) functional relevance.

All fractions obtained during the isolation procedure were analysed for DPP IV-activity (Tab. 1). In kidney, an enrichment factor of 160 with a recovery of almost 8% of the DPP IV activity was achieved. Simple calculations indicate that about 45 mU of the DPP IV activity correspond to 1 μg protein. Since 285 mU/mg protein were measured in the homogenate, DPP IV represents about 0.2% of all renal proteins. With respect to the limited localisation of DPP IV in kidney (glomeruli and brush border of proximal tubules), DPP IV should be present at these subcellular sites in high concentrations (at least 5% of all proteins as estimated). This seems to be a fairly high proportion, especially for an enzyme with so far not fully characterised functions in kidney.

Table 1. Purification of DPP IV from one rat kidney

Purification fraction	Total protein (mg)	Total activity (U)	DPP IV recovery (%)	Specific activity (mU/mg)	Enrichment factor
Homogenate	1170.0	333.6	100	285	1.0
Crude membranes	729.0	257.0	77	353	1.2
Solubilisate	555.0	219.8	66	396	1.4
Con A-eluate	35.6	162.4	49	4556	16.0
Immunoaffinity eluate	0.6	25.5	8	45365	160.0

DPP IV activity was determined with glycyl-prolyl-p-nitroanilid as substrate.

Eluted fractions of the immunoaffinity chromatography step were also used for the analysis of biochemical properties of DPP IV (Fig. 1). In electrophoresis under denaturing and reducing conditions DPP IV runs as 105 kDa and 60 kDa bands (corresponding to the complete molecule and the fragment). If the electrophoresis was performed under non-reducing and denaturing conditions, DPP IV was found as 110 kDa and 50 kDa bands. These results indicate that the destruction of the disulphide bridges changes the apparent molecular mass of both the complete DPP IV-molecule and the fragment only minimally as also suggested by chemical titration[7]. If the electrophoresis was performed under non-denaturing conditions (either in the presence or absence of reducing agents), a 150 kDa protein was observed instead of the 105 kDa protein which exhibited DPP IV-activity in an overlay assay with the coupling reagent Fast Garnet Blue. A 290 kDa band expressed also DPP IV-activity, whereas the 60 kDa fragment was inactive. How do we interpret these results? We suggest that the 150 kDa protein represents the active monomer and the 290 kDa protein the active dimer. The inactive dimer is demonstrable in SDS-PAGE only after chemical crosslinking with agents as disuccinimidylsuberate[13].

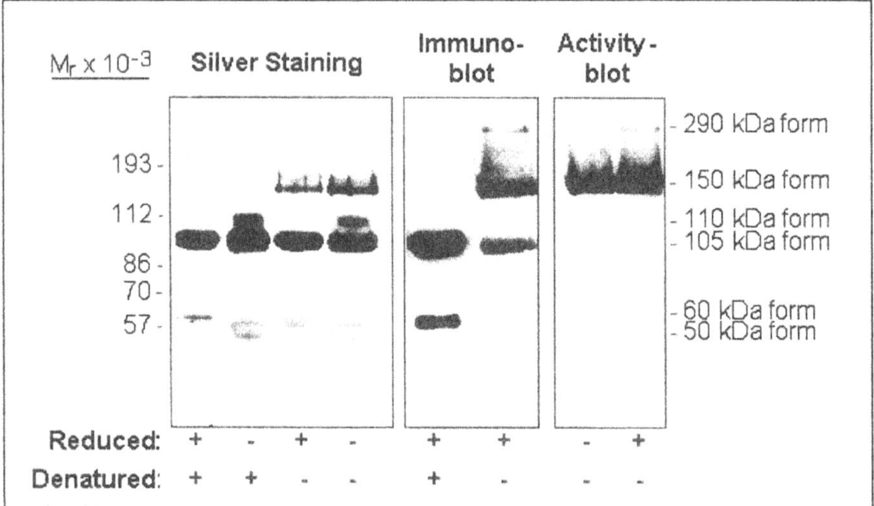

Figure 1. Characterisation of immunopurified DPP IV under various electrophoresis conditions (reduced/non-reduced; active/denatured). The electrophoretic separation was performed on a 7.5% SDS polyacrylamide gel. The immunoblot was developed with the mAb 13.4 against rat DPP IV and the enzyme activity blot with gly-pro-methoxy-beta-naphthylamid/Fast Garnet GBC salt.

Other studies[7,10] implicated without direct evidence that the 150 kDa protein observed in electrophoresis under non-denaturing conditions represents the active dimer of DPP IV. These studies propose no molecular mechanisms how a (pH-, temperature- and salt-labile) 150 kDa protein could bear two 105 kDa monomers and, additionally, cannot explain the existence of the 290 kDa band. We, therefore, prefer the explanation that the 150 kDa form turns after denaturation by an uncharacterised intramolecular unfolding into the apparently 45 kDa smaller denatured form and suggest that dimerisation and catalytic activity are two independent properties of DPP IV.

3. DPP IV EXHIBITS GELATINASE ACTIVITY

Immunoaffinity-purified DPP IV was subjected to gelatin zymography to directly demonstrate the gelatinase (and, correspondingly, endopeptidase) activity of DPP IV. Therefore, polyacrylamide gels were prepared containing co-polymerised heat-denatured collagen (gelatin). The subsequent electrophoresis of DPP IV was performed at 4°C to inhibit early activities of native DPP IV. The gels were then incubated at 37°C to allow DPP IV to degrade gelatin and, then, stained with Coomassie Blue (Fig. 2A). Negative staining of a transparent band in an otherwise stained gel revealed gelatinase activity of DPP IV since both the gelatin fragments and DPP IV had diffused out of the gel during the incubation while the collagen chains were still integrated in the gel. The zone of proteolytic activity became greater with increasing concentrations of DPP IV and incubation time.

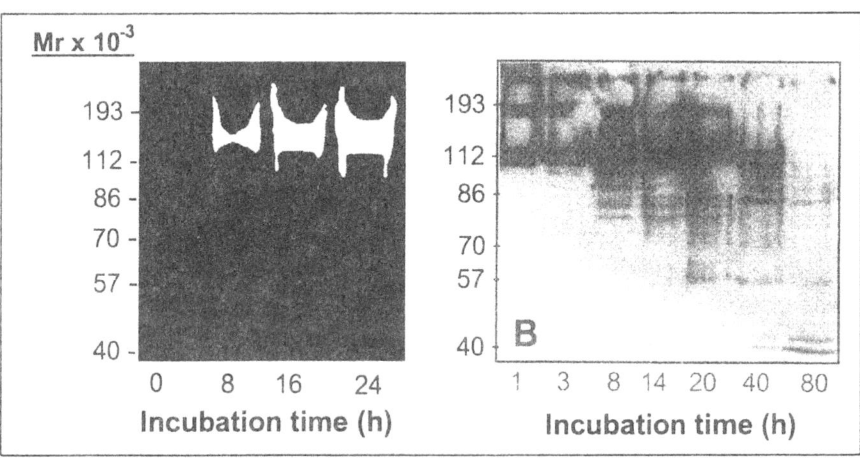

Figure 2. Demonstration of gelatinase activity of DPP IV in gelatin zymography (A) and the soluble proteolytic assay (B). A: 3.2 μg of DPP IV were incubated in a gelatin-containing polyacrylamide gel for time points as indicated. B: Immunoblot with anti-collagen-antibodies demonstrated the presence of collagen fragments generated by gelatinase activity incubated with native DPP IV for time points as indicated.

A soluble proteolytic assay was used to characterise the gelatinase activity of DPP IV[11]. For this purpose, immunopurified active DPP IV was incubated with denatured collagens at 37°C. Aliquots of the proteolytic mixture were separated by SDS-PAGE and stained with Coomassie Blue. Due to degradation of gelatin by DPP IV, the amount of detectable collagen gradually decreased during the incubation. Cleavage products were not detected with Coomassie Blue staining, but were visualised as multiple peptide bands in a stepladder pattern on immunoblots (Fig. 2B). These results suggest that DPP IV recognises multiple cleavage sites within the collagen chains.

The decrease of collagen alpha 1-chains was densitometrically quantified. 1 mg of DPP IV digested approximately 7 x 10^{-6} μmol denatured type I collagen per min, that is approximately 3 mg in 80 hours. This finding suggests that DPP IV is exhibiting a weak endopeptidase activity compared to other gelatinases or DPP IV exopeptidase activity. Optimal pH and temperature conditions were similar for exo- and endopeptidase activity. Experiments investigating the substrate specificity revealed, that DPP IV does not digest native collagen, albumin, fibronectin or the enzyme itself but only denatured collagens. Collagens types I, II, III and V were cleaved more rapidly than the basement membrane collagen type IV.

Experiments with peptidase inhibitors showed a similar inhibition profile for endo- and exopeptidase acitivity of DPP IV. Both endo- and exopeptidase activity were equally inhibited by the serine peptidase inhibitors DFP and PMSF as well as by the tripeptides Diprotin A and B. Diprotin A and B are regarded as competitive inhibitors for the DPP IV exopeptidase activity. This similar inhibition profile suggests that both activities of DPP IV reside within a single active site.

To test the biological relevance of our *in vitro* findings, we performed histochemical investigations on DPP IV-deficient Fischer rats in comparison to Wistar rats. Catalytic histochemistry revealed a high amount of active DPP IV in glomeruli and proximal tubules of the kidney as well as in liver and tendon in Wistar rats. As expected, DPP IV-activity was not detected in Fischer rats. The immunohistochemical analysis showed similar distribution patterns of collagen in kidney and tendon of both Fischer and Wistar rats (Fig. 3). In contrast, liver parenchym of Fischer rats contained fibrillar structures that were not seen in Wistar rats. These fibrillar structures presumably represent reticular fibers. We suggest that DPP IV-deficient Fischer rats accumulate collagen and its fragments in the space of Disse. This deficit might be compensated by other peptidases in renal and tendon tissue, where these fibrillar structures were not detected. However, similar histochemical investigations should be performed on substrains of Fischer rats expressing DPP IV (supplied e.g. by an American distributor) as controls to evaluate theses results.

4. CONCLUSION

We propose that the exopeptidase activity of DPP IV does not depend on its appearance as homodimer while other studies suggested a strong connection between dimerisation and catalytic activity[10]. Further investigations (e.g. crystallisation data) should help to clarify these apparently contradictory points of view.

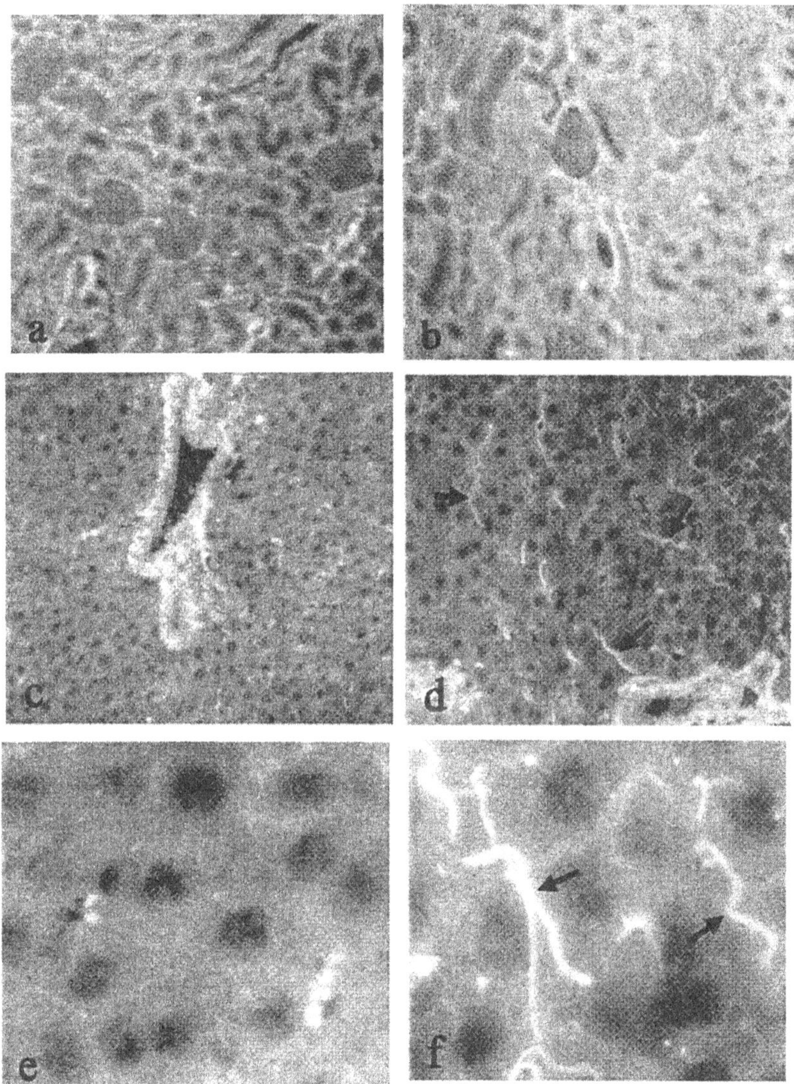

Figure 3. Localisation of collagen type-1 in Wistar (a,c,e) and Fischer 344 (b,d,f) rats in kidney (a,b) and liver (c-f). Immunohistochemistry was performed with polyclonal anti-collagen-antibodies (kindly provided by Prof. H.-J. Merker, Institut für Anatomie, Freie Universität Berlin) and Cy3-coupled secondary antibodies. Arrows in d and f mark collagenous fibrils present in the space of Disse in the hepatic parenchym. Magnification in a,b 10-fold, in c,d 20-fold and in e,f 100-fold.

A tripeptidase activity of DPP IV has not been described so far, and a contamination of the immunopurified fractions with other gelatinases seems rather unlikely (inhibitor profile; exact correspondence of the transparent areas in the zymograms with the molecular mass of active DPP IV). Therefore, the gelatinase activity described here should actually reflect the endopeptidase activity of DPP IV.

At this point, we can only speculate on the possible biological implications of the relatively low gelatinase activity of DPP IV with specificity for denatured fibrillar collagens endopeptidase activity. Collagenous tissue is usually not degraded by a single enzyme, but rather through a complex interaction of multiple peptidases, including matrixmetalloproteases, gelatinases, other serine proteases and cysteine proteases. Therefore, DPP IV might be involved as gelatinase in biological processes that demand complex patterns of collagen cleavage as during the absorption of nutrients as well as cell adhesion and translocation processes of cells, e.g., in wound healing, inflammation, tumor invasion and metastasis. This hypothesis is supported by recent findings of Ghersi *et al.*[14] found DPP IV and FAP to form protein complexes on the cell surface of fibroblasts that elicit both gelatin binding and gelatinase activities localised at invadopodia of cells migrating on collagenous fibers.

ACKNOWLEDGEMENTS

This work was supported by the Deutsche Forschungsgemeinschaft, Bonn (SFB 366, Teilprojekt C4) and the Sonnenfeld-Stiftung, Berlin. Helpful discussions with Dr. Klemens Löster and the excellent technical assistance of Werner Hofmann are gratefully acknowledged. We would also like to thank Dr. Christoph Weise (Insitut für Biochemie der FU Berlin) for N-terminal sequencing of the 60 kDa DPP IV-fragment. O.B. and F.B. contributed equally to this work.

REFERENCES

1. Mentlein, R., 1999, Dipeptidyl-peptidase IV (CD26)-role in the inactivation of regulatory peptides. *Regul. Pept.* **85**: 9-24.
2. Reutter, W., Baum, O., Löster, K., Fan, H., Bork, J.P., Bernt, K., Hanski, C., and Tauber, R., 1995, Functional aspects of the three extracellular domains of dipeptidyl peptidase IV: characterization of glycosylation events, of the collagen-binding site and of endopeptidase activity. In *Dipeptidyl peptidase IV (CD26) in metabolism and the immune response* (B. Fleischer, ed.), Springer-Verlag, Heidelberg, pp.55-78.

3. Fujiwara, T., Tsuji, E., Misumi, Y., Takami, N., and Ikehara, Y., 1992, Selective cell-surface expression of dipeptidyl peptidase IV with mutations at the active site sequence. *Biochem. Biophys. Res. Commun.* **185**: 776-784.
4. David, F., Baricault, L., Sapin, C., Gallet, X., Marguet, D., Thomas-Soumarmon, A., and Trugnan, G., 1996, Reduced cell surface expression of a mutated dipeptidyl peptidase IV (DPP IV/CD26) correlates with the generation of a beta strand in its C-terminal domain. *Biochem. Biophys. Res. Commun.* **222**: 833-838.
5. Fan, H., Meng, W., Kilian, C., Grams, S., and Reutter, W., 1997, Domain-specific N-glycosylation of the membrane glycoprotein dipeptidylpeptidase IV (CD26) influences its subcellular trafficking, biological stability, enzyme activity and protein folding. *Eur. J. Biochem.* **246**: 243-251.
6. Abbott, C.A., McCaughan, G.W., and Gorrell M.D., 1999, Two highly conserved glutamic acid residues in the predicted beta propeller domain of dipeptidyl peptidase IV are required for its enzyme activity. *FEBS Lett.* **458**: 278-284.
7. Dobers, J., Grams, S., Reutter, W., and Fan, H., 2000, Roles of cysteines in rat dipeptidyl peptidase IV/CD26 in processing and proteolytic activity. *Eur. J. Biochem.* **267**: 5093-5100.
8. Hopsu-Havu, V.K., and Glenner, G.G., 1966, A new naphthylamidase hydrolysing glycyl-prolyl-beta-naphthylamide. *Histochemie* **7**: 197-201.
9. Löster, K., Zeilinger, K., Schuppan, D., and Reutter, W., 1995b, The cysteine-rich region of dipeptidyl peptidase IV (CD26) is the collagen-binding site. *Biochem. Biophys. Res. Commun.* **217**: 341-348.
10. Pineiro-Sanchez, M.L., Goldstein, L.A., Dodt, J., Howard, L., Yeh, Y., and Chen, W.T., 1997, Identification of the 170-kDa melanoma membrane-bound gelatinase (seprase) as a serine integral membrane protease. *J. Biol. Chem.* **272:** 7595-7601.
11. Bermpohl, F., Löster, K., Reutter, W., and Baum O., 1998, Rat dipeptidyl peptidase IV (DPP IV) exhibits endopeptidase activity with specificity for denatured fibrillar collagens. *FEBS Lett.* **428**: 152-156.
12. Iwaki-Egawa, S., Watanabe, Y., and Fuimoto Y., 1993, N-terminal amino acid sequence of the 60-kD protein of rat kidney dipeptidyl peptidase IV. *Biol. Chem. Hoppe-Seyler* **374**: 973-975.
13. Löster, K., Baum, O., Hofmann, W., and Reutter, W., 1995a, Characterization of molecular aggregates of $\alpha 1\beta 1$-integrin and other rat liver membrane proteins by combination of size-exclusion chromatography and chemical cross-linking. *J. Chromat.* **711**: 187-199.
14. Ghersi, G., Dong, H., Goldstein, L.A., Yeh, Y., Hakkinen, L., Larjava, H.S., and Chen, W.T., 2002, Regulation of fibroblast migration on collagenous matrix by a cell surface peptidase complex. *J. Biol. Chem.* **277**: 29231-29241.

Exploration of the Active Site of Dipeptidyl Peptidase IV From *Porphyromonas gingivalis*

Comparison with the human enzyme

ANNE-MARIE LAMBEIR*, DEAN REA#, VILMOS FÜLÖP#, YUMI KUMAGAI§, KOEN AUGUSTYNS*, INGRID DE MEESTER*, ACHIEL HAEMERS*, and SIMON SCHARPÉ*

** Department of Pharmaceutical Sciences, University of Antwerp, Universiteitsplein 1, Wilrijk, Belgium; # Department of Biological Sciences, University of Warwick, Gibbet Hill Road, Coventry, UK; § Department of Microbiology, Nippon Dental University, Tokyo, Japan*

1. INTRODUCTION

Porphyromonas gingivalis is a pathogen associated with adult periodontitis. These microorganisms produce dipeptidyl-peptidase IV (DPP IV) and other extracellular proteases, which cause the destruction of periodontal tissue[1-3]. Mutant strains were constructed lacking DPP IV activity[3]. The decreased virulence of these mutants compared to the parent strains suggests that DPP IV is involved in the pathogenesis of *P. gingivalis*[1,3].

Approximately 30 % of the amino acids in the *P. gingivalis* DPP IV sequence are identical with their counterparts in human DPP IV[4]. This is sufficient to conclude that both enzymes have a similar overall structure and catalytic mechanism. However, there may be subtle differences that could be exploited for the development of selective inhibitors, or that could shed some light on the catalytic action of DPP IV in general. The biochemical characterization of the purified *P. gingivalis* DPP IV and the recombinant protein expressed in *E. coli* revealed significant similarity with the enzymatic properties of human DPP IV[2,3].

In this study we explored differences in selectivity mainly originating from interactions with amino acid residues preceding the scissile bond. For

Dipeptidyl Aminopeptidases in Health and Disease, Edited by Hildebrandt et al.
Kluwer Academic/Plenum Publishers, New York, 2003

this purpose we used dipeptide derived chromogenic substrates and a selection of inhibitors that were previously reported to interact with human DPP IV[5-7].

2. MATERIAL AND METHODS

The *p*-nitroanilide substrates Gly-Pro-*p*NA, Ala-Pro-*p*NA and Lys-Ala-*p*NA were purchased from Sigma or Bachem. Other substrates were synthesized following standard procedures. Inhibitors were synthesized as previously reported[5-7].

The plasmid with the *P. gingivalis* DPP IV gene was provided by Dr Y. Kumagai[3]. The extracellular part of the protein (starting at Gln20) was cloned in an expression vector and produced in *E. coli* as a fusion protein with an N-terminal histidine tag. The recombinant protein was purified by metal affinity chromatography following standard procedures. The protein concentration was determined from the absorbance at 278 nm using the calculated extinction coefficient, $A^{0.1\%} = 1.545$. The specific activity of the preparation was 7.5 U/mg.

Soluble human DPP IV was isolated from seminal plasma as described[8]. The specific activity was 35 U/mg.

One unit DPP IV is defined as the amount of enzyme that converts 1 μmole of substrate per minute in presence of 0.5 mM Gly-Pro-*p*NA and 50 mM Tris-buffer, pH 8.3 at 37 °C.

All kinetic experiments were performed in 50 mM Tris-buffer, pH 8.3, at 37 °C in a final volume of 200 μl using a Spectramax340 (Molecular Devices) microtiterplate reader. Kinetic parameters (k_{cat} and K_m) were determined by varying the substrate concentration between 10 μM and 10 mM and fitting the data with the Michaelis-Menten equation (using Grafit). The IC_{50} value is defined as the inhibitor concentration required for 50 % reduction of the initial rate measured at a substrate concentration equal to the K_m (i.e. 400 μM Gly-Pro-*p*NA for *P. gingivalis* DPP IV and 100 μM for human DPP IV). The IC_{50} was determined by varying the inhibitor concentration between $IC_{50}/10$ and $IC_{50}*10$ with a maximal concentration of 10 mM. The data were fitted with the following equation: $v_i/v_0 = 1 / \{1 + ([I]/IC_{50})^S\}$, where v_i and v_0 are the initial rates in presence and absence of inhibitor, [I] is the inhibitor concentration and s is the slope factor.

Progress curves of inactivation were determined by measuring the absorbance at 405 nm in function of time in presence of 400 μM Gly-Pro-*p*NA and fitting the data with the following equation:

$$A_t = A_0 + (v_0 - v_b)*(1 - \exp(-k_{obs}*t))/k_{obs} + v_b*t$$

Where A_t and A_0 are the absorbance at time t and 0, respectively, v_0 is the initial rate at time 0, v_b is the rate of spontaneous hydrolysis of the substrate and k_{obs} is the observed pseudo-first order rate constant for inhibitor binding. k_{obs} is a function of both the substrate and the inhibitor concentration. A plot of k_{obs} *versus* the inhibitor concentration yields a value for the apparent second order rate constant, which is dependent on the substrate concentration as the inhibitor and the substrate compete for the same binding site[9].

3. RESULTS AND DISCUSSION

Catalytic parameters were determined for *P. gingivalis* or human DPP IV and a series of Xaa-Pro-*p*NA and Xaa-Ala-*p*NA substrates. The selectivity of the substrates for both enzymes is represented in Figure 1 using k_{cat}/K_m as a relevant parameter.

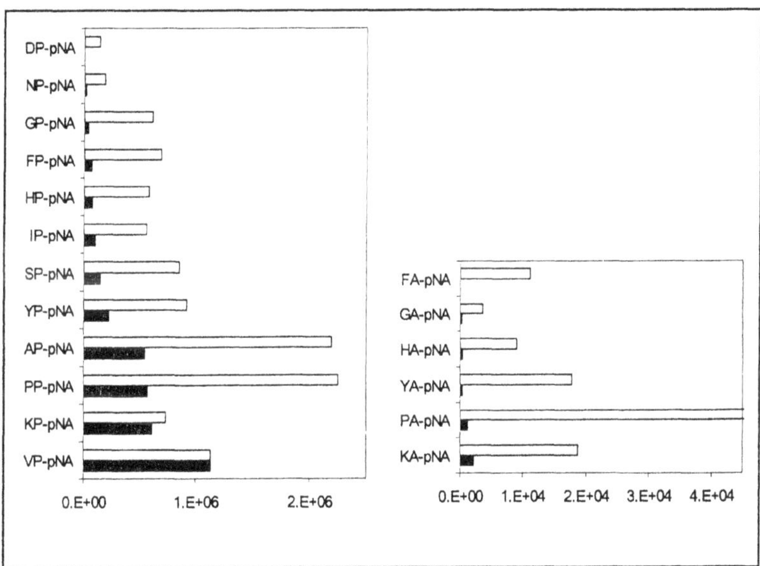

Figure 1. Differences in selectivity constants ($M^{-1}.s^{-1}$) between *P. gingivalis* (black bars) and human DPP IV (white bars) for dipeptide-derived chromogenic substrates.

Whereas the selectivity constants for *P. gingivalis* DPP IV are in the same range as those of the human enzyme, the effect of varying the P_2 residue is much larger. Moreover, the preferred P_2 residues are different, Val for the *P. gingivalis* DPP IV and Pro/Ala for the human enzyme. For the Xaa-Pro-*p*NA substrates the differences originate mainly from the K_m. In the case of the Xaa-Ala-*p*NA substrates, the k_{cat}/K_m values are 10 to 100 times lower for the bacterial than for the human enzyme. For this type of substrates the differences originate both from a reduction in k_{cat} and a relative increase of K_m. The preference for the P_2 residues appears to be independent of the type of P_1 residue (Pro or Ala).

In the past we reported on the synthesis and structure-activity relationships of pyrrolidides as DPP IV inhibitors[5,7]. These were also tested with *P. gingivalis* DPP IV. Because they lack a functional group interacting with the catalytic serine, they can be used to probe the S_1 and S_2 binding pockets. The results are shown in Table 1.

Table 1. Selectivity of reversible inhibitors for *P. gingivalis* DPP IV

Inhibitor	IC_{50} (μM)		Selectivity
	P. gingivalis	Human	
1-(L-Isoleucyl)pyrrolidine	27	4	0.15
1-(L-Isoleucyl)-3(R,S)-chloropyrrolidine	120	66	0.55
1-(L-Isoleucyl)-3(R,S)-hydroxypyrrolidine	420	93	0.22
1-(L-Isoleucyl)-3(R,S)-azidopyrrolidine	70	107	1.5
1-(L-Isoleucyl)-3-pyrroline	110	18	0.16
1-(L-Isoleucyl)-1,2,5,6-tetrahydropyridine	90	45	0.50
1-(L-Isoleucyl)piperidine	120	67	0.56
1-(L-Isoleucyl)hexamethyleneimine	260	370	1.4
1-(L-Valyl)pyrrolidine	29	4	0.14
1-(L-Lysyl)pyrrolidine	60	39	0.65
1-(L-Lysyl)piperidine	110	247	2.2
1-(L-Prolyl)pyrrolidine	260	15	0.06
1-(L-Tyrosyl)pyrrolidine	360	14	0.04
1-(L-Alanyl)pyrrolidine	380	41	0.11
1-(L-cyclohexylalanyl)pyrrolidine	420	17	0.04
1-(L-cyclohexylalanyl)piperidine	430	190	0.44
1-(L-Phenylalanyl)pyrrolidine	440	21	0.05

Substitution of the hydrogen atom on the 3-position of 1-(isoleucyl)pyrrolidine by –Cl, –OH, or $–N_3$ caused an increase in IC_{50} for both enzymes but the *P. gingivalis* DPP IV accepts the substituents relatively better. Similarly, replacing the pyrrolidine moiety with six and seven-membered rings was more easily accepted by the *P. gingivalis* than the human DPP IV. The preference for the P_2 residue is different for both enzymes. Within the group of compounds tested, 1-(lysyl)piperidine was the most discriminating compound in favour of *P. gingivalis* DPP IV.

Dipeptide-derived diaryl phosphonate esters cause inactivation of DPP IV by forming a covalent adduct with the catalytic serine[6]. The rate of inactivation is influenced by the interactions with the enzyme, activating the phosphorous atom, and the electron withdrawing properties of the substituted phenyl groups[9]. The kinetics of inactivation of *P. gingivalis* DPP IV were determined for bis(4-acetamidophenyl) 1-(*(S)*-prolyl) pyrrolidine-

2*(R,S)*-phosphonate (AB192) and bis{4-[(ethoxycarbonyl)-methylamino-carbonyl]-phenyl} 1-(*(S)*-prolyl)pyrrolidine-2*(R,S)*-phosphonate (AB207).

The results are shown in Figure 2.

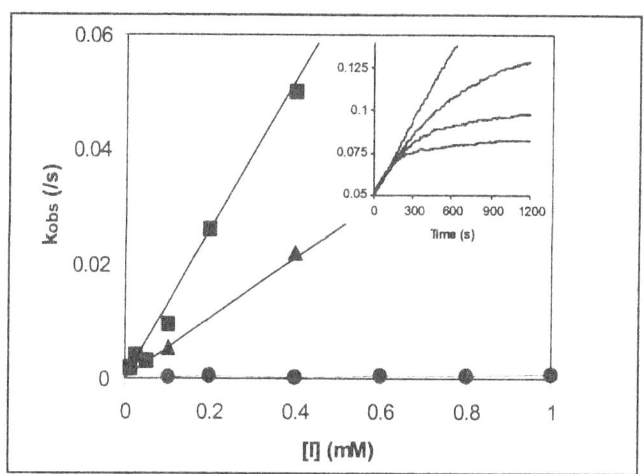

Figure 2. Rate constants for binding of AB192 (circles) and AB207 (squares) to P. gingivalis DPP IV and AB192 to the human enzyme (triangles). Inset: progress curves for AB207 at 0, 12.5, 25 and 100 μM in presence of 400 μM Gly-Pro-pNA.

The apparent second order rate constants obtained from this experiment (0.6 and 130 $M^{-1}.s^{-1}$) are at least 100 times lower than the values obtained with human DPP IV measured in parallel. These differences do not follow any of the catalytic parameters of Pro-Pro-*p*NA since the k_{cat} of the *P. gingivalis* enzyme is similar to that of human DPP IV and the K_m is only 3 times higher. The effect appears to be caused by differences in the active sites of the enzymes since the correlation with the electron withdrawing properties of the substituents is maintained.

In conclusion: This comparative study revealed both subtle and striking differences between two highly homologous enzymes. The results may provide a starting point for the development of selective inhibitors. They also raise some questions concerning the catalytic machinery of DPP IV that deserve more attention.

ACKNOWLEDGEMENTS

This work was supported by a research grant from the University of Antwerp and by the National Fund for Scientific Research Flanders. We thank Nicole Lamoen for her technical assistance.

REFERENCES

1. Yagishita, H., Kumagai, Y., Konishi, K., Takahashi, Y., Aoba, T., and Yoshikawa, M., 2001, Histopathological studies on virulence of dipeptidyl aminopeptidase IV (DPPIV) of Porphyromonas gingivalis in a mouse abscess model: use of a DPPIV-deficient mutant. *Infect. Immun.* **69**:7159-7161.
2. Banbula, A., Bugno, M., Goldstein, J., Yen, J., Nelson, D., Travis, J., and Potempa, J., 2000, Emerging family of proline-specific peptidases of Porphyromonas gingivalis: purification and characterization of serine dipeptidyl peptidase, a structural and functional homologue of mammalian prolyl dipeptidyl peptidase IV. *Infect. Immun.* **68**:1176-1182.
3. Kumagai, Y., Konishi, K., Gomi, T., Yagishita, H., Yajima, A., and Yoshikawa, M., 2000, Enzymatic properties of dipeptidyl aminopeptidase IV produced by the periodontal pathogen Porphyromonas gingivalis and its participation in virulence. *Infect. Immun.* **68**:716-724.
4. Kiyama, M., Hayakawa, M., Shiroza, T., Nakamura, S., Takeuchi, A., Masamoto, Y., and Abiko, Y., 1998, Sequence analysis of the Porphyromonas gingivalis dipeptidyl peptidase IV gene. *Biochim. Biophys. Acta.* **1396**:39-46.
5. Augustyns, K.J.L., Lambeir, A.M., Borloo, M., De Meester, I., Vedernikova, I., Vanhoof, G., Hendriks, D., Scharpé, S., and Haemers, A., 1997, Pyrrolidides: synthesis and structure-activity relationships as inhibitors of dipeptidyl peptidase IV. *Eur. J. Med. Chem.* **32**:301-309.
6. Belyaev, A., Zhang, X., Augustyns, K., Lambeir, A.M., De Meester, I., Vedernikova, I., Scharpé, S., and Haemers, A., 1999, Structure-activity relationship of diaryl phosphonate esters as potent irreversible dipeptidyl peptidase IV inhibitors. *J. Med. Chem.* **42**:1041-1052.
7. Senten, K., Van der Veken, P., Bal, G., De Meester, I., Lambeir, A.M., Scharpé, S., Bauvois, B., Haemers, A., and Augustyns, K., 2002, Development of potent and selective dipeptidyl peptidase II inhibitors. *Bioorg. Med. Chem. Lett.*, in press.
8. De Meester, I., Vanhoof, G., Lambeir, A.M., and Scharpé, S., 1996, Use of immobilized adenosine deaminase (EC 3.5.4.4) for the rapid purification of native human CD26/dipeptidyl peptidase IV (EC 3.4.14.5). *J. Immunol. Methods.* **189**:99-105.
9. Lambeir, A.M., Borloo, M., De Meester, I., Belyaev, A., Augustyns, K., Hendriks, D., Scharpé, S., and Haemers, A., 1996, Dipeptide-derived diphenyl phosphonate esters: mechanism-based inhibitors of dipeptidyl peptidase IV. *Biochim. Biophys. Acta* **1290**:76-82.
10. Fujiwara, T., Tsuji, E., Misumi, Y., Takami, N., and Ikehara, Y., 1992, Selective cell-surface expression of dipeptidyl peptidase IV with mutations at the active site sequence. *Biochem. Biophys. Res. Commun.* **185**: 776-784.
11. Hopsu-Havu, V.K., and Glenner, G.G., 1966, A new naphthylamidase hydrolysing glycyl-prolyl-beta-naphthylamide. *Histochemie* **7**: 197-201.
12. Ghersi, G., Dong, H., Goldstein, L.A., Yeh, Y., Hakkinen, L., Larjava, H.S., and Chen, W.T., 2002, Regulation of fibroblast migration on collagenous matrix by a cell surface peptidase complex. *J. Biol. Chem.* **277**: 29231-29241.
13. Iwaki-Egawa, S., Watanabe, Y., and Fuimoto Y., 1993, N-terminal amino acid sequence of the 60-kD protein of rat kidney dipeptidyl peptidase IV. *Biol. Chem. Hoppe-Seyler* **374**: 973-975.
14. Löster, K., Baum, O., Hofmann, W., and Reutter, W., 1995a, Characterization of molecular aggregates of $\alpha 1\beta 1$-integrin and other rat liver membrane proteins by

combination of size-exclusion chromatography and chemical cross-linking. *J. Chromat.* **711**: 187-199.

15. Löster, K., Zeilinger, K., Schuppan, D., and Reutter, W., 1995b, The cysteine-rich region of dipeptidyl peptidase IV (CD26) is the collagen-binding site. *Biochem. Biophys. Res. Commun.* **217**: 341-348.
16. Mentlein, R., 1999, Dipeptidyl-peptidase IV (CD26)-role in the inactivation of regulatory peptides. *Regul. Pept.* **85**: 9-24.
17. Pineiro-Sanchez, M.L., Goldstein, L.A., Dodt, J., Howard, L., Yeh, Y., and Chen, W.T., 1997, Identification of the 170-kDa melanoma membrane-bound gelatinase (seprase) as a serine integral membrane protease. *J. Biol. Chem.* **272:** 7595-7601.
18. Reutter, W., Baum, O., Löster, K., Fan, H., Bork, J.P., Bernt, K., Hanski, C., and Tauber, R., 1995, Functional aspects of the three extracellular domains of dipeptidyl peptidase IV: characterization of glycosylation events, of the collagen-binding site and of endopeptidase activity. In *Dipeptidyl peptidase IV (CD26) in metabolism and the immune response* (B. Fleischer, ed.), Springer-Verlag, Heidelberg, pp.55-78.

Modification of the Biological Activity of Chemokines by Dipeptidyl Peptidase IV – a Side Effect in the Use of Inhibitors?

ROLF MENTLEIN*, FLORIAN SCHIEMANN#, ANDREAS LUDWIG#, and ERNST BRANDT#
*Department of Anatomy, University of Kiel, Olshausenstr. 40, 24098 Kiel, Germany; #Research Center Borstel, Parkalle 22, 23845 Borstel, Germany

1. INTRODUCTION

Dipeptidyl peptidase IV (DPP IV) is a multifunctional enzyme as well as a binding protein or co-activator[1-3]. As a regulatory protease, it can cleave and thereby inactivate or modify the activity of various biologically active peptides[4-6]. This includes the

Inactivation of circulating peptide hormones:
Termination of incretin action (GLP-1, GIP) in vitro and in vivo, and that of other hormones (NPY, PYY, GRH) in vitro[7-10].

Inactivation of neuropeptides:
Differential inactivation of NPY in vitro[7]
Inactivation of chemokines:
Termination of chemotaxis for leukocytes in vitro (RANTES[11,12], SDF-1[13,14], eotaxin[15], I-TAC[16], MDC[17], LD78beta[17])

However, most of these effects have only been shown in vitro and their physiological significance remains questionable. The exception is the inactivation of incretins, in particular of GLP-1. Here, the pivotal role of

Dipeptidyl Aminopeptidases in Health and Disease, Edited by Hildebrandt et al.
Kluwer Academic/Plenum Publishers, New York, 2003

DPP IV has been clearly documented in vivo, and DPP IV inhibitors as anti-diabetic drugs for humans are under development.

To evaluate or understand potential side effects of DPP IV inhibitors used as drugs, it is important to know what relevance have other DPP IV targets described so far. We addressed this question on chemokines, a group of chemotactic cytokines, several of which are DPP IV substrates.

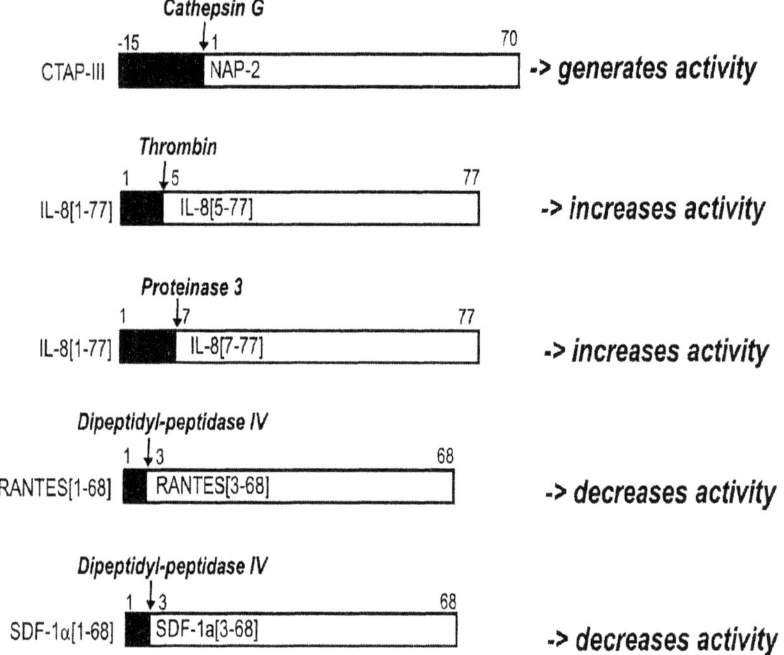

Figure 1. N-terminal cleavage of chemokines modulates their biologic activity. The cleavage of chemokines by several proteases is schematically depicted. N-terminal truncation may result in an activation or inactivation depending on the chemokine.

1.1 Chemokines as protease substrates

Regulation of the biological activity of chemokines by N-terminal proteolytic truncation is not a phenomenon restricted to DPP IV. Other proteases like cathepsin G[19], thrombin[20] or proteinase-3[21] have been shown to generate or increase the biological activity of some chemokines (Fig. 1). But is DPP IV a physiological regulator of chemokine activity of similar importance? And furthermore, would DPP IV-inhibitors strongly affect chemokine functions?

2. I-TAC AS A MODEL FOR DPP IV TRUNCATION OF CHEMOKINES

To investigate the importance of DPP IV as a physiological regulator of chemokine activity, we chose I-TAC (interferon-γ-inducible T cell α-chemoattractant; CXCL11) as a model substrate. I-TAC is a member of the CXC chemokine subfamily. It is expressed upon stimulation with INF-γ in a variety of cells, including monocytes, neutrophils, fibroblasts and especially endothelial cells. I-TAC shares its receptor CXCR3, a G protein-coupled 7-TMD-receptor, with the chemokines IP-10 (CXCL10) and Mig (CXCL9). I-TAC induces chemotaxis in CXCR3-expressing cells, e.g. natural killer cells and activated T cells.

Two main features of the chemokine argued for using I-TAC: first, it turned out that I-TAC, as compared to most other chemokines, is a relatively good substrate for DPP IV[22]. Second, and most important, the major target cells for I-TAC, namely activated T cells, co-express I-TAC-receptor CXCR3 and DPP IV. These are optimal conditions which should allow to answer the question: Do DPP IV-expressing cells exhibit sufficient enzymatic activity to affect the I-TAC-induced cellular response? And, have DPP IV inhibitors direct effect on I-TAC-induced chemotaxis?

3. DPP IV-MEDIATED CLEAVAGE OF I-TAC

I-TAC consists of 73 amino acids and has a proline in the N-terminal penultimate position (Fig 2) – it has therefore been considered as a potential substrate for DPP IV.

Indeed, co-incubation of DPP IV purified form human placenta and I-TAC leads to the formation of a degraded molecule lacking the N-terminal dipeptide FP as shown by mass spectrometrical analysis. No further cleavage was observed[16]. Moreover, it turned out that I-TAC represents an even better substrate than the two other chemokines that target the CXCR3 receptor, IP-10 and Mig. These molecules, having also a penultimate proline, were cleaved with considerably lower rates[22].

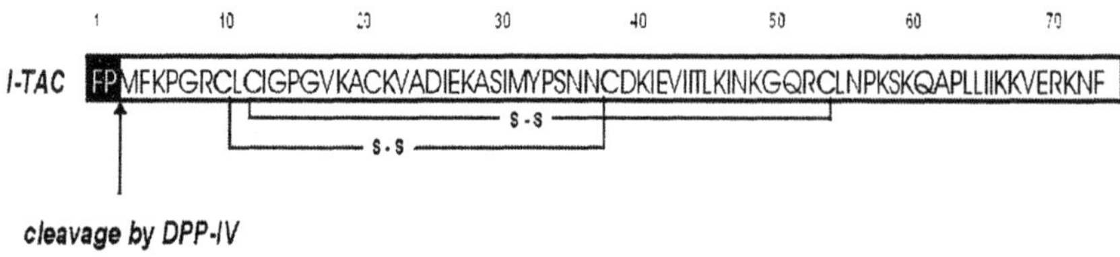

Figure 2. Representation of the primary structure of I-TAC, including the positions of disulfide bridges. The cleavage site for DPP-IV and the N-terminal residues becoming removed are indicated.

These observations raised the question whether DPP IV expressed on activated T cells would also efficiently cleave I-TAC.

To elucidate this, we incubated these cells with I-TAC in the absence or presence of DPP IV inhibitors. Indeed, cleavage resulted in the formation of N-terminally truncated I-TAC as found with purified DPP IV. No proteolytic cleavage products other than I-TAC truncated by the N-terminal dipeptide FP were observed. Moreover, degradation of I-TAC by activated T cells was substantially inhibited by specific DPP IV inhibitors, e. g. Lys-pyrollidide. This experiment shows that activated T cells are not only targets for I-TAC, but can cleave this chemokine.

4. MODIFICATION OF THE BIOLOGICAL ACTIVITY OF I-TAC BY DPP IV

In the following experiments we investigated whether DPP IV-mediated cleavage of I-TAC would modulate the chemokine's activity towards activated T cells. First, we examined potential modulation of its binding activity.

As seen in competition binding assays using intact iodine-125-labeled I-TAC as a tracer, both the intact as well as the DPP IV-truncated chemokine were able to displace labeled I-TAC from its specific binding sites on activated T cells. However, the binding potency of truncated I-TAC was about 8-fold lower than that of the full-size chemokine. This experiment shows that I-TAC binding activity is significantly decreased by DPP IV truncation.

To confirm and extend this observation, we investigated whether truncated I-TAC would also affect receptor dynamics. Upon ligand binding chemokine receptors rapidly undergo internalization. Thus, we measured the surface expression of CXCR3 by flow cytometric analysis on activated T cells after 30 min stimulation with the I-TAC variants. Interestingly, we observed that truncated I-TAC had an about 8-fold reduced potency to induce CXCR3 downregulation from the cell surface (Fig.3). These results indicated that it might also have a reduced ability to affect biological activity.

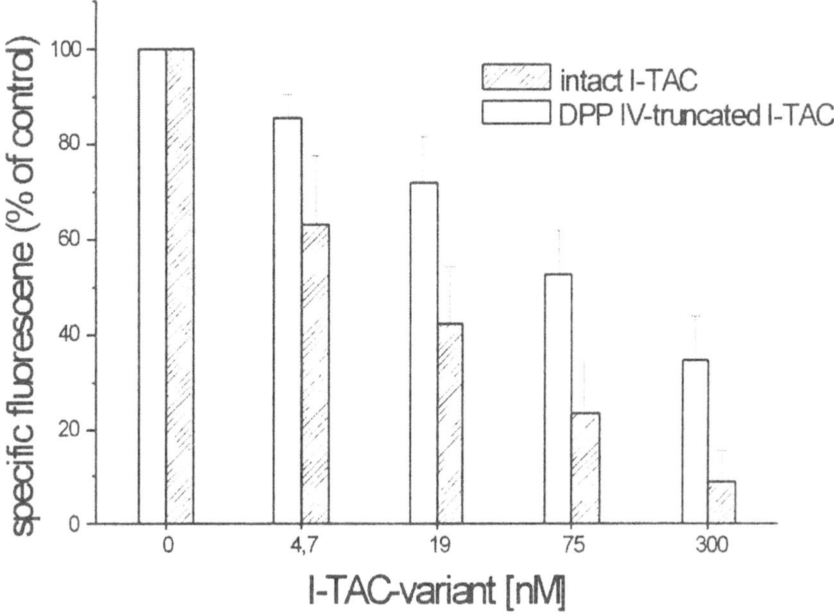

Figure 3. Downregulation of CXCR3 surface expression on activated human T cells (stimulated with 5 µg/ml phytohemagglutinin (PHA) and 100 U/ml interleukin-2 for 3 days) by I-TAC and DPP IV-truncated I-TAC (flow cytometric analysis after 30 min exposure at 37°C). DPP IV cleavage reduces the potency of the chemokine about 8-fold.

To examine this, we checked two prominent biological functions that become elicited by I-TAC in activated T cells, (i) the induction of a transient increase in intracellular calcium, that can be measured by an increase from the fluorescence an intracellular label, Fura-2 that complexes calcium, and (ii) the activation of the cells to perform chemotactic migration in response to the chemokine.

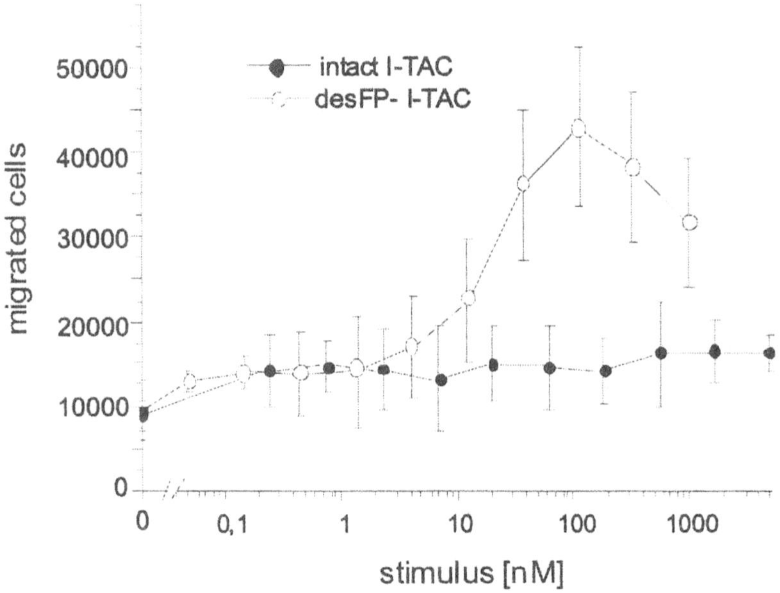

Figure 4. Chemotaxis of activated human T cells (stimulated with 5 μg/ml phytohemagglutinin (PHA) and 100 U/ml interleukin-2 for 3 days) by I-TAC and DPP IV-truncated I-TAC (Boyden chamber assays). DPP IV cleavage abolishes the chemotactic activity of the chemokine.

Whereas we found full size I-TAC to be able to elevate intracellular calcium in a dose-dependent manner, DPP IV-truncated I-TAC was ineffective even at concentrations up to 16-fold higher than that required for a minimal effect with the intact chemokine. Corresponding results were found for chemotactic activity of the I-TAC variants, as measured in a Boyden chamber-assay (Fig.4). Here, an even about 500-fold excess of truncated I-TAC over that of the full-size chemokine was not sufficient to stimulate a minimal chemotactic response in activated T cells. As these experiments clearly show, cleavage by DPP IV abolishes the capability of I-TAC to stimulate cell functions.

However, the latter results appeared inconsistent with respect to the observation that truncated I-TAC was still able to interact with its receptor but did not stimulate biological activity. As it is known from many examples, receptor binding and downregulation does not necessarily correlate with cell activation but may also result in functional desensitization of the target cell.

To exploit this possibility we examined whether truncated I-TAC would desensitize the T cell migratory response. For this we measured chemotaxis of activated T cells exposed to increasing concentrations of truncated I-TAC towards a fixed concentration of intact I-TAC. Under these conditions a 30-fold excess of truncated I-TAC was sufficient to completely abrogate the chemotactic response towards the full-size molecule, demonstrating that truncated I-TAC may act as a desensitizing agent.

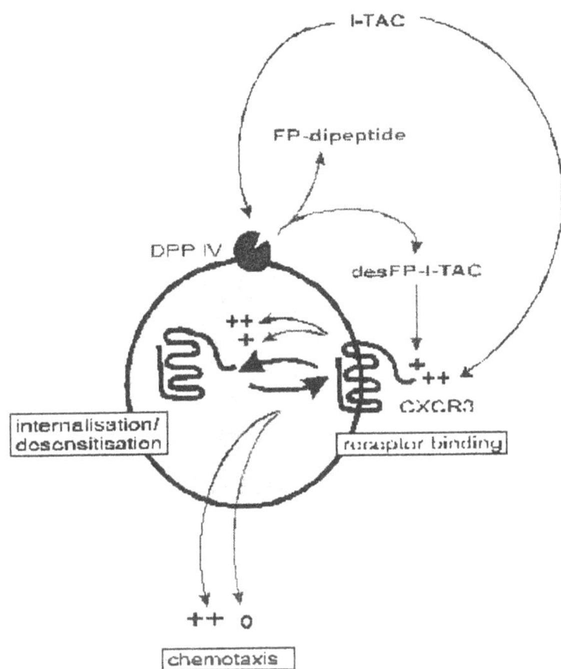

Figure 5. Model for T cell mediated cleavage of I-TAC through DPP IV and its functional consequences. ++: strong effect, + weak effect, 0: no effect.

The above results taken together demonstate, that DPP IV (Fig. 5)

(1) in its purified as well as in its T cell-associated form can cleave I-TAC,

(2) converts I-TAC into a truncated molecule that exhibits significantly reduced capacity for receptor binding and internalization.

(3) cleaves I-TAC to generate a molecule without any chemotactic activity for T cells, but that has still the capacity to functionally desensitize the cells for activation through the intact chemokine.

5. EFFECT OF DPP IV INHIBITORS ON I-TAC-INDUCED CHEMOTAXIS

The crucial question is now: Is the activity of DPP IV on T cells strong enough to degrade I-TAC during its chemotactic action? In other words: Do DPP IV inhibitors affect the I-TAC-induced chemotaxis of activated T cells?

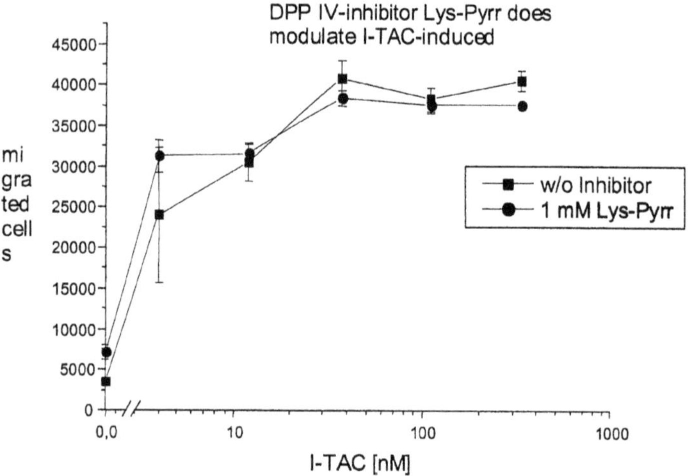

Figure 6. Chemotaxis of activated human T cells (stimulated with 5 μg/ml phytohemagglutinin (PHA) and 100 U/ml interleukin-2 for 3 days) by I-TAC is not changed by the presence of the DPP IV inhibitor Lys-pyrollidide. DPP IV cleavage is not rapid enough to disturb the chemotactic activity of I-TAC.

The answer shown in Fig. 6 is clearly: no. Chemotaxis assays at various concentrations were not disturbed by the addition of Lys-pyrollidide a specific and - at the concentration applied – a 100% inhibitor of DPP IV. Truncation of I-TAC – and probably – other chemokines by DPP IV appear to be long term effects: They may only be active where an accumulation of chemokines is responsible for their effects.

6. CONCLUSION

We can summarize our investigation with respect to the initial question:

Despite the receptor and the inactivation enzyme are directly neighboured, DPP IV does not affect immediate chemotactic effects. Since I-TAC is a relatively good chemokine substrate, this conclusion should be valid also for other chemokines. Furthermore, many chemokines are redundant, and not all of them are substrates for DPP IV.

Therefore, from the pharmacological view we think that DPP IV inhibitors have only little or no side effects on chemokines. Such side effects would be the prolongation or increase of inflammatory responses which have, to our knowledge, not yet been reported after in vivo applications in humans.

ACKNOWLEDGEMENTS

This experimental work in this study was supported by the Hensel Foundation of the University of Kiel, Germany.

REFERENCES

1. Gorrell, M. D., Gysbers, G., and McCaughan, G. W., 2001, CD26: A multifunctional integral membrane and secreted protein of activated lymphocytes. *Scand. J. Immunol.* **54**: 249-264.
2. Audustyns, K., Bal, G., Thonus, G., Belyaev, A., Zhang, X. M., Bollaert, W., Lambeir, A. M., Durinx, C., Goossens, F., and Haemers, A., 1999, The unique properties of dipeptidyl peptidase IV (DPP IV / CD26) and the therapeutic potential of DPP IV inhibitors. *Current Med. Chem.* **6**: 311-327.
3. Kähne, T., Lendeckel, U., Wrenger, S., Neubert, K., Ansorge, S., and Reinhold, D., 1999, Dipeptidyl peptidase IV: A cell surface peptidase involved in regulating T cell growth. *Int. J. Mol. Med.* **4**: 3-15.
4. Mentlein, R., 1999, Dipeptidyl-peptidase IV (CD26) – role in the inactivation of regulatory peptides. *Regul. Peptides* **85**: 9-24.
5. De Meester, I., Korom, S., Van Damme, J., and Scharpé, S., 1999, CD26, let it cut or cut it down. *Immunoloy Today* **20**: 367-375.
6. Scharpé, S., and De Meester, I., 2001, Peptide truncation by dipeptidyl peptidase IV: A new pathway for drug discovery? *Verhandelingen van de Koninklijke Academie voor Geneeskunde van Belgie* **LXIII**: 5-33.
7. Mentlein, R., Dahms, P., Grandt, D., and Krüger, R., 1993, Proteolytic processing of neuropeptide Y and peptide YY by dipeptidyl peptidase IV. *Regul. Peptides* 49: 133-144.
8. Mentlein, R., Gallwitz, B., and Schmidt, W. E., 1993, Dipeptidyl peptidase IV hydrolyses gastric inhibitory polypeptide, glucagon-like peptide-1(7-36) amide, peptide histidine methionine and is responsible for their degradation in human serum. *Eur. J. Biochem.* 214: 829-835.

9. Deacon, C. F., Hughes, T. E., and Holst, J. J., 1998, Dipeptidyl peptidase IV inhibition potentiates the insulinotropic effect of glucagon-like peptide 1 in the anesthetized pig. *Diabetes* **47**: 764-769.
10. Pederson, R. A., White, H. A., Schlenzig, D., Pauly, R. P., McIntosh, C. H. S., and Demuth, H.-U., 1998, Improved tolerance in Zucker fatty rats by oral administration of the dipeptidyl peptidase IV inhibitor isoleucine thiazolidide. *Diabetes* **47**: 1253-1258.
11. Oravecz, T., Pall, M., Rodriquez, G., Gorrell, M. D., Ditto, M., Nguyen, N. Y., Boykins, R., Unsworth, E., and Norcross, M. A. 1997, Regulation of the receptor specificty and function of the chemokine RANTES (regulated on activation, normal T cell expressed and secreted) by dipeptidyl peptidase IV (CD26)-mediated cleavage. *J. Exp. Med.* **186**: 1865-1872.
12. Proost, P., De Meester, I., Scholz, D., Struyf, S., Lambeir, A.M., Wuyts, A., Opdenakker, G., De Clerq, E., Scharpe, S., Van Damme, J., 1998, Aminoterminal truncation of chemokines by CD26/dipeptidyl-peptidase IV. Conversion of RANTES into a potent inhibitor of monocyte chemotaxis and HIV infection. *J. Biol. Chem.* **273**: 7222-7227.
13. Proost, P., Struyf, S., Schols, D., Durinx, C., Wuyts, A., Lenaerts, J. P., De Clerq, E., De Meester, I., and Van Damme, J., 1998, Processing by CD26/dipeptidyl-peptidase IV reduces the chemotactic and anti-HIV-1 activity of stromal-cell-derived factor-1alpha. *FEBS Lett.* **432**: 73-76.
14. Shioda, T., Kato, H., Ohnishi, Y., Tashiro, K., Ikegawa, M., Nakayama, E.E., Hu, H., Kato, A., Sakai, Y., Liu, H., Honjo, T., Nomot, A., Iwamoto,A., Morimoto, C., Nagai, Y., 1998, Anti-HIV-1 and chemotactic activities of human stromal cell-derived factor 1alpha (SDF-1alpha) and SDF-beta are abolished by CD26/dipeptidyl peptidase IV-mediated cleavage. *Proc. Natl. Acad. Sci.* USA **95**: 6331-6336.
15. Struyf, S., Proost, P., Schols, D., De Clerq, E., Opdenakker, G.,. Lenaerts, J.-P., Detheux, M., Parmentier, M., De Meester, I., Scharpé, S., Van Damme, J., 1999, CD26/Dipeptidyl-peptidase IV down regulates the eosinophil chemotactic potency, but not the anti-HIV activity of human eotaxin by affecting its interaction with CC chemokine receptor 3. *J. Immunol.* **162**: 4903-4909.
16. Ludwig, A., Schiemann, F., Mentlein, R., Lindner, B., Brandt, E., 2002, Dipeptidyl peptidase IV (CD26) on T cells cleaves the CXC chemokine CXCL11 (I-TAC) and abolishes the stimulating but not the desensitizing potential of the chemokine. *J. Leukoc. Biol.* **72**: 183-191
17. Proost, P., Struyf, S., Schols, D., Opdenakker, G., Sozzani, S., Allavena, P., Mantovani, A., Augustyns, K., Bal, G., Haemers, A., Lambeir, A.M., Scharpe, S., Van Damme, J. De Meester, I., 1999, Truncation of macrophage-derived chemokine by CD26/dipeptidyl peptidase IV beyond its predicted cleavage site affects chemotactic activity and CC chemokine receptor 4 interaction. *J. Biol. Chem.* **274**: 3988-3993.
18. Proost, P., Menten, P., Struyf, S., Schutyser, E., De Meester, I., and Van Damme, J., 2000, Cleavage by CD26/dipeptidyl peptidase IV converts the chemokine LD78beta into a most efficient monocyte attractant and CCR1 agonist. *Blood* **96**: 1674-1680.
19. Brandt, E., Van Damme, J., Flad, H.D., 1991, Neutrophils can generate their activator neutrophil-activating peptide 2 by proteolytic cleavage of platelet-derived connective tissue-activating peptide III. *Cytokine* **3**: 311-321.
20. Hebert, C. A., Luscinskas, F. W., Kiely, J. M., Luis, E. A., Darbonne, W. C., Bennett, G. L., Liu, C. C., Obin, M. S., Gimbrone, M. A. Jr., Baker, J. B., 1990, Endothelial and leukocyte forms of IL-8. Conversion by thrombin and interactions with neutrophils. *J. Immunol.* **145**: 3033-3040.

21. Padrines, M.., Wolf, M.., Walz, A.., Baggiolini, M., 1994, Interleukin-8 processing by neutrophil elastase, cathepsin G and proteinase-3. *FEBS Lett.* **352**: 231-235.
22. Lambeir, A. M., Proost, P., Durinx, C., Bal, G., Senten, K., Augustyns, K., Scharpé, S., Van Damme, J., De Meester, I., 2001, Kinetic investigation of chemokine truncation by CD26/dipeptidyl peptidase IV reveals a striking selectivity within the chemokine family. *J. Biol. Chem.* **10**: 29839-29845.

Molecular Chimeras and Mutational Analysis in the Prolyl Oligopeptidase Gene Family

KATERINA AJAMI, CATHERINE A. ABBOTT[1], VANESSA GYSBERS, THILO KÄHNE[2], GEOFFREY W. MCCAUGHAN, and MARK D. GORRELL
A.W. Morrow Gastroenterology and Liver Centre, Royal Prince Alfred Hospital, Centenary Institute of Cancer Medicine and Cell Biology and The University of Sydney, New South Wales, Australia. [2]*Research Center of Immunology, Institute of Experimental Internal Medicine, University of Magdeburg, Magdeburg, Germany.*

[1] Present address: *School of Biological Sciences, Flinders University of South Australia, Adelaide, Australia.*

1. INTRODUCTION

CD26/DPIV (E. C. 3.4.14.5), DP8, DP9, fibroblast activation protein (FAP), DP-like1 (DPL1/DP6/DPX) and DPL2 form the CD26 gene family and those with DP activity (CD26, DP8, DP9 and FAP) form the S9b peptidase family[1-4]. Members of this family exhibit 27% to 60% amino acid identity with each other. The three dimensional structure of the related S9a peptidase prolyl oligopeptidase (POP; E. C. 3.4.21.26; Protein Data Base code 1qfm) has been solved[5] and has been used as a template for a model of CD26[1,3]. This model predicts that the extracellular 738 residues of this 766-residue protein form an α/β hydrolase fold consisting of residues 29-132 and 502-766 and a seven blade β propeller fold of residues 133-501[3] (Fig 1). The propeller covers the catalytic cleft of the hydrolase domain and in POP is a gating filter of substrates[6].

In the absence of a crystal structure of CD26 some understanding of its structure – function relationships has been made using point mutations and truncations of CD26, rat-human chimeric CD26 proteins, a model of CD26 and characterisation of the related molecules FAP and DP8 (reviewed by Gorrell et al[3]). N-terminal truncations of up to 39 residues, which remove the cytoplasmic and transmembrane domains and produce a soluble form, retain both enzyme and ADA binding activities[1,7,8]. We were unable to align the sequences of POP and CD26 for building a model of the CD26 region N-terminal to residue 133. Rather, we predicted by analogy with POP that residues of CD26 between the β propeller and transmembrane domains form part of the α/β hydrolase fold. We reasoned that deletion of this portion of CD26 would either show it to be essential for peptidase activity or suggest the possibility of a truncated form of CD26 that is active. The existence of such a form would indicate a potential to solve the structure of a CD26 fragment.

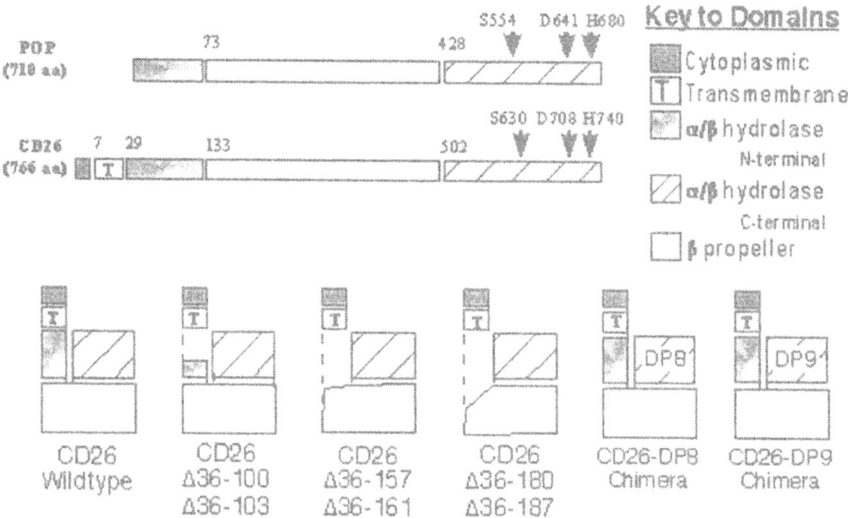

Figure 1. Schematic of the POP and CD26 proteins and of the CD26 constructs examined in this study. The propeller domain is the adenosine deaminase binding domain. Not to scale

Carbohydrate moieties can reduce the propensity of a glycoprotein crystal to yield useful diffraction data and CD26 has nine potential N-linked glycosylation sites and is about 30% carbohydrate of variable composition[9]. Therefore, a further purpose in deleting residues and making chimeras with the less glycosylated CD26 relatives DP8 and DP9 was to remove some glycosylation sites. The CD26 glycosylation sites at positions 85, 92, 150, 520 and 685 were targeted.

2. MATERIALS AND METHODS

The antibodies and methodologies have been described previously[1, 9-12]. The wild type cDNAs used were CD26[13], DP8[12] and DP9[4] which have GenBank accession numbers M80536, AF221634 and AF542510 respectively.

2.1 Preparation of mutants

Table 1. Primers

	primer sequence (5' ⇒ 3')
FORWARD Primers	
CD26_ClaI_Δ36-100	GGC CGG ATC GAT TCT ATC AAT GAT TAT TC
CD26_ClaI_Δ36-103	GGC CGG ATC GAT GAT TAT TCA ATA TCT CCT G
CD26_ClaI_Δ36-157	GGC CGG ATC GAT TCA CCA GTG GGT CAT AAA TTG GC
CD26_ClaI_Δ36-161	GGC CGG ATC GAT CAT AAA TTG GCA TAT GTT TG
CD26_ClaI_Δ36-180	GGC CGG ATC GAT GCT TCA TAT GAC ATT TAT G
CD26_ClaI_Δ36-187	GGC CGG ATC GAT CCA AGT TAC AGA ATC ACA TGG
DP8E259K_forward	CCT TTG TTC TCC AAA AGG AAT TTG ATA G
DP8S739A_forward	CCA CGG CTG GGC CTA TGG AGG ATA C
DP8D817A_forward	CAT GGT TTC CTG GCT GAG AAT GTC C
DP8H849A_forward	CCT CAG GAG AGA GCC AGC ATA AGA G
DP8_ClaI_Forward	GGG CGG ATC GAT TCA GCA GGT CCT CTT C
DP9_ClaI_Forward	CGC CCC ATC GAT GCA GCC AGC TG
REVERSE Primers	
CD26_ApaI_Reverse	GCC GCG ATC CTA GAG GGC CCT TCG AA
DP8E259K_reverse	CTA TCA AAT TCC TTT TGG AGA ACA AAG G
DP8S739A_reverse	GTA TCC TCC ATA GGC CCA GCC GTG G
DP8D817A_reverse	GGA CAT TCT CAG CCA GGA AAC CAT G
DP8H849A_reverse	CTC TTA TGC TGG CTC TCT CCT GAG G
DP8_XbaI_Reverse	GCG GCC TCT AGA TTA TAT CAC TTT TAG AGC AG
DP9_XbaI_Reverse	GGC GCG TCT AGA GGG CGG GAC AAA GTG CCT CAC TGG

CD26 deletion mutants were prepared by PCR using primers that introduced a ClaI restriction site and an ApaI restriction site at the 5' and 3' ends, respectively, of the fragment that was subsequently excised. To clone the CD26-DP8 and CD26-DP9 chimeric molecules, point mutations were engineered into the wild-type CD26 cDNA to introduce ClaI and XbaI unique restriction sites. The ClaI site at nucleotide position 2482 ends the CD26 portion of each chimera at amino acid position 501. DP8 and DP9 3' ends encoding the C-terminal portion of each hydrolase domain were prepared by PCR using primers designed to introduce a ClaI restriction site and an XbaI restriction site at the 5' and 3' ends respectively. The natural stop codons of CD26, DP8 and DP9 were retained in the pcDNA3.1 expression vector.

3. RESULTS

3.1 CD26 Deletion Mutations

The deletion mutants were transfected into COS cells and assessed by flow cytometry, immunocytochemistry and enzyme assay. The two mutants that retained the cytoplasmic and transmembrane domains but lacked N-terminal portions of extracellular CD26 (Δ36-100, Δ36-103) were poorly cell surface expressed, reflected in their limited ability to bind mAb 2A6, mAb TA5.9 or ADA (Table 1). MAb EF6/B10 recognises a conformationally dependent epitope formed by CD26 residues 117 to 187[14] so the inabilities of the deletion mutants to either bind mAb B10 or exhibit detectable peptidase activity indicated disruption of the structure. Residues 133-501 were predicted to form a β propeller domain (Fig 1). Deletion mutants that lacked part of this domain coincided with loss of both ADA binding and antibody binding (Δ36-157; Table 1), indicating that the region 104 to 156 is essential for biosynthesis of the CD26 molecule.

Table 2. Deletion mutants of CD26.

Name of Mutation	Enzyme Activity[a]	MAb binding[b]			ADA binding[c]
		EF6/B10	2A6	TA 5.9	
CD26 wild-type	+++[d]	+++	+++	+++	+++
CD26_Δ36-100	None	None	+	++	+
CD26_Δ36-103	None	None	+	++	+
CD26_Δ36-157	None	None	None	None	None
CD26_Δ36-161	None	None	None	None	None
CD26_Δ36-180	None	None	None	None	None
CD26_Δ36-187	None	None	None	None	None

[a]Enzyme cytochemistry
[b]Flow cytometry and immunocytochemistry using mAbs to CD26
[c]Flow cytometry with biotinylated ADA
[d]Staining intensity indicated by +

3.2 CD26-DP8 and CD26-DP9 Chimeras

The level of structural and functional homology between CD26 and DP8 and DP9 was investigated by making chimeric molecules in which the C-terminal portion of the CD26 α/β hydrolase domain (residues 502 to 766; see Fig 1) was replaced by the equivalent residues of DP8 or DP9 (Figure 1). The CD26-DP8 and CD26-DP9 chimeric constructs encoded 776 and 767 amino acid proteins respectively, which exhibited mobilities of about 90 kDa and 95 kDa, respectively, on SDS–PAGE (Fig 2). Unlike CD26, the chimeras did not dimerise. All chimeric proteins lacked peptidase activity.

Flow cytometry and immunocytochemistry using mAb B10 showed that, like CD26, the chimeric proteins were expressed intact and in abundance on the surface of transfected 293T cells. Immunostaining with mAb TA5.9 showed that the TA5.9 epitope does not contain CD26 residues 502 to 552.

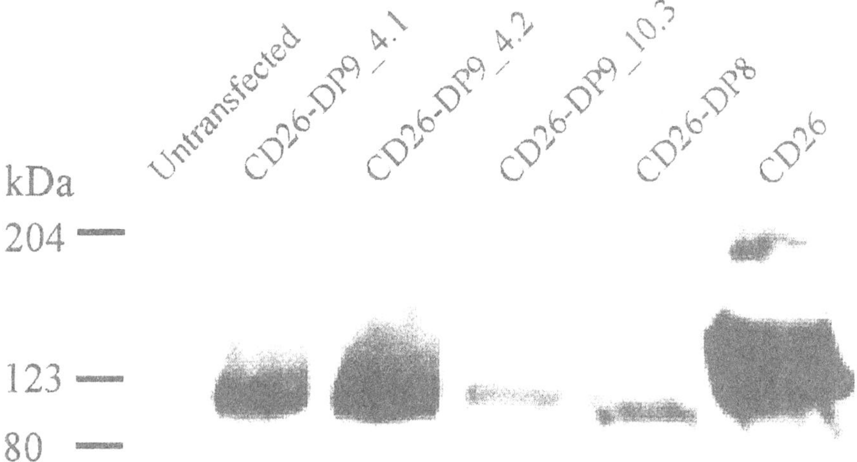

Figure 2. Western blot of 8-16 % SDS-PAGE using anti-CD26 mAb 2A6 and membrane preparations of 293T cells. CD26-DP8 chimera Mr was about 90 kDa. Mr of all three DNA clones of the CD26-DP9 chimera (4.1, 4.2 and 10.3) was about 95 kDa. Wild-type CD26 ran at the expected mobilities of about 170 kDa (dimer) and 105 to 110 kDa (monomer).

4. DISCUSSION

Deletions and chimeras were used here to improve our understanding of CD26 and the POP gene family. Both N-terminal and C-terminal portions of the α/β hydrolase domain of CD26, residues 37 to 103 and 502 to 766, were shown to be essential for catalytic activity and intact structure but not for the TA5.9 epitope. Furthermore, residues 502 to 766, which is the C-terminal portion of the hydrolase domain, was found to be essential for CD26 dimerisation.

Human CD26 has nine potential N-linked glycosylation sites. The POP family lacks potential sites for other types of glycosylation. About 25% of the mass of CD26 is carbohydrate. POP and DP8 lack glycosylation and DP9 has one potential glycosylation site in the peptidase domain, 211 residues from the C-terminus, and a second site in the propeller domain. Therefore, our observation that the Mr of the CD26 monomer was greater than those of the CD26-DP8 and CD26-DP9 chimeras suggests that the potential glycosylation sites in the C-terminal portion of CD26, Asn520 and Asn685, are glycosylated. Furthermore, our observation that the Mr of the CD26-DP9 chimera was intermediate between those of CD26 and the CD26-DP8 chimera suggests that the potential glycosylation site at Asn652 in DP9 is glycosylated.

Our CD26 model predicts that the Δ36-101 and Δ36-103 mutants lacked most of the N-terminal portion (residues 29 to 132) of the α/β hydrolase domain. The presence of this additional portion of the α/β hydrolase fold is unique to the POP gene family and is distant from the catalytic site. For these reasons it might not directly contribute to catalysis but rather provide stability to the circular structure of the propeller by covalently linking it to the hydrolase domain[5]. The absence of enzyme activity coinciding with little cell surface expression from these mutants supports this concept. Concordantly, the additional deletion of a small N-terminal part of the propeller, led to failure to express cell surface CD26 protein, indicating that the first propeller blade is essential for the biosynthesis of CD26. Truncation of only 20 residues from the CD26 C-terminus ablates enzyme activity[1], so the entire α/β hydrolase domain is required for expression of peptidase activity.

The physical data reported here concords with our CD26 model. Further investigations are needed into the mechanisms of dimerisation and catalysis. The current understanding of CD26 structure-function indicates that it will be necessary to produce the entire extracellular portion of this glycoprotein rather than isolated domains or fragments for deriving a crystal structure.

ACKNOWLEDGEMENTS

The authors are grateful for antibodies from Dr Ingrid De Meester and Professor Sibrand Poppema and cDNA from Dr Chikao Morimoto.

REFERENCES

1. Abbott, C. A., McCaughan, G. W., Levy, M. T., Church, W. B. and Gorrell, M. D., 1999, Binding to human dipeptidyl peptidase IV by adenosine deaminase and antibodies that inhibit ligand binding involves overlapping, discontinuous sites on a predicted beta propeller domain. *Eur. J. Biochem.* **266:** 798-810.
2. Barrett, A. J., Rawlings, N. D. and O'Brien, E. A., 2001, The MEROPS database as a protease information system. *J. Struct. Biol.* **134:** 95-102.
3. Gorrell, M. D., Gysbers, V. and McCaughan, G. W., 2001, CD26: A multifunctional integral membrane and secreted protein of activated lymphocytes. *Scand J Immunol* **54:** 249-64.

4. Abbott, C. A. and Gorrell, M. D., 2002, The family of CD26/DPIV and related ectopeptidases. *In Ectopeptidases: CD13/Aminopeptidase N and CD26/Dipeptidylpeptidase IV in Medicine and Biology* (J. Langner and S. Ansorge ed.), Vol. ISBN 0-306-46788-7 Kluwer/Plenum, NY, p. 171-95.
5. Fülop, V., Bocskei, Z. and Polgar, L., 1998, Prolyl oligopeptidase - an unusual beta-propeller domain regulates proteolysis. *Cell* **94:** 161-70.
6. Fülop, V., Szeltner, Z. and Polgar, L., 2000, Catalysis of serine oligopeptidases is controlled by a gating filter mechanism. *EMBO Reports* **1:** 277-81.
7. Ogata, S., Misumi, Y. and Ikehara, Y., 1989, Primary structure of rat liver dipeptidyl peptidase IV deduced from its cDNA and identification of the NH2-terminal signal sequence as the membrane-anchoring domain. *J. Biol. Chem.* **264:** 3596-601.
8. Durinx, C., Lambeir, A. M., Bosmans, E., Falmagne, J. B., Berghmans, R., Haemers, A., Scharpe, S. and De Meester, I., 2000, Molecular characterization of dipeptidyl peptidase activity in serum - Soluble CD26/dipeptidyl peptidase IV is responsible for the release of X-Pro dipeptides. *Eur. J. Biochem.* **267:** 5608-13.
9. Kähne, T., Kroning, H., Thiel, U., Ulmer, A. J., Flad, H. D. and Ansorge, S., 1996, Alterations in structure and cellular localization of molecular forms of DP IV/CD26 during T cell activation. *Cell. Immunol.* **170:** 63-70.
10. De Meester, I., Vanham, G., Kestens, L., Vanhoof, G., Bosmans, E., Gigase, P. and Scharpé, S., 1994, Binding of adenosine deaminase to the lymphocyte surface via CD26. *Eur. J. Immunol.* **24:** 566-70.
11. Abbott, C. A., McCaughan, G. W. and Gorrell, M. D., 1999, Two highly conserved glutamic acid residues in the predicted beta propeller domain of dipeptidyl peptidase IV are required for its enzyme activity. *FEBS Lett.* **458:** 278-84.
12. Abbott, C. A., Yu, D. M. T., Woollatt, E., Sutherland, G. R., McCaughan, G. W. and Gorrell, M. D., 2000, Cloning, expression and chromosomal localization of a novel human dipeptidyl peptidase (DPP) IV homolog, DPP8. *Eur. J. Biochem.* **267:** 6140-50.
13. Tanaka, T., Camerini, D., Seed, B., Torimoto, Y., Dang, N. H., Kameoka, J., Dahlberg, H. N., Schlossman, S. F. and Morimoto, C., 1992, Cloning and functional expression of the T cell activation antigen CD26. *J. Immunol.* **149:** 481-6.
14. Hühn, J., Olek, S., Fleischer, B. and von Bonin, A., 1999, The adenosine deaminase-binding region is distinct from major anti-CD26 mAb epitopes on the human dipeptidyl peptidase IV(CD26) molecule. *Cell. Immunol.* **192:** 33-40.

The Specificity of DP IV for Natural Substrates is Peptide Structure Determined

KERSTIN KÜHN-WACHE, TORSTEN HOFFMANN, SUSANNE MANHART, W. BRANDT*, and HANS-ULRICH DEMUTH
*Probiodrug AG, Weinbergweg 22, D-06120 Halle, Germany, *Institute of Plant Biochemistry, Leibniz-Institute Halle, Weinberg 3, D-06120 Halle, Germany*

1. INTRODUCTION

Dipeptidyl peptidase IV (DP IV, EC 3.4.14.5, CD26) is known as a highly specific dipeptidyl aminopeptidase cleaving its oligopeptide substrates N-terminally after a penultimate proline or alanine residue. The catalytic properties of short artificial substrates of the enzyme are well characterised, but there are only limited data concerning the kinetic properties of DP IV cleaving natural peptide substrates[1]. Although the biological role of DP IV in mammalian systems has not been completely established, it is believed that DP IV could play a key regulator function in physiological processes as immune response, energy homeostasis and glucose homeostasis by modulating the bioactivity of such peptides *in vivo* [2].

Previously, it was demonstrated that for glucagon, glucagon-like peptide-1 (GLP-1) and glucose-dependent insulinotropic polypeptide (GIP), DP IV is the main convertase in the circulation[3, 4]. These peptide hormones are involved in opposite processes, e.g., GLP-1 and GIP and stimulate the insulin secretion while the substrate glucagon suppresses insulin release. The control of activity of such counterregulating hormones by one enzyme seems to be conflicting at the first view but the different specificity of the enzyme towards their substrates in combination with their actual physiological concentrations could be an explanation for a fine-tuned regulation network. While GLP-1 and GIP act in a strong limited time slot (in the postprandial

Dipeptidyl Aminopeptidases in Health and Disease, Edited by Hildebrandt et al.
Kluwer Academic/Plenum Publishers, New York, 2003

phase, if the glucose concentration is high) both hormones undergo a fast biodegradation by DP IV. Intact glucagon has obviously a longer lasting effect on the glucose level in the organism[3, 4] and it shows a slow inactivation by DP IV.

2. SUBSTRATE ENZYME INTERACTIONS

Previously Lambeir et al. could show that DP IV hydrolyses the related peptides PACAP27 and PACAP38 with different rates. The C-terminal extension of PACAP38 improves the PACAP binding and the turnover rate [5]. These results support the hypothesis of secondary interactions which facilitate the binding of substrates to DP IV or stabilise transition states of productive proteolysis.

2.1 Hydrolysis of GIP-analogues and GIP-fragments

Hence, the substrate structure of peptides determines their life-time and also their bioactivity. Ala in P_1-position contributes to the excellent substrate properties of the two incretins (GLP-1 and GIP) for DP IV *in vitro* [6] and *in vivo*[3, 7]. In contrast, glucagon having a serine in P_1-position is rather slowly degraded by DP IV[1,8]. Stabilised analogues of GIP could enhance the insulinotropic action of the hormone and so they are of interest for therapeutic use in type 2 diabetes. Therefore, we have synthesised and characterised various GIP-analogues and -fragments. In general, it was observed that substitution of Ala in P_1-position by Gly, Val or Ser results in an improved stability against DP IV-mediated degradation (see table 1).

The binding constant K_D describes the affinity of the peptide to the enzyme. Regarding the K_D-values it could be clearly shown that Ala in P_1-position of the substrate is optimised for binding in the active site. Nevertheless the enzyme also accept unusual amino acid residues in P_1-position but the hydrolysis rate is drastically decreased. In order to develop stabilised GIP-analogues as therapeutics such slow degradation is desired because the action of the hormone on the one hand is prolonged and on the other hand the drug can be renal cleared by the organism and so its action is limited. The stability of the analogues is dependent on the amino acid in P_1-position and increases in the order Ala$\ll$Ser$<$Gly$<$Val.

Whereas all fragments with 30 amino acid residues were cleaved by DP IV the N-terminal hexapeptide analogues exhibit different properties. Only the Ser^2-fragment was hydrolysed by DP IV, the Val^2- and Gly^2-analogues were not accepted as substrates by DP IV (see figure 1and 2).

Table 1. Kinetic properties of DP IV catalysed hydrolysis of various GIP-analogues and fragments determined using MALDI-TOF mass spectrometry and biomolecular interaction analysis.

GIP/GIP-fragments	$t_{1/2}$ [min]	DP IV binding	
		K_D [μM]	χ^2
GIP_{1-42}	2.56 ± 00.12	1.74	0.10
GIP_{3-42}	stable	3.15	0.33
GIP_{1-30}	2.68 ± 00.16	9.15	1.15
$[Ser^2]GIP_{1-30}$	137.14 ± 12.3	9.89	1.90
$[Gly^2]GIP_{1-30}$	150.02 ± 27.3	19.61	1.96
$[Val^2]GIP_{1-30}$	298.04 ± 92.2	35.22	1.70
GIP_{1-6}	>7.5 min	not detectable	not detectable
$[Ser^2]GIP_{1-6}$	79.04 ± 12.1	not detectable	not detectable
$[Gly^2]GIP_{1-6}$	stable	not detectable	not detectable
$[Val^2]GIP_{1-6}$	stable	not detectable	not detectable

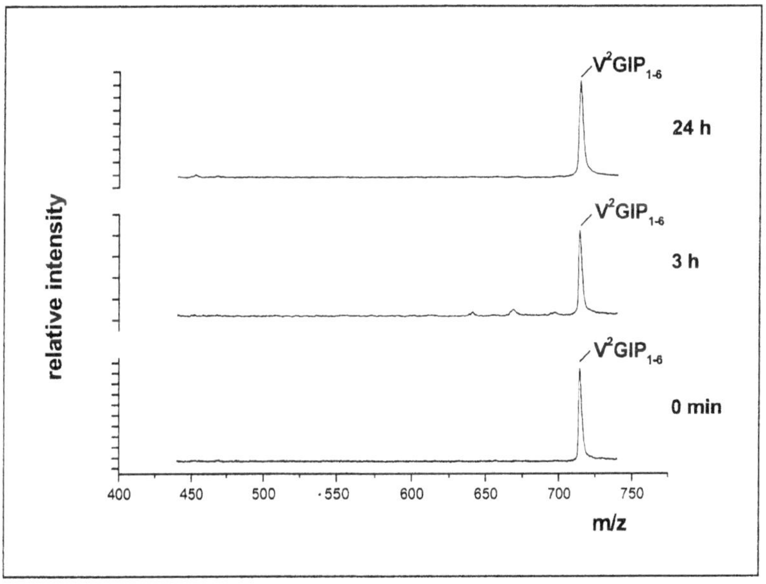

Figure 1. Mass spectra of V2GIP1-6, time dependent recorded during the incubation of 25μM peptide with 40mU porcine DP IV at 30°C, pH 7.6.

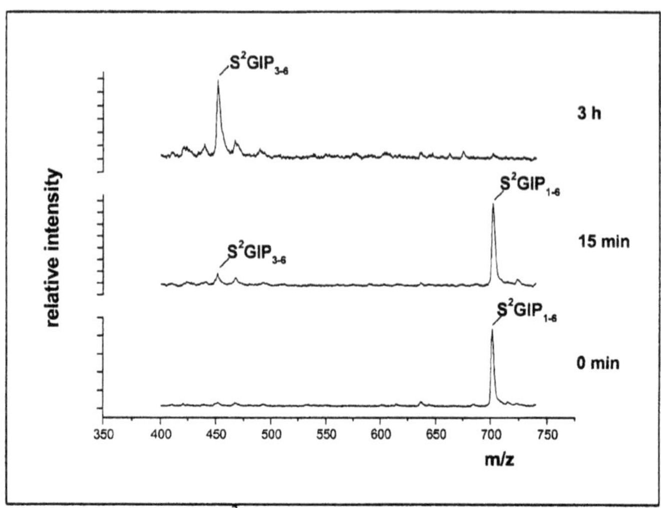

Figure 2. Mass spectra of S^2GIP_{1-6}, time dependent recorded during the incubation 25μM peptide with 40mU porcine DP IV at 30°, pH7.6

This observation indicates the existence of one or more secondary binding sites which allow a productive binding of non-proline and non-alanine substrates in the catalytic site. In order to proof this hypothesis based on the X-ray structure of prolyl oligopeptidase we developed a computational DP IV 3D-structure model.

2.2 Prediction of secondary binding sites using a 3D-structure DP IV-model

Molecular dynamic simulations on the DP IV model suggest a multitude of interactions between substrates and enzyme. Regarding the binding of Ser^2-GIP_{1-6} it has been shown that H-bonds between the Glu^3 of the substrate and Gln^{553}-residue of enzyme and Thr^5(substrate) and His^{740} (enzyme) as well as a salt bridge at the C-terminal end of the substrate (Phe^6 and Arg^{560}) facilitate the binding of the substrate in a proper orientation. The active serine is in a optimal position to attack nucleophilic the peptide bond (see figure 3).

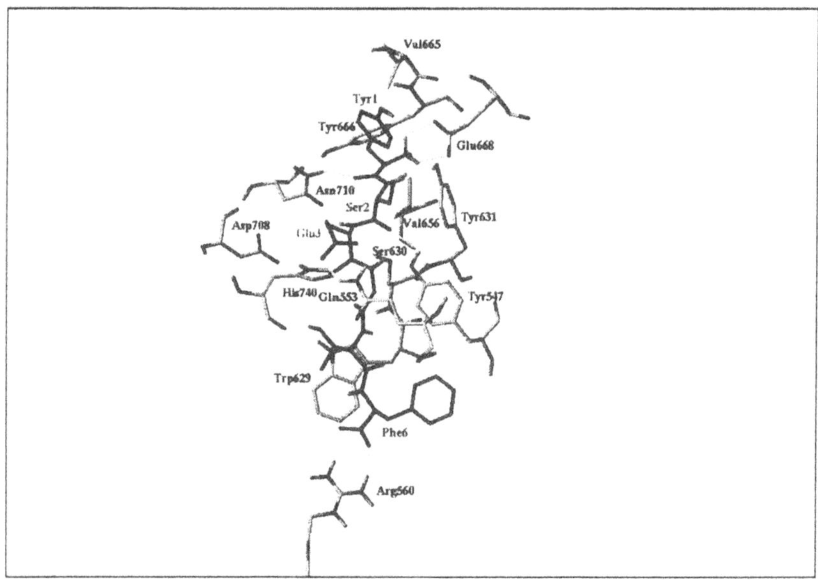

Figure 3. Potential interaction sites between Ser2GIP1-6 and the active site of DP IV derived from molecular dynamic simulations

In contrast, Val^2GIP_{1-6} can not be stabilised by interaction of Val^2 and Tyr^{666}. From the model one can calculate a distance of 5.8 Å between oxygen atom of the hydroxyl group of the active serine and the carbonyl group of Val^2 (see figure 4). Therefore a nucleophilic attack from the active serine is impossible.

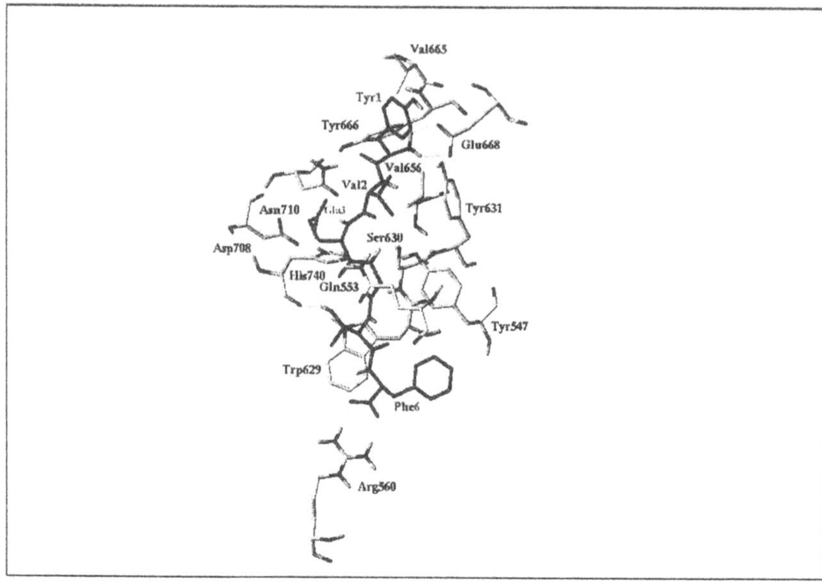

Figure 4. Potential interaction sites between Val2GIP1-6 and the active site of DP IV derived from molecular dynamic simulations.

3. CONCLUSION

Our results indicate that the substrate properties of peptides are encoded by their own structure. That means, that substrate characteristics depend not only on the primary structure around the catalytic site rather C-terminal located secondary interactions strongly influence the binding and catalysis of the substrates. Such interaction sites seem to force the ligand in a proper orientation to the active site of DP IV. As result of these relations the hydrolysis of peptides with non-proline and non-alanine residues in P_1-position (Ser, Val, Gly) becomes possible in longer peptides.

Such specific secondary interactions opens the opportunity for development of new inhibitors.

REFERENCES

1. Bongers, J., Lambros, T., Ahmad, M. and Heimer, E.P.; 1992, Kinetics of dipeptidyl peptidase IV proteolysis of growth hormone-releasing factor and analogs. *Biochim. Biophys. Acta:* 1122-147.

2. DeMeester, I., Korom, S., Van Damme, J. and Scharpé, S., 1999, CD26, let it cut or cut it down. *Immunol Today* **20**: 367-375.
3. Kieffer, T.J., McIntosh, C.H. and Pederson, R.A., 1995, Degradation of glucose-dependent insulinotropic polypeptide and truncated glucagon-like peptide 1 in vitro and in vivo by dipeptidyl peptidase IV. *Endocrinology* **136 (8)**: 3585-3596.
4. Hinke, S.A., Pospisilik, J.A., Demuth, H.-U., Mannhart, S., Kühn-Wache, K., Hoffmann, T., Nishimura, E., Pederson, R.A. and McIntosh, C.H., 2000, Dipeptidyl peptidase IV (DPIV/CD26) degradation of glucagon. Characterization of glucagon degradation products and DPIV-resistant analogs. *J Biol Chem* **275**:3827-3834.
5. Lambeir, A.M., Durinx, C., Proost, P., Van Damme, J., Scharpé, S. and DeMeester, I., 2001, Kinetic study of the processing by dipeptidyl-peptidase IV/CD26 of neuropeptides involved in pancreatic insulin secretion. *FEBS Lett.* **507**:327-330.
6. Mentlein, R., Gallwitz, B. and Schmidt, W.E., 1993, Dipeptidyl-peptidase IV hydrolyses gastric inhibitory polypeptide, glucagon-like peptide-1(7-36)amide, peptide histidine methionine and is responsible for their degradation in human serum. *Eur J Biochem* **214**:829-835.
7. Deacon, C.F., Johnsen, A.H. and Holst, J.J., 1995, Degradation of glucagon-like peptide-1 by human plasma in vitro yields an N-terminally truncated peptide that is a major endogenous metabolite in vivo. *J.Clin.Endocrinol.Metab* **80**:952-957.
8. Pospisilik, J.A., Hinke, S.A., Pederson, R.A., Hoffmann, T., Rosche, F., Schlenzig, D., Glund, K., Heiser, U., McIntosh, C.H. and Demuth, H.-U., 2001, Metabolism of Glucagon by Dipeptidyl Peptidase IV (CD26). *Regul Pept* **93(3)**:133-141.

New Results on the Conformations of Potent DP IV (CD26) Inhibitors bearing the N-terminal *MWP* Structural Motif

CARMEN MRESTANI-KLAUS[*], WOLFGANG BRANDT[#], JÜRGEN FAUST[*], SABINE WRENGER[†], DIRK REINHOLD[†], SIEGFRIED ANSORGE[‡], and KLAUS NEUBERT[*]
[*]Department of Biochemistry/Biotechnology, Martin-Luther-University Halle-Wittenberg, Kurt-Mothes-Strasse 3, Halle; [#]Plant Biochemistry Institute, Leibniz Institute Halle, Weinberg 3, Halle; [†]Institute of Immunology, Otto-von-Guericke-University Magdeburg, Leipziger Strasse 44, Magdeburg; [‡]IMTM, Magdeburg, Germany

1. INTRODUCTION

The cell surface ectopeptidase dipeptidyl peptidase IV, also known as CD26, plays an important role in the activation and proliferation of T lymphocytes, besides a number of other biological functions. We have shown previously that N-terminal HIV-1 Tat peptides such as Tat(1-9) inhibit DP IV mediating the immunosuppressive effect of the HIV-1 Tat protein via inhibition of DP IV[1]. Replacement of Asp in the second position leads to analogues with enhanced inhibition. In particular, tryptophan in position 2 strongly improves DP IV inhibition as could be shown for Trp^2-Tat(1-9), $K_i = (2.19 \pm 0.07)\ 10^{-6}$ M, and for TXA2-R(1-9), $K_i = (5.06 \pm 0.04)\ 10^{-6}$ M, the N-terminal nonapeptide of the thromboxane A2 receptor[2].

An understanding of the conformations of these peptides containing the N-terminal *MWP* motif may shed some light on their binding to DP IV and could help to design more potent and specific peptide-based inhibitors.

Dipeptidyl Aminopeptidases in Health and Disease, Edited by Hildebrandt et al.
Kluwer Academic/Plenum Publishers, New York, 2003

For this purpose, we present here the results of conformational investigations of both DP IV/CD26 inhibitors Trp^2-Tat(1-9) (MWPVDPNIE) and TXA2-R(1-9) (MWPNGSSLG) by using NMR spectroscopy and molecular modeling.

2. RESULTS AND DISCUSSION

The solution conformations of Trp^2-Tat(1-9) and TXA2-R(1-9) in water (H_2O/D_2O, 90:10) were studied by 1- and 2-dimensional ^{1}H NMR techniques and molecular modeling. For both peptides, two major sets of signals could be identified and unambiguously assigned using COSY, TOCSY, NOESY and ROESY spectra. One isomer (44% for Trp^2-Tat(1-9), 55% for TXA2-R(1-9)) adopts an *all-trans* conformation. The second signal set (about 56% for Trp^2-Tat(1-9), 45% for TXA2-R(1-9)) belongs to the *cis* isomer characterized by one Trp^2-Pro^3 *cis* bond based on the corresponding cross peaks in the 2D ROESY spectra. Several ROEs were observed between backbone $d_{NN}(i,i+1)$ and $d_{\alpha N}(i,i+1)$ for both peptides. The lack of medium and long range ROEs, together with averaged $^3J(NH,\alpha H)$ coupling constants could be interpreted as lack of a defined secondary structure. However, αH shifts of a few residues of both peptides had substantial deviations from the tabulated random-coil values, indicating that the backbones are not fully extended or random. After molecular dynamics (MD) simulations using distance restraints (18 interresidue distances for *trans* and 10 interresidue distances for *cis*) and energy minimization several solution conformations could be determined suggesting similar overall backbone conformations for both isomers of the Trp^2 analogue as it was already reported[2,3] (Figure 1).

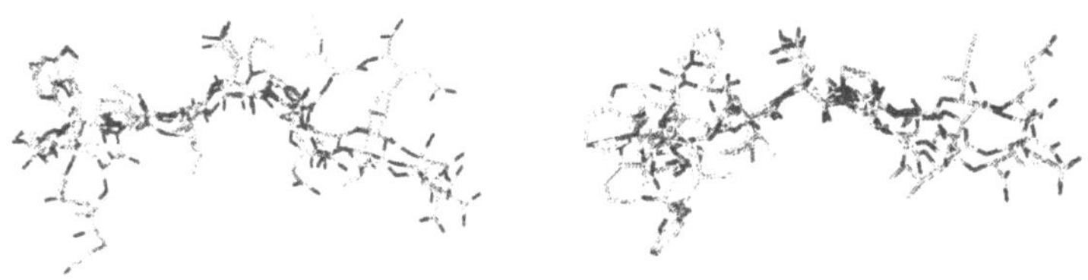

Figure 1. Bundle of 8 best Trp^2-Tat(1-9) structures of lowest energy obtained by MD calculations. These structures fit best the NMR data such as ROEs, torsion angles ϕ and temperature dependence studies. Left: *trans* Trp^2-Tat(1-9). Right: *cis* Trp^2-Tat(1-9).

The substitution of Asp^2 with the hydrophobic tryptophan does not cause a significant rearrangement of the backbone structures of Trp^2-Tat(1-9) compared to the parent peptide Tat(1-9)[2,3]. The structures of Trp^2-Tat(1-9) are flexible with the propensity to form fairly rigid conformations along the residues Pro^3 to Pro^6. In fact, the close inspection of the ϕ, ψ values indicates that this peptide adopts a left-handed polyproline II helix (PPII) around the mid segment region covering Trp^2 to Pro^6.

The solution conformations of TXA2-R(1-9) calculated by a simulated annealing approach using distance constraints (altogether 44, 17 of that interresidue) turned out to be highly flexible, although the NMR data show certain similarities to those of Trp^2-Tat(1-9), particularly at the N-terminus. The inherent flexibility of this molecule makes definitive conformational conclusions difficult (Figure 2).

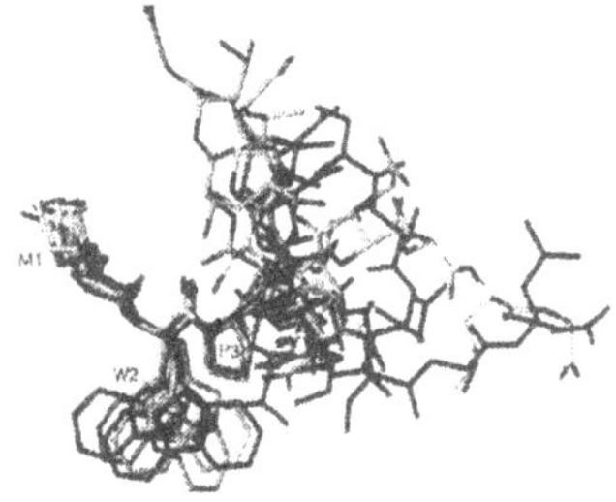

Figure 2. Representation of the superposition of an ensemble of 8 low-energy structures of *trans* TXA2-R(1-9). In an attempt to calculate *cis* TXA2-R(1-9) using the program X-PLOR the ROE constraints were not fulfilled. Therefore, the structures of this conformer could not be shown.

3. CONCLUSION

Conformational analysis by NMR spectroscopy and molecular modeling revealed a left-handed PPII helix-like structure for Trp^2-Tat(1-9) (*cis* and *trans*) and an even more flexible structure for TXA2-R(1-9).

PPII helices form a well-defined structural class comparable with the other structures defined in proteins and are characterized by exposed, mobile structures with 4-8 residues, mostly found on the protein surface. Polyproline II helices are mainly identified by their torsion angles of $\phi \sim -75°$ and $\psi \sim 145°$. They do not form regular interchain hydrogen bonds, but are hydrogen bonded with water molecules. PPII helices have a strong preference for the amino acid proline, although it is not necessarily present. These features were also reported for the parent peptide Tat(1-9)[4] as well as for the well known DP IV substrates neuropeptide Y and pancreatic polypeptide[5] suggesting that PPII-like helical structures represent a favored structural class for the interaction with DP IV.

Thus, the considerable enhancement of the inhibition capacity of both Trp^2-Tat(1-9) and TXA2-R(1-9) compared to the moderate inhibitor Tat(1-9)[2], $K_i = 2.68 \pm 0.01\ 10^{-4}$ M, can only be due to tryptophan in the second position suggesting that its side chain is favored to exhibit attractive hydrophobic interactions with DP IV compared with aspartic acid.

On the other hand, we could show recently that Tat(1-9) and its analogues as well as TXA2-R(1-9) inhibit DP IV according to different inhibition mechanisms (Lorey et al., manuscript submitted). One possible explanation for these findings might be enzyme-ligand interactions relying on multiple weak binding sites as described for PPII helices[5] rather than specific lock and key binding. Certainly, only an X-ray structure of DP IV would help to understand the interaction of DP IV with inhibitors.

ACKNOWLEDGEMENTS

Financial support was obtained from the Deutsche Forschungsgemeinschaft, SFB 387 and NE 501/2-1, and is gratefully acknowledged.

REFERENCES

1. Wrenger, S., Reinhold, D., Hoffmann, T., Kraft., M., Frank, R., Faust, J., Neubert, K., and Ansorge, S., 1996, The N-terminal X-X-Pro sequence of the HIV-1 Tat protein is important for the inhibition of dipeptidyl peptidase IV (DP IV/CD26) and the suppression of mitogen-induced proliferation of human T cells. *FEBS Lett.* **383**: 145-149.
2. Wrenger, S., Faust, J., Mrestani-Klaus, C., Fengler, A., Stöckel-Maschek, A., Lorey, S., Kähne, T., Brandt, W., Neubert, K., Ansorge, S., and Reinhold, D., 2000, Down-regulation of T cell activation following inhibition of dipeptidyl peptidase IV/CD26 by the N-terminal part of the thromboxane A2 receptor. *J. Biol. Chem.* **275**: 22180-22186.
3. Mrestani-Klaus, C., Fengler, A., Faust, J., Brandt, W., Wrenger, S., Reinhold, D., Ansorge, S., and Neubert, K., 2001, Conformational study of a tryptophan containing Tat(1-9) analogue as potent inhibitor of dipeptidyl peptidase IV. In *Peptides 2000* (J. Martinez and J. A. Fehrentz, eds.), Editions EDK, Paris, pp.511-512.
4. Kanyalkar, M., Srivastava, S., and Coutinho, E., 2001, Conformation of N-terminal HIV-1 Tat (fragment 1-9) peptide by NMR and MD simulations. *J. Peptide Sci.* **7**: 579-587.
5. Williamson, M. P., 1994, The structure and function of proline-rich regions in proteins. *Biochem. J.* **297**: 249-260.

Different Inhibition Mechanisms of Dipeptidyl Peptidase IV by Tryptophan Containing Peptides and Amides

ANGELA STÖCKEL-MASCHEK[*], BEATE STIEBITZ[*], JÜRGEN FAUST[*], ILONA BORN[*], THILO KÄHNE[#], MARK D. GORRELL[§], KLAUS NEUBERT[*]

[*]*Department of Biochemistry and Biotechnology, Martin-Luther-University Halle-Wittenberg, Kurt-Mothes-Strasse 3, Halle, Germany,* [#]*Department of Internal Medicine, Otto-von-Guericke-University Magdeburg, Leipziger Str. 44, Magdeburg, Germany,* [§]*A W Morrow Gastroenterology and Liver Center, Royal Prince Alfred Hospital and University of Sydney, NSW 2042, Australia.*

1. INTRODUCTION

The dipeptidyl peptidase IV (DPP IV, EC 3.4.14.5) is a proline specific serine protease which cleaves Xaa-Pro dipeptides from the N-terminus of oligo- and polypeptides[1].

During recent years our attention was directed to the development of DPP IV inhibitors. It is well-known, that Xaa-Pro dipeptides that are products of substrate hydrolysis are competitive inhibitors of DPP IV. Furthermore, the product analogous amino acid pyrrolidides (Pyrr) and thiazolidides (Thia) are also known as potent competitive inhibitors[2,3]. On the other side, inhibition studies with peptides containing a N-terminal Xaa-Yaa-Pro sequence, e.g. Tat(1-9), TXA2-R(1-9) and analogous peptides, showed that these are also DPP IV inhibitors. In contrast to the first mentioned compounds, these peptides are not competitive inhibitors of DPP IV[4]. The enzymekinetic studies indicated that Trp in front of the Pro residue may have a great influence on the inhibition mechanism and the inhibition constants. Therefore, we investigated generally the influence of Trp in front

Dipeptidyl Aminopeptidases in Health and Disease, Edited by Hildebrandt et al.
Kluwer Academic/Plenum Publishers, New York, 2003

of Pro or Pro analogues on the inhibition mechanism and the inhibition constants of human recombinant and pig kidney DPP IV.

2. RESULTS

Amino acid pyrrolidides and thiazolidides are known as potent product analogous inhibitors of DPP IV. Generally, these compounds inhibit DPP IV competitively[2], but our investigations demonstrate that Trp-Pyrr, Trp-Thia and related compounds inhibit DPP IV in many cases according to other inhibition mechanisms.

Table 1. Inhibition of pig kidney DPP IV and human recombinant DPP IV by different product analogous compounds

compound	pig kidney DPP IV			human recombinant DPP IV		
	K_i (μM)		mechanism	K_i (μM)		mechanism
Ile-Pro	14.9		competitive	29.2		competitive
Lys(ZNO$_2$)-Pro	12.0		competitive	14.7		competitive
Trp-Pro	57.4	α=0.8	linear mixed-type	185	α=0.3	linear mixed-type
Ile-Pyrr	0.218		competitive	0.547		competitive
Lys(ZNO$_2$)-Pyrr	0.271		competitive	0.422		competitive
Trp-Pyrr	19.9	α=6.8	linear mixed-type	18.6		competitive
Ile-Thia	0.126		competitive	0.216		competitive
Lys(ZNO$_2$)-Thia	0.0923		competitive	0.136		competitive
Trp-Thia	8.01		competitive	7.87		competitive
Pro-Pro$^{(P)}$[OEt]$_2$ diastereomeric mixture	1230		competitive	n.d.		
Phe-Pro$^{(P)}$[OEt]$_2$ diastereomeric mixture	76.2		competitive	136		competitive
Trp-Pro$^{(P)}$[OEt]$_2$ diastereomere 1	44.4	α=0.7 ß=0.06	hyperbolic mixed-type	n.d.		
Trp-Pro$^{(P)}$[OEt]$_2$ diastereomere 2	no inhibition (1 mM)			n.d.		

n.d. not determined

The data in table 1 show, that DPP IV is inhibited by Trp-Pro, Trp-Pyrr and Trp-Thia up to two orders of magnitude weaker than by dipeptides containing the amino acids Ile or N^{ε}-nitrobenzoyloxycarbonyl-Lys (Lys(ZNO$_2$)) instead of Trp. Furthermore, the Trp-containing inhibitors exhibit also linear mixed-type and hyperbolic mixed-type inhibition in addition to competitive inhibition. For instance, Trp-Pyrr inhibits pig kidney DPP IV according to a linear mixed-type mechanism (Fig. 1).

In case of amino acid phosphonates, two separate diastereomeres of Trp-Pro$^{(P)}$[OEt]$_2$ as well as the diastereomeric mixtures of Pro-Pro$^{(P)}$[OEt]$_2$ and Phe-Pro$^{(P)}$[OEt]$_2$ were used for kinetic investigations. As expected, Pro-Pro$^{(P)}$[OEt]$_2$ and Phe-Pro$^{(P)}$[OEt]$_2$ inhibit DPP IV competitively, but diastereomere 1 of Trp-Pro$^{(P)}$[OEt]$_2$ is a hyperbolic mixed-type inhibitor of the enzyme. Diastereomere 2 was not able to inhibit DPP IV.

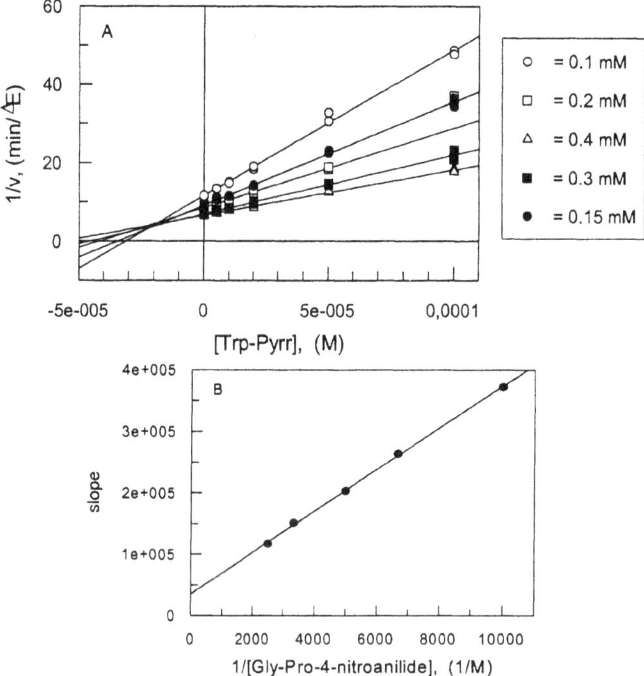

Figure 1. Inhibition of pig kidney DPP IV by Trp-Pyrr. The data were obtained by monitoring the hydrolysis of Gly-Pro-4-nitroanilide in 40 mM Tris buffer pH 7.6 containing NaCl (I=0.125) and different concentrations of Trp-Pyrr at 30°C. The final concentration of DPP IV amounts to 1.4 nM. A) Dixon plot 1/v vs. [I]. B) Replot of the slopes of the Dixon plot vs. 1/[S].

3. DISCUSSION

The results of our investigation indicate that Trp-Pro dipeptide and related compounds containing Trp in P_2-position inhibit DPP IV in many cases according to a linear mixed-type inhibition with K_i values in the micromolar range. Therefore, the inhibitor and the enzyme do not only form an EI complex, furthermore an EIS complex exists[5]. This EIS complex is not catalytically active. In contrast to the above mentioned compounds, diastereomere 1 of Trp-Pro$^{(P)}$[OEt]$_2$ inhibits DPP IV according to the hyperbolic mixed-type mechanism. In this case, the EIS complex is catalytically active. This separated diastereomere is a better inhibitor of DPP IV as the diastereomeric mixtures of the other amino acid phosphonates.

Pig kidney DPP IV is inhibited with slightly more potency than human recombinant DPP IV. In most cases, both enzymes show the same inhibition mechanism. Only in the case of Trp-Pyrr the inhibition mechanism differs. Whereas pig DPP IV is inhibited according to the linear mixed-type mechanism, the human enzyme is inhibited competitively.

4. CONCLUSION

Xaa-Pro dipeptides and amino acid pyrrolidides as well as thiazolidides are well-known competitive inhibitors of DPP IV. If these compounds contain the amino acid Trp the compounds are in many cases linear mixed-type or hyperbolic mixed-type inhibitors.

ACKNOWLEDGEMENTS

The Deutsche Forschungsgemeinschaft, SFB 387 supported this work.

REFERENCES

1. Fleischer, B. 1995, Molecular Biology Intelligence Unit. *Dipeptidyl Peptidase IV (CD26) in Metabolism and the Immune Response*, Austin Texas.
2. Born, I., Faust, J., Heins, J., Barth, A. and Neubert, K., 1994, Potent Inhibitors of Dipeptidyl Peptidase IV. *Eur. Cell Biol. Suppl.* **40:**83.
3. Stöckel-Maschek, A., Stiebitz, B., Born, I., Faust, J., Mögelin, W.and Neubert, K., 2000, Potent Inhibitors of Dipeptidyl Peptidase IV and Their Mechanisms of Inhibition. *Adv. Exp. Med. Biol.* **477:** 117-123.
4. Lorey, S., Stöckel-Maschek, A., Faust, J., Brandt, W., Stiebitz, B., Gorrell, M.D., Kähne, T., Mrestani-Klaus, C., Wrenger, S., Reinhold, D., Ansorge, S., Neubert, K., 2002, Different Modes of Dipeptidyl Peptidase IV (CD26) Inhibition by Oligopeptides derived from the N-terminus of HIV-1 Tat indicate at least two inhibitor binding sites. *Submitted.*
5. Segel, I.H., 1993, *Enzyme Kinetics*. John Wileys & Sons, Inc. New York.

Re-Uptake Mechanisms of Peptide Fragments after DPP IV-Mediated Proteolysis in the Peripheral Nervous System

Q.THAI DINH, CHRISTIAN PEISER, AXEL FISCHER, AND DAVID A. GRONEBERG

Dept. of Internal Medicine, Psychosomatics/Psychotherapie, Humboldt-University, Charite Campus Mitte, Liusenstraße13A, 10117 Berlin. Dept.of Pediatric Pneumology and Immunology, Charite Campus Wirchow,Augustenburger Platz 1, 13353 Berlin, Germany.

1. INTRODUCTION

DPP IV and other plasma membrane ectoenzymes such as aminopeptidase N have been identified in glial and vascular structures of the peripheral nervous system in recent years. Following DPP IV-mediated proteolysis, peptide fragments may be re-utilized via uptake mechanisms. In contrast to the detailed knowledge on the functional and molecular properties of these DPP IV-products, only little is known about the metabolism of these products after their inactivation. Apart from single amino acids, a large number of di- and tripeptides may arise from proteolysis and there is a growing interest in the fate of these short chain peptides[1]. Within the central nervous system, a specific pH-dependent transport for dipeptides such as carnosine has been characterized functionally[2], which was demonstrated to be localized in glial cells[3].

So far, two pH-dependent mammalian peptide transporters have been cloned from various species, which possess 12 membrane spanning domains and share an identity of approximate 47% at the protein level[4]. The transporter mediate the electrogenic uphill peptide transport of the proteolysed products by coupling the substrate translocation to the movement of H^+/H_3O^+ with the transmembrane electrochemical proton gradient serving as the driving force[4]. In addition to DPP IV products such as di- and tripeptides, both carrier proteins transport several peptidomimetic drugs such as cephalosporins, bestatin, or selected ACE inhibitors.

In the present study, the question was addressed, if the protein of the high-affinity, low capacity peptide transporter is expressed in guinea pig dorsal root ganglia.

2. ESTABLISHMENT OF TISSUE HARVESTING AND IMMUNOHISTOCHEMISTRY

Adult male and female guinea-pigs (300–500 g body weight) were anesthetized with pentobarbital sodium (40 mg/kg body weight i.p.) and perfused retrogradely through the aorta with freshly prepared 4% paraformaldehyde (PFA) in phosphate-buffered saline (PBS) at pH 7.4 for 5 min. The dorsal root ganglia were then removed and thoroughly washed in PBS. Cryoprotection was performend by immersion in sucrose-PBS solution, which was adjusted to 800 mosmol/kg. After shock-freezing in liquid nitrogen, the ganglia were processed to 8-10 μm sections using a cryostate. For immunohistochemistry, the sections were thawed onto gelatine-chromalum-coated glass slides and air dried for 1 h. The sections were then washed in PBS 3 x 5 min and preincubated for 1h at room temperature with 2% low fat milk powder in PBS, pH 7.4. The sections were then incubated with polyclonal anti-rabbit-anti high-affinity low capacity peptide transporter-serum[5], which was diluted 1:1000 in the preincubation solution overnight. As the secondary antibody an anti-rabbit indocarbocyanin (Cy3)- antibody (1:1000, Dianova, Germany) was used. The slides were finally coverslipped in carbonate-buffered glycerol (pH 8.6) and viewed using an epifluorescence microscope.

3. PROTEIN EXPRESSION WITH GUINEA PIG DORSAL ROOT GANGLIA

Immunohistochemistry for the protein of the high-affinity, low-capacity peptide transporter PEPT2 was carried out using cryostat sections of guinea pig dorsal root ganglia. And a polyclonal specific antibody raised against the C-terminal region of the carrier protein.

The incubation resulted in positive staining for transporter immunoreactivity in cells which were in close contact to the ganglionic neurons. The immune-signal was of a non-granular type and localised in the cytoplasma. The positive cells were identified as glial cells by their typical shape and close connection to the neurons. No immunostaining signals were found in neuronal cells, vascular structures or connective tissue cells (Fig. 1).

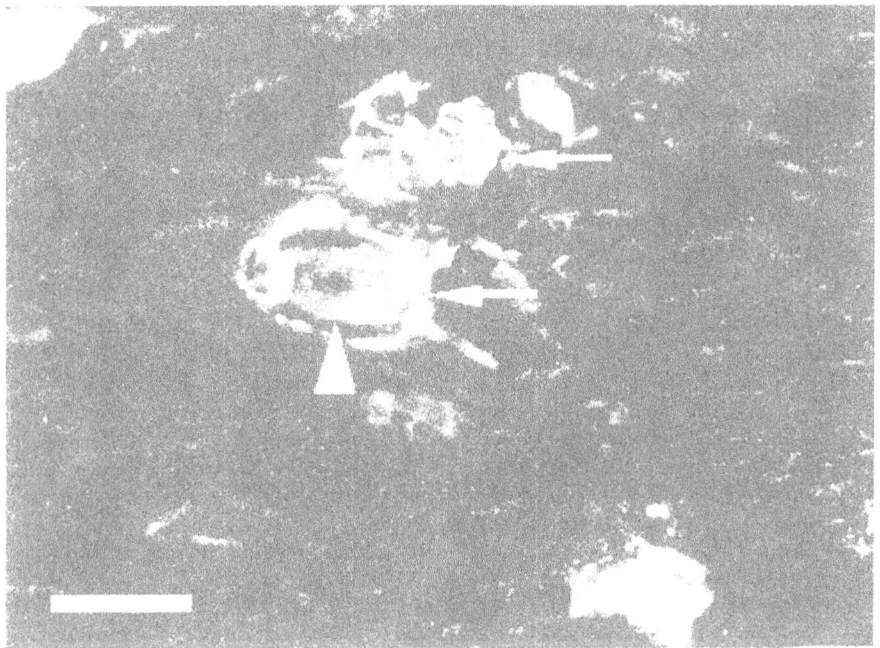

Fig 1. Immunostaining for the high-capacity, low affinity transporter protein in adult guinea pig dorsal root ganglia. The immunoreactivity is localized to the satellite cells (arrows) surrounding the non-reactive ganglionic neurons (arrowshead). Scale Bar = 30 μm

4. CONCLUSION

In the present study we demonstrated the presence of the high-affinity, low capacity peptide transporter PEPT2 at the translational level in dorsal root ganglia of the species guinea pig by employing immunohistochemistry. This finding provides new insights into the fate of peptide fragments, following DPP IV-mediated proteolysis. The results indicate, that the fragments may be re-utilized via uptake mechanisms in the peripheral nervous system of the guinea pig. For the rat nervous system, several studies demonstrated the presence of the transporter mRNA in nervous system astrocytes, satellite cells, subependymal cells, ependymal cells and cells of the choroid plexus[6,7].

The expression of the transporter in glial cells of the guinea pig peripheral nervous system (PNS) suggests a role of the transporter within the intraganglionic neuropeptide metabolism. Apart from a role of the transporter as a scavenging system for DPP IV products, the carrier may also catalyze the uptake of biologically active short chain peptides such as carnosine[8], or the neuropeptide N-acetylaspartateglutamate (NAAG [9]) and thyreotropin releasing hormone (TRH, Glu-His-Pro[10]).

ACKNOWLEDGEMENTS

The work was supported by a grant from the DFG, Zi 110/22-1.

REFERENCES

1. Cameron R.S., Rakic P., 1991: Glial cell lineage in the cerebral cortex: a review and synthesis, *Glia.* **4:** 124-1372. Abraham D., Pisano J.J., Udenfried S., 1964: Uptake of carnosine and homocarnosine by rat brain slices, *Arch. Biochem. Biophys.* **104:**160-165
3. Tom Dieck S., Heuer H., Ehrchen J., Otto C., Bauer K., 1999: The peptide transporter in Pept2 is expressed in rat brain and mediates the accumulation of the fluorescent derivative β-Ala-Lys-N_ε-AMCA in Astrocytes, *Glia.* **25:** 10-20
4. Daniel H., Herget M., 1997: Cellular and molecular mechanisms of renal peptide transport, *Am. J. Physiol.* **273**: F1-F8
5. Groneberg D.A., Döring F., Theis S., Nickolaus M., Fischer A., Daniel H., 2002: Peptide transport in the mammary gland: Expression and distribution of the peptide transporter PEPT2 mRNA and protein, Am. J. Physiol. Endocrinol. Metab. **282:** E1172-E1179
6. Berger U.V., Hediger M.A., 1999: Distribution of peptide transporter PEPT2 mRNA in the rat nervous system, *Anat. Embryol.*, **199:** 439-449
7. Groneberg D.A., Döring F., Nickolaus M., Daniel H., Fischer A, 2001: Expression of peptide transporter PEPT2-protein and -mRNA in rat dorsal root ganglia, *Neurosci. Lett.* **304:** 181-184
8. Hoffmann A.M., Bakardijev A., Bauer K., 1996: Carnosine-synthesis in cultures of rat glial cells is restricted to oligodendrocytes and carnosine uptake to astrocytes, *Neurosci. Lett.* **215:** 29-32
9. Cassidy M., Neale J.H., 1993: Localization and transport of N-acetylaspartylglutamate in cells of whole murine brain in primary culture. *J. Neurochem.* **60:** 1631-1638
10. Pacheco M.F., Woodward D.J., McKelvy, J.F., Griffin, W.S., 1981: Trh in the rat cerebellum: II. Uptake by cerebellar slices, *Peptides.* **2:** 282-288

2

DPP IV-LIKE ENZYMES

Dipeptidyl Peptidase IV Gene Family

The DPIV family

TONG CHEN, KATERINA AJAMI[1], GEOFFREY W. MCCAUGHAN[1], MARK D. GORRELL[1], and CATHERINE A. ABBOTT

School of Biological Sciences, Flinders University of South Australia, Adelaide, Australia. 1AW Morrow Gastroenterology and Liver Centre, Royal Prince Alfred Hospital, Centenary Institute of Cancer Medicine and Cell Biology and The University of Sydney, New South Wales, Australia.

1. INTRODUCTION

DPIV, FAP/seprase and DPL1 (DPX/DP6) are members of the DPIV gene family. DPIV and FAP also belong to the prolyl oligopeptidase (POP)/S9 enzyme family. This family is characterized by homology in the last 200 C-terminal amino acid residues which contain a catalytic triad. The order of the catalytic triad is the reverse of the classical serine proteases, the nucleophilic Ser residue preceeds the Asp and His residues. DPIV and FAP are post-proline dipeptidyl amino peptidases which cleave after a penultimate proline near the amino terminus of their peptide substrates. In contrast, while DPL1 shares structural homology with members of this family it lacks the serine residue of the triad required for catalytic activity. This report focuses on three new members of the DPIV gene family, the novel enzymes DP8 and DP9 and a non-enzyme homologue DPL2 whose characterisation has been aided by data-base mining approaches.

Dipeptidyl Aminopeptidases in Health and Disease, Edited by Hildebrandt et al.
Kluwer Academic/Plenum Publishers, New York, 2003

Table 1. Comparison of physical characteristics of DPIV and its related proteins

Physical Characteristics	DPIV CD26 ADAbp	FAP Seprase	DP8	DP9	DPL-1 DP6 DPX	DPL-2
Accession Number (DNA)	M 80536	U09278	AF221634	AF542510	M96859 M96860	Tong *et al* BC030832
Number of amino acids	766	760	882	969 (partial)	865 803	789 796
DNA identity with DPIV		52%	27%	27%	33%	36%
Number of exons	26	26	20	>19	26	26
Gene size(kb)	81.8	72.8	71	>21	>911	1402
Gene location	2q24.2	2q24.3	15q22.32	19p13.3	7q36.2	2q14.1
Catalytic Ser-Asp-His triad	✓	✓	✓	✓	✗	✗
Intron near Ser motif	✓	✓	✗	✗	✓	✓
Transmembrane Domain	✓	✓	✗	✗	✓	✓
Potential N-linked Glycoslyation Sites	9	5	0	2	7	7
Monomer mobility on SDS page	110kDa	95 kDa	100kDa	?	?	100kDa
Dimer mobility on SDS -PAGE	150kDa	180kDa	None	None	?	None
DP activity	✓	✓	✓	✓	None	None

Key: √- yes; ×- no; ?- not known

2. NOVEL MEMBERS OF THE S9B FAMILY

Database searches for novel proteins related to DPIV and FAP were initially performed on the expressed sequence tags (EST) databases at the National Centre for Biotechnology Information. Both DP8 and DPL2 were identified using this approach. The discovery of the full-length DP8 cDNA, together with the completion of draft sequence of Human Chromosome 19 led to the search for the cDNA for DP9. While structural features are highly conserved in all six family members, the DPP-IV gene family has split to include four post-proline cleaving enzymes and two non-enzymes which appear to be mostly expressed in the brain (Table 1).

2.1 Novel enzymes in the DPIV gene family

While the DP8 and DP9 genes are encoded on different chromosomes their amino acid sequences share 60% amino acid identity and 77% amino acid similarity[1].

2.1.1 DP8

Overall DP8 shares 27% amino acid identity and 51% amino acid similarity with the protein sequences of DPIV and FAP, this increases to 35% amino acid identity and 57% amino acid similarity in the hydrolase domain[1] (Table1). Several features that are only observed in the S9b family are also evident in the primary structure of DP8, such as: the conserved pair of glutamates, Glu205Glu206, in blade 2 of DPIV that are essential to enzyme for enzyme activity; and the strong similarity (43%) between propeller blades 1 and 2[1]. It has been found that the Glu259 of DP8 that aligns with Glu205 of DPIV is essential for enzyme activity[2]. Besides this, unlike other members of S9b family, DP8 is a soluble cytoplasmic protein. Just like prolyl endopeptidase, it is active as a monomer and lacks N-linked and O-linked glycosylation sites. DP8 hydrolyses the prolyl bond after a penultimate proline, as do DPIV and FAP. Despite finding DP8 in the trans golgi, which is part of the secretion pathway, there is no evidence of secretion of DP8 by transfected COS cells[1]. However, it is possible that DP8 is secreted under circumstances not yet identified.

2.1.2 DP9

DP9 is the newest peptidase member of DPIV gene family. It was found that DP8 cDNA had high homology with two overlapping cosmids when a BLASTn search into GenBank was performed[3]. These cosmids encode a region of the human chromosome at 19p13.3 with a gene size greater than 47.5 kb and it also contains the residues required for DP enzyme activity (Table 1). The partial DP9 sequence (Genbank accession number AF542510) has no transmembrane domain, contains two N-linked glycosylation sites and an Arg-Gly-Asp (RGD) potential cell attachment sequence. The RGD motif is one of the best characterized integrin binding motifs[4], and it has therefore been hypothesized that if DP9 is externally expressed by a cell it may help them to mediate cell-cell adhesion through the binding of the motif to various integrin receptors. Northern blot hybridization showed ubiquitous DP9 mRNA tissue expression, similar to that of DP8 and DPIV (Ajami *et al*, in preparation).

2.2 Non-enzymes in the DPIV-gene family

While DPL1 contains two of the catalytic triad residues in its C terminal (Asp, His) it lacks the Ser residue in the appropriate motif (GKDYGG instead of GWSYGG). The nucleophilic serine is essential for catalysis thus DPL1 lacks DP activity. DPL1 shares structural homology with DPIV as post-proline activity can be reinstated simply by mutating the Asp residue in the recognition site to a Ser residue[5]. The cloning of a second human DPIV paralog that lacks the catalytic serine (see below), has led us to use the names Dipeptidyl Peptidase - Like (DPL) 1 and 2 for these proteins.

2.2.1 DPL1/DP6/DPX

DPL1 has at least two forms encoded by alternate splicing of exon 1. The N-terminal cytoplasmic tail of DPL1-long (DPL1-L) is 90 amino acids in length while DPL1-short (DPL1-S) is only 32 amino acids. DPL1-L has 33% amino acid identity with DPIV (Table 1). The relevance of these different cytoplasmic tails to the respective functions of each protein is still not understood. In rats, DPL1-L mRNA is expressed only in brain while DPL1-S is also expressed in prostate, kidney, ovary and testis and other organs but not in liver, spleen or heart[6,7].

Despite the absence of DP activity, DPL1 exerts an important developmental function. The mouse rump white mutation, which lacks expression of the DPL1 gene, is embryonic lethal in homozygotes and causes a pigmentation defect in heterozygotes[8].

2.2.2 DPL2

An expressed sequence tag with homology to DPIV, accession number H11799, derived from Soares infant brain, was identified using database mining. Using a combination of 5' rapid amplification of cDNA ends and reverse transcriptase polymerase chain reaction from human brain tissue a cDNA encoding a 789 aa protein was obtained. This protein has greatest homology (53 % amino acid identity and 70% amino acid similarity) with human DPL1-L. Like DPL1, it lacks the catalytic serine residue and is thus a non-peptidase member of the S9b family. Therefore, it was named dipeptidyl peptidase like protein 2 (DPL2). DPL2 contains a transmembrane domain and 7 potential N-linked glycosylation sites just like DPL1 (Table 1). The DPL2 gene is also alternatively spliced as recently another cDNA has been sequenced (GenBank accession number BC030832) which encodes a 796 aa protein which arises also from alternate use of exon 1. Interestingly, the DPL2 gene is located on chromosome 2 at 2q14.1, proximal to the DPIV and FAP genes, and is more closely related to DPIV and FAP than is DPL1 (Table 1).

A Master RNA Blot of 50 human tissues (Clontech, Palo Alto, USA) revealed high levels of DPL2 mRNA expression in both adult and fetal brain, adrenal gland and pancreas. Very weak mRNA expression was observed in the pituitary gland and kidney. DPL2 was not expressed in any other tissues examined (Tong *et al,* in preparation). The Master RNA Blot also showed different expression levels of DPL2 mRNA in different areas of brain. High mRNA expression levels were observed in the cortex while lower expression levels were observed in the cerebellum, which is similar to the expression pattern of DPL1-L [6,7].

The Master RNA blot results were confirmed by a Multiple Tissue Northern Blot (Clontech, Palo Alto, USA). Interestingly, six DPL2 mRNA transcripts were observed in brain tissue ranging from 3.8 to 7.4 kb in size. In contrast, in the pancreas, only two DPL2 mRNA transcripts were observed, one 4.0 kb and the other 4.4 kb. We do not know whether all these six forms of DPL2 mRNA will generate different isoforms of DPL2 protein with different sequence and different functions or whether the six forms will just contain varying amounts of 5' and 3' untranslated sequence.

We have expressed full-length DPL2 as a fusion protein using the cloning vector pcDNA3.1V5His. Recombinant DPL2 has a similar mobility to DP8 by SDS-PAGE and runs as a 100kDa monomer (Figure 1). The functions of DPL2 are so far unknown, but because it has high identity and similarity to DPL1, we suggest that it might have similar functions to DPL1 in embryonic development.

3. EVOLUTION OF DPIV GENE FAMILY

Current examinations of both the human and mouse draft genome sequences[9] (http://genome.ucsc.edu) lead us to believe that all members of the DPIV gene family have been identified. A mouse ortholog has been identified for each of the six members of DPIV gene family (Table 2). Examination of the genomic arrangement of these genes suggests that the ancestral gene of the DPIV gene family was similar to DP8 or DP9 (Tables 1 and 2). DP8 and DP9 have the smallest gene size and the fewest exons. In addition, the serine recognition site is contained in one exon for both of these genes but is split over two exons in the other family members.

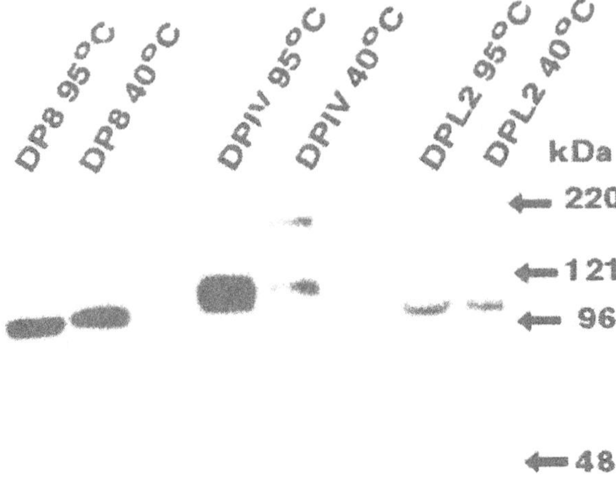

Figure 1. Western Blot of 8% SDS-PAGE using preparations of 293T cells transfected with DP8, DPIV and DPL2 fusion constructs and the anti-V5 mAb (Invitrogen, The Netherlands) to detect recombinant proteins. Recombinant proteins are expressed as C- terminal fusion proteins using the pcDNA3.1V5-His expression vector. Unboiled cell preparations demonstrate that only DPIV forms dimers.

4. SUMMARY

We have identified three novel members of the DPIV gene family using database mining approaches. Recombinant DP8 shares a post-proline dipeptidyl aminopeptidase activity with the closely related enzymes DPIV and FAP. The similarities between DP8, DP9 and DPIV in tissue expression pattern suggest a potential role for DP8 and DP9 in liver disease, T cell activation and immune function. The role of the two novel enzymes DP8 and DP9 and the other non-enzyme member DPL2 in human disease will be the focus of further studies.

Table 2. Comparison of human and mouse DPIV-gene family members

Physical Characteristics	DPIV CD26 ADAbp	FAP seprase	DP8	DP9	DPL-1 DP6 DPX	DPL-2
Number of amino acids						
Human	766	760	882	969(partial)	865 803	789 796
Mouse	760	761	883	869(partial)	✓ 804	✓
Number of exons						
Human	26	26	20	>22	26	26
Mouse	26	26	21	>19	26	✓
Gene size(kb)						
Human	81.8	72.8	71	>47.3	>911	1402
Mouse	85	73.1	>47	>25	905	✓
Gene location						
Human	2q24.2	2q24.3	15q22.32	19p13.3	7q36.2	2q14.1
Mouse	2	2	9	17	5	1

✓ Gene is present but not enough data to predict number of amino acids, exons and gene size.

ACKNOWLEDGEMENTS

The authors are grateful for DPIV cDNA from Dr Chikao Morimoto and tissue obtained from the NHMRC Brain Bank of South Australia.

REFERENCES

1. Abbott, C. A., G. W. McCaughan and M. D. Gorrell (1999). Two highly conserved glutamic acid residues in the predicted beta propeller domain of dipeptidyl peptidase IV are required for its enzyme activity. *FEBS Lett.* **458**(3): 278-284.

2. Abbott, C. A., D. M. T. Yu, E. Woollatt, G. R. Sutherland, G. W. McCaughan and M. D. Gorrell (2000). Cloning, expression and chromosomal localization of a novel human dipeptidyl peptidase (DPP) IV homolog, DPP8. *Eur. J. Biochem.* **267**(20): 6140-6150.
3. Abbott, C. A. and M. D. Gorrell (2002). The family of CD26/DPIV related ectopeptidases. *Ectopeptidases: CD13/Aminopeptidase N and CD26/Dipeptidylpeptidase IV in Medicine and Biology*. J. Langner and S. Ansorge. New York, Kluwer/Plenum. ISBN 0-306-46788-7: 171-95.
4. Dubljevic, V., A. Sali and J. W. Goding (1999). A conserved RGD (Arg-Gly-Asp) motif in the transferrin receptor is required for binding to transferrin. *Biochem. J.* **341**(Pt 1): 11-4.
5. Yokotani, N., K. Doi, R. J. Wenthold and K. Wada (1993). Non-conservation of a catalytic residue in a dipeptidyl aminopeptidase IV-related protein encoded by a gene on human chromosome 7. *Hum. Mol. Genet.* **2**(7): 1037-9.
6. Wada, K., N. Yokotani, C. Hunter, K. Doi, R. J. Wenthold and S. Shimasaki (1992). Differential expression of two distinct forms of mRNA encoding members of a dipeptidyl aminopeptidase family. *Proc. Natl. Acad. Sci. USA* **89**(1): 197-201.
7. de Lecea, L., E. Soriano, J. R. Criado, S. C. Steffensen, S. J. Henriksen and J. G. Sutcliffe (1994). Transcripts encoding a neural membrane CD26 peptidase-like protein are stimulated by synaptic activity. *Brain Res Mol Brain Res* **25**(3-4): 286-96.
8. Hough, R. B., A. Lengeling, V. Bedian, C. Lo and M. Bucan (1998). *Rump white* inversion in the mouse disrupts dipeptidyl aminopeptidase-like protein 6 and causes dysregulation of *Kit* expression. *Proc. Natl. Acad. Sci. USA* **95**: 13800-13805.
9. McPherson, J. D., *et al.* (2001). A physical map of the human genome. *Nature* **409**(6822): 934-41.

Seprase-DPPIV Association and Prolyl Peptidase and Gelatinase Activities of the Protease Complex

G. GHERSI [1], H. DONG [2], L.A. GOLDSTEIN [2], Y. YEH [2], L. HAKKINEN [3], H.S. LARJAVA[3] and W-T. CHEN [2].
[1] *Department of Cellular and Developmantal Biology, University of Palermo, Viale delle scienze 90138 Palermo, Italy.* [2] *Department of Medicine/Medical Oncology, State University of New York, Stony Brook New York 11794-8154, USA.* [3]*University of British Columbia, Division of Periodontics, 2199 Westbrook Mall, Vancouver, BC V6T 1Z3, CANADA.*

1. INTRODUCTION

Tissue repair requires remodeling of the extracellular matrix (ECM) by migratory cells[1,2]. Such cellular activities occur on membrane protrusions invadopodia[3], which exhibit dynamic membrane mobility, cell adhesion molecules and proteolytic enzymes. Collagenolytic activities were found to be involved in these processes[1]. One of the matrix metalloproteases, type I collagenase is rapidly induced in human skin at the wound-edge after acute injury, its activity persists during healing and stops at wound-closure[4]. Moreover, cells in healing wounds express more proteases, including other matrix metalloproteases[4] and neutrophil elastase[5], than their quiescent counterparts.

Recent evidence has demonstrated the involvement of serine-integral membrane proteases (SIMP), including dipeptidyl peptidase IV (DPPIV)/CD26 and seprase, in cell surface proteolysis and cell migration[6]. SIMP members are type II transmembrane proteins, with a short cytoplasmic tail of 6 amino acids (a.a.) followed by a 20 a.a. (seprase) or 22 a.a. (DPPIV) transmembrane domain at the N-terminus and a stretch of 200 a.a. at the C-terminus containing the catalytic region with the catalytic triad in a non-classical orientation[7,8]. DPPIV specifically removes N-terminal dipeptides

Dipeptidyl Aminopeptidases in Health and Disease, Edited by Hildebrandt et al.
Kluwer Academic/Plenum Publishers, New York, 2003

from oligo-peptides, which include Neuro-Peptide Y and other peptide hormones, with either L-proline, L-hydroxyproline, or L-alanine at the penultimate position [9,10]. DPPIV has been shown to be an adhesion receptor for collagen[11] or fibronectin[12]. In addition, a recent report showed that DPPIV also possesses a seprase-like gelatinase activity and therefore endopeptidase activity[13], suggesting its involvement in collagen degradation. DPPIV is expressed constitutively on brush border membranes of intestine and kidney epithelial cells (14). However, its transient expression on T-cells has been implicated as a marker for T-cell activation[14].

Seprase, originally identified as a 170 kDa membrane-bound gelatinase is expressed on invadopodia of highly aggressive LOX human melanoma cells[15,16]. The active enzyme is a homodimer of 97 kDa subunits, which are proteolytically inactive[8]. Analysis of the deduced amino acid sequence from a cDNA that encodes the 97 kDa subunit[7] revealed that it is homologous to DPPIV, and is essentially identical to fibroblast activation protein α (FAPα), which is expressed on reactive stromal fibroblasts of epithelial cancers and healing wounds[17].

To define the role of seprase and DPPIV in the tissue invasive phenotype, we investigated the functional expression of these proteases on surfaces of migratory connective tissue cells. We have identified an invadopodia-specific protease complex consisting of DPPIV and seprase, which is activated on human connective tissue cells in response to wounding. The novel protease complex having both prolyl peptidase and gelatinase activities exhibits synergistic action with matrix metalloproteases and contributes to the local degradation of type I collagen and cellular migration, necessary for tissue repair[18].

2. RESULTS

2.1 Immuno-isolation of the seprase-DPPIV complex

To isolate the seprase-DPPIV complex, monoclonal antibodies (mAbs) against seprase and DPPIV were used. Immunoprecipitation using mAbs against seprase or DPPIV were performed on plasma membrane components labeled with biotin; it has identified two major similar intensity bands in

SDS gels, the top band at 200 kDa was identified by immunoblotting as DPPIV, and the lower band migrating at 170 kDa as seprase. However, when cell surface components were cross-linked using BS^3 crosslinker (Peirce) before SDS solubilization, the heteromeric aggregate dissociated into two stable dimers of 200 kDa DPPIV and 170 kDa seprase, In three independent experinnents involving RIPA cell extracts, a stable association of seprase and DPPIV was detected using mAbs against seprase and DPPIV.

2.2 Gelatinolytic activity of the seprase-DPPIV complex

To determine the proteolytic activities of the immuno-isolated complex, antigens were isolated from RIPA extracts WI38 human lung embryonic fibroblasts by affinity purification using mAbs that recognize either seprase or DPPIV. The eluates were analyzed for a 170 kDa (seprase) gelatinase, 200 kDa (DPPIV) proline-specific dipeptidyl-aminopeptidase, and denatured type I collagen-degrading activity. Gelatin zymography immunoprecipitates a 170 kDa gelatinase activity. As previously shown by zymography[8], that DPPIV dimer exhibited no gelatinase activity, the 170 kDa band on the gelatin zymogram identified with the DPPIV antibody represented the presence of seprase in the protease complex. Similarly, substrate overlay assay detected a 200 kDa proline-specific dipeptidyl-aminopeptidase activity in immunoprecipitates of anti-seprase mAb or anti-DPPIV mAb. No 170 kDa gelatinase or DPPIV activity could be observed for immunoprecipitates using antibodies against αv, α2, α6 or β3 integrin or control IgG. Previous studies using zymography and substrate overlay assay involved SDS PAGE separation of proteins, showed that seprase was active as an 170 kDa dimer and degraded gelatin in the 170 kDa band of the SDS gel; DPPIV was also active as a 200 kDa dimer and cleaved prolyl dipeptides in the 200 kDa band of the substrate overlay membrane. To determine proteolytic activity of the seprease-DPPIV complex, a soluble collegen-degrading assay was used, in which the release of peptide fragments from biotinylated type I collagen gel by immuno-isolated protease complexes was measured. Both seprase and DPPIV immuno-isolates exhibit collagen-degrading activities but integrin complexes (β1 or β3 integrins) or control immuno-isolates do not. Importantly, uncomplexed seprase derived from LOX melanoma cells and uncomplexed DPPIV isolated from bovine kidney brush border membranes by mAb affinity chromatography did not show collagen degrading activity.

2.3 Prolyl peptidase activity of the seprase-DPPIV complex

As described above, a soluble enzymatic assay using a classical color substrate Gly-Pro-pNA was used to evaluate prolyl dipeptidase activity of the protease complex. The complex was purified from MDA-MB-436 cells using different mAbs against DPPIV or seprase, seprase was obtained from Lox cells and recombinant DPPIV from DPPIV transfected COS-1 cells. All three proteases examined exhibited the peptidase activity against the Gly-Pro-pNA substrate, confirming the highly homologous catalytic sites of individual dipeptidases, i.e. seprase and DPPIV.

2.4 Involvement of the seprase-DPPIV complex in the local collagen degradation and cell migration

To determine the role of the seprase-DPPIV complex in the local collagen degradation and cell migration, we overlaid a thin layer of type I collagen on monolayers to establish an in vitro wound model for morphological examination. Cell migration and associated local collagen removal were measured by counting the area of cell migration/collagen removal using image analysis (NIH Image 1.62b4/fat program) in conjunction with phase contrast and fluorescence microscopy on live cells. We observed that in a collagen gel, WI38 cells at the wound edge migrated into the gel and close the wound within 2 days. An evenly intensive red fluorescent collagen gel that covered the cell layer and glass surface were seen at the beginning, however, local collagen degradation and extensive cell migration occurred from the wound edge within 18 h. Addition of the inhibitory mAbs against DPPIV into the wound-closure model blocked cell migration and local collagen removal, while class matched mAbs against other cell surface antigens did not. There was an increase in inhibition with increasing amounts of inhibitory mAbs and the antibody inhibitory effect could be reversed by removing it from the culture. Furthermore, the local collagen degradation by activated cells was quantified by counting fluorescent peptides released from fluorescent collagen fibers by WI38 cells in a sparse culture using spectrofluorometry. Cells in sparse culture are known to be migratory due to less "contact inhibition of migration"[3]. Migratory WI38 cells showed time-dependent collagen degradation within 4 days, and specific inhibitory mAb against DPPIV inhibited collagen degradation by migratory cells while the control mAb against glycoprotein 90 did not. These data demonstrate the involvement of the seprase-DPPIV complex of wound-activated fibroblasts in collagen degradation and cellular migration.

2.5 Immuno-localization of the seprase-DPPIV complex on invadopodia

To examine the localization of the seprase - DPPIV complex on a specialized plasma membrane domain, invadopodia, immunofluorescence experiments were performed on activated fibroblast in collagen gels. We found that the local degradation of collagen fibers occurred at regions immediately adjacent to invadopodia of the cell at the wound edge. Coincidently, invadopodia of the cell migrating from the wound edge were stained positively with FITC-mAb against seprase and TRITC-mAb against DPPIV. Labeled type-I collagen fibers were removed from the region immediately adjacent to invadopodia where seprase and DPPIV were accumulated. Furthermore, mAbs directed against DPPIV and seprase demonstrated co-localization of these molecules in the same invadopodia. Such labeling should be specific as FITC-mAb against β_1 integrin stained large portion of plasma membrane and only co-localized with seprase in sites where the local collagen degradation was observed.

2.6 Induction of seprase and DPPIV expression in stromal fibroblasts during wound closure *in vivo*

To analyze the presence of seprase and DPPIV in stromal fibroblasts during wound closure *in vivo,* we localized these molecules in human oral mucosal wounds by using immunohistochemistry. A high expression of both proteases was seen in connective tissue cells adjacent to wound within day 3 after wounding. No immuno-reaction was seen in the fibrin clot area, epithelium or in the unwounded normal mucosa. After 7 days, seprase and DPPIV was expressed by cells in the middle of granulation tissue. Their expression was down regulated and no immunoractivity for seprase or DPPIV was observed in 14- or 28-day-old wounds.

3. CONCLUSION

We have observed that the seprase-DPPIV complex becomes localized on cell surface invadopodia of migrating fibroblasts to degrade surrounding connective tissue components locally. It has been reported that DPPIV is expressed constitutively on apical plasma membranes of epithelial cell types. The DPPIV dimer may act on epithelial surfaces for processing collagen-like macromolecules. The complex formation of DPPIV with seprase appears to be important for other cell types that are activated to express cellular invasiveness. Supporting this hypothesis, we observed the transient expression of the seprase-DPPIV complex to invadopodia of connective tissue cells that are activated in response to wounding. Invadopodial localization of the protease complex may provide a driving force for cell migration occurring during wound closure. In previous studies, we suggested that seprase was a biomarker for cell invasiveness of melanoma and other carcinoma[15,16]. Other studies also demonstrated the involvement of seprase and matrix metalloproteases in breast cancer cell invasion[19,20]. It is also apparent that the seprase-DPPIV complex is an invasiveness marker for connective tissue cells and an activation marker for cell migration as well. However, molecular inducers for expression of the seprase-DPPIV complex in these tissue cells remain to be elucidated.

Collagen fibers in 50 μm-zone adjacent to the leading edge of the cell were removed by migrating cells that may be due to the mobility and degradative activity of invadopodia in 3-D collagen fibers. This differs from that of crosslinked gelatin films by transformed cells where degradation spots on planar substratum were observed[21]. Our observations provide several lines of evidence supporting the role of the seprase-DPPIV complex in the local collagen degradation and cell migration. These included the demonstration of peptidase and gelatinase activities of the protease complex , inhibition of the local collagen degradation and cellular migration by specific mAbs against the seprase-DPPIV complex. In addition, we found that the seprase-DPPIV complex could act synergistically with matrix metalloproteases in the local collagen degradation and cell migration. Membrane type-1 matrix metalloprotease and active matrix metalloprotease-2 can degrade cell-associated collagen immediately adjacent to invadopodia, and their inhibitors block the collagen-degrading activity[22]. However, it is still not clear how collagenases of the matrix metalloprotease family are involved in these processes.

REFERENCES

1. Birkedal-Hansen, H., Moore, W.G.I., Bodden, M.K., Windsor, L.J., Birkedal-Hansen, B., DeCarlo, A., and J.A.Engler, 1993, Matrix Metalloproteinases: A Review, *Critical Reviews in Oral Biology and Medicine* **4:** 197.
2. Martin, P., 1997, Wound healing - Aiming for perfect skin regeneration, *Science* **276:** 75.
3. Chen, W.-T., 1979, Induction of spreading during fibroblast movement, *J.Cell Biol* **81:** 684.
4. Young, P.K., and Grinnell, F., 1994, Metalloproteinase activation cascade after burn injury: A longitudinal analysis of the human wound environment, *J.Invest.Dermatol.* **103:** 660.
5. Grinnell, F., and Zhu, M., 1994, Identification of neutrophil elastase as the proteinase in burn wound fluid responsible for degradation of fibronectin, *J.Invest.Dermatol.* **103:** 155.
6. Chen, W.T., 1996, Proteases associated with invadapodia, and their role in degradation of extracellular matrix., *Enzyme Protein* **49:** 59.
7. Goldstein, L.A., Ghersi, G., Piñeiro-Sánchez, M.L., Salamone, M., Yeh, Y.Y., Flessate, D., and Chen, W.-T., 1997, Molecular cloning of seprase: A serine integral membrane protease from human melanoma, *Biochimica et Biophysica Acta:* Molecular Basis of Disease **1361:** 11.
8. Pineiro-Sanchez, M.L., Goldstein, L.A., Dodt, J., Howard, L., Yeh, Y., Tran, H., Argraves, W.S., and Chen, W.-T., 1997, Identification of the 170-kDa melanoma membrane-bound gelatinase (seprase) as a serine integral membrane protease. *J.Biol.Chem.* **272:** 7595.
9. Ghersi, G., Chen, W., Lee, E.W., and Zukowska, Z., 2001, Critical role of dipeptidyl peptidase IV in neuropeptide Y-mediated endothelial cell migration in response to wounding, *Peptides* **22:** 453.
10. Heins, J., Welker, P., Schonlein, C., Born, J., Hartrodt, B., Neubert, K., Tsuru, D., and Barth, A., 1988, Mechanism of proline-specific proteinases: (I) substrate specificity of dipeptidyl peptidase IV from pig kidney and proline-specific endopeptidase from Flavobacterium meningosepticum., *Biochim.Biophys.Acta* **954:** 161.
11. Löster, K., Zeilinger, K., Schuppan, D., and Reutter, W., 1995, The cysteine-rich region of dipeptidyl peptidase IV (CD 26) is the collagen-binding site, *Biochem.Biophys.Res.Commun.* **217:** 341.
12. Abdel-Ghany, M., Cheng, H., Levine, R.A., and Pauli, B.U., 1998, Truncated dipeptidyl peptidase IV is a potent anti-adhesion and anti- metastasis peptide for rat breast cancer cells, *Invasion Metastasis* **18:** 35.
13. Bermpohl, F., Löster, K., Reutter, W., and Baum, O., 1998, Rat dipeptidyl peptidase IV (DPP IV) exhibits endopeptidase activity with specificity for denatured fibrillar collagens, *FEBS Lett* **428:** 152.
14. Morimoto, C., and Schlossman, S.F., 1994, CD26: A key costimulatory molecule on CD4 memory T cells, *The Immunologist* **2:** 4.

15. Mueller, S.C., Ghersi, G., Akiyama, S.K., Sang, Q.X., Howard, L., Pineiro-Sanchez, M., Nakahara, H., Yeh, Y., and Chen, W.-T., 1999, A novel protease-docking function of integrin at invadopodia, *J.Biol.Chem.* **274:** 24947.
16. Monsky, W.L., Lin, C.-Y., Aoyama, A., Kelly, T., Mueller, S.C., Akiyama, S.K., and Chen, W.-T., 1994, A potential marker protease of invasiveness, seprase, is localized on invadopodia of human malignant melanoma cells, *Cancer Res.* **54:** 5702.
17. Garin-Chesa, P., Old, L.J., and Rettig, W.J., 1990, Cell surface glycoprotein of reactive stromal fibroblasts as a potential antibody target in human epithelial cancers, *Proc.Natl.Acad.Sci.USA* **87:** 7235.
18. Ghersi, G., Dong, H., Goldstein, L.A., Yeh, Y., Hakkinen, L., Larjava, H.S., and Chen, W.-T., 2002, Regulation of fibroblast migration on collagenous matrix by a cell surface peptidase complex, *J.Biol.Chem.* **277:** 29231.
19. Kelly, T., Yan, Y., Osborne, R.L., Athota, A.B., Rozypal, T.L., Colclasure, J.C., and Chu, W.S., 1998, Proteolysis of extracellular matrix by invadopodia facilitates human breast cancer cell invasion and is mediated by matrix metalloproteinases, *Clinical & Experimental Metastasis* **16:** 501.
20. Kelly, T., Kechelava, S., Rozypal, T.L., West, K.W., and Korourian, S., 1998, Seprase, a membrane-bound protease, is overexpressed by invasive ductal carcinoma cells of human breast cancers, *Modern Pathology* **11:** 855.
21. Chen, W.-T., 1989, Proteolytic activity of specialized surface protrusions formed at rosette contact sites of transformed cells, *J.Exp.Zool.* **251:** 167.
22. Nakahara, H., Howard, L., Thompson, E.W., Sato, H., Seiki, M., Yeh, Y., and Chen, W.-T., 1997, Transmembrane/cytoplasmic domain-mediated membrane type 1-matrix metalloprotease docking to invadopodia is required for cell invasion, *Proc.Natl.Acad.Sci.U.S.A* **94:** 7959.

Dipeptidyl Peptidase-IV Activity and/or Structure Homologues (DASH) in Transformed Neuroectodermal Cells

[1,2]RADEK MALÍK, [1]PETR BUŠEK, [1]VLADISLAV MAREŠ, [1]JAN ŠEVČÍK, [1]ZDENĚK KLEIBL, and [1]ALEKSI ŠEDO

[1]Joint Laboratory of Cancer Cell Biology of the Institute of Biochemistry and Experimental Oncology, 1st Faculty of Medicine of Charles University Prague and Institute of Physiology of Academy of Sciences of Czech Republic, U Nemocnice 5, Prague 2, 12853, Czech Republic; [2]Department of Oncology of the First Faculty of Medicine and General Teaching Hospital, U Nemocnice 2, 12808, Prague 2, Czech Republic

1. INTRODUCTION

Dipeptidyl peptidase-IV (DPP-IV, EC 3.4.14.5, CD26) was for many years believed to be a unique cell membrane protease cleaving X-Pro dipeptides from the N-terminal end of peptides and proteins. DPP-IV has been implicated in numerous physiological functions, although the precise underlying mechanisms await further clarification[1].

Subsequently, a number of other molecules exhibiting DPP-IV-like enzyme activity, thus possibly sharing similar set of catalytic partners, were discovered. A term "DPP-IV activity and/or structure-homologues" (DASH) has been coined for the group comprising DPP-IV, Fibroblast activation protein-α, DPP6, 8, 9, Attractin, N-acetylated-γ-linked-acidic dipeptidases I, II and L, Quiescent cell proline dipeptidase, Prolyl carboxypeptidase, Thymus-specific serine protease, DPP-IV-β etc. Furthermore, biological functions mediated by DASH independently on their hydrolytic potential[1] likewise "on-demand" expression of enzymatically active or inactive form of DASH were suggested[2,3]. Molecular complexity of DASH could explain some seeming contradictions of formerly hypothesized DPP-IV roles[1].

Dipeptidyl Aminopeptidases in Health and Disease, Edited by Hildebrandt et al.
Kluwer Academic/Plenum Publishers, New York, 2003

Functional studies of most of DASH are still lacking and particular biological functions of these molecules as yet remain mostly speculative rather than proven. However, the possible ability of DASH to complement and/or functionally substitute DPP-IV/CD26 – on the level of its enzymatic activity – implies that DASH may be involved in analogous processes like DPP-IV. Similar substrate specifity combined with differences in other enzymatic properties of particular DASH may be the reason why a subtle difference in their expression pattern might be of biological relevance.

The objective of our work is to analyse complexity of DPP-IV-like enzymatically active DASH molecules in transformed neuroectodermal cells.

2. METHODS

Panel of neuroectodermal cell lines of different growth rate and degree of malignity was used (Tab.1). Cells were cultured under the standard conditions according to provider (ETCC) instructions. Proliferation rate was quantified in exponentially growing cells by DNA flow cytometry, expression of glial differentiation marker GFAP was determined by immunohistochemistry.

Table 1. Characteristics of the neuroectodermal cell lines studied

Cell line	U87	U138	T98G	U373	Hs683	SK-MEL-1	SK-MEL-28	Hs294T
Origin*/grade	G/IV	G/IV	G/IV	G/III	G/I	M	M	M
Proliferation rate (S phase%)	18	23	31	20	25	n.d.	19	19
GFAP	+++	-	-	+++	-	n.d.	n.d.	n.d.

*G – glioma, M – melanoma; n.d. – not determined

In experiments with quiescent cells, culture media were devoid of foetal calf serum.

Expression of particular DASH mRNA was assayed using reverse transcription PCR, the specificity of PCR products was confirmed by cyclic sequencing using the same PCR primers. Enzyme activities were measured by continuous fluorimetric assay. Further characterization of enzymatic and molecular properties of individual DASH was performed by inhibition studies, subcellular localization, gel filtration and gradient polyacrylamide gel electrophoresis (GPAGE) with fluorescent contact print enzyme activity detection. Abovementioned methods were used as we described before[2].

3. RESULTS

3.1 Expression of DASH mRNA

As shown in Fig.1, mRNA of DPP8, QPP, Attractin, DPP-IV and FAP-α is expressed in most cell lines studied, cultured both under standard or serum-deprived conditions ("proliferating" and "quiescent" cells respectively). Negative results were observed in case of DPP-IV/CD26 in SK-MEL-28 and T98G cells. Moreover, FAP-α was absent in SK-MEL-28 cells and downregulated by withdrawing of serum in T98G cells.

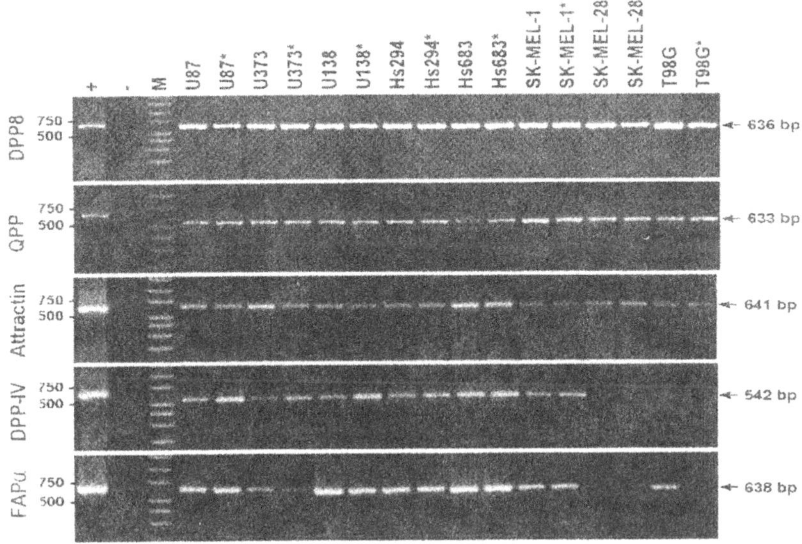

Figure 1. RT-PCR analysis of DASH expression in neuroectodermal cell lines in proliferating and *quiescent populations

3.2 Characterization of DASH enzymatic and molecular properties

There are basically two patterns of DPP-IV-like enzyme activity after the gel chromatography separation of the cell line extracts. They comprise 2 or 3 peaks of Glycyl-Prolyl-7-amino-4-methycoumarin (G-P-NHMec) cleaving enzyme activity, differing in the molecular weight and pH preference. Two peaks (2 and 3 in Fig.2A) were found in U373, T98G, SK-MEL-28 and HS294T, whereas three peaks (1-3 in Fig.2B) in U87, U138 and Hs683 cells. Enzyme activity in fractions from peaks 1 and 3, in contrast with the ones from peak 2, was trypsin treatment resistant.

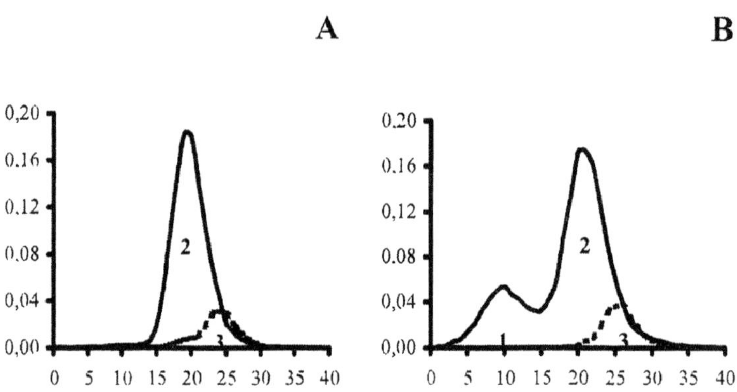

Figure 2. Distribution of DPP-IV-like activity from neuroectodermal cell lines after Sephacryl S-300 gel filtration. Vertical axes: relative DPP-IV-like enzyme activity measured with GP-NHMec as a substrate; horizontal axes: number of fraction. Solid line: pH 7.4, dashed line: pH 5.5. 669, 440, 140 (kDa) MW markers were eluted in fractions 7, 13, 25 respectively. Panel A: U373, T98G, SK-MEL-28, HS294T Panel B:U87, U138, Hs683.

Inhibition studies (Tab.2) showed some preferences toward DPP-IV-like activity from peaks 1-3: Ala-Pyr-CN 1>2>>3; Diprotin A: 1>2>>3; β-mercaptoethanol inhibits preferentially enzyme activity of peak 3, whereas Bestatin is a potent inhibitor of DPP-IV-like activity of peak 1, with almost no effect toward the enzyme activity from peaks 2 and 3. Moreover, DPP-IV-like activity of peak 1 was trypsin resistant (not shown). The efficiency

Table 2. IC_{50} of selected inhibitors of DPP-IV-like enzyme activity in peaks from gel filtration

Peak No	**1**	**2**	**3**
Ala-Pyrr-CN	0.01 μM	0.1 μM	> 10 μM
Arg(PMC)-Pyrr-CN	0.005 μM	0.05 μM	1 μM
Ile-Thia	0.5 μM	2 μM	> 20 μM
Ile-Pyr	0.5 μM	2 μM	> 20 μM
Diprotin-A	1 μM	5 μM	20 μM
Diprotin-B	5 μM	5 μM	> 30 μM
Bestatin	10 μM	> 50 μM	> 50 μM
β-mercaptoethanol	2%	1%	< 1%

of all inhibitors was very similar in all cell lines tested. Considering all these observations, we assume the DPP-IV-like enzyme activity might be, eventhough not exclusively, carried by Attractin in peak 1. Indeed, immunoreactivity of Attractin has been observed in peak 1 from U87 cells previously[2]. Acidic pH optimum suggests presence of QPP in peak 3. Molecular species carrying DPP-IV-like enzyme activity in peak 2 could only be speculated.

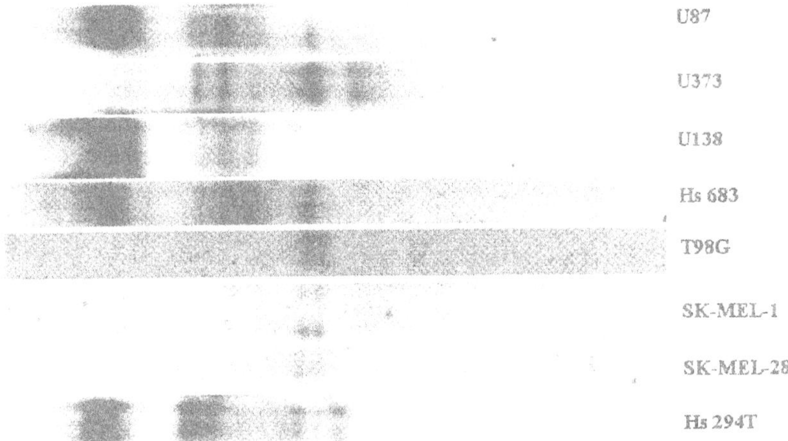

Figure 3. Non-denaturing GPAGE of DPP-IV-like enzyme activity from neuroectodermal cell lines.

Non-denaturing GPAGE argues for further cell line specific DASH heterogeneity. (Fig.3). This heterogeneity may be caused by presence of more DASH, differences in posttranslational processing, existence of more splice variants of particular DASH as well as by the presence of naturally existing heterodimers[4] or artificial complexes of some DASH molecules. However, absence of slowly migrating bands observed in GPAGE combined with no detectable levels of DPP-IV and FAPα mRNA (Fig.1) in material from T98G and SK-MEL-28 cells should be noticed.

Taken together, decisive identification of cell line specific patterns of DPP-IV-like enzyme activity bearing molecules of different MW is still not possible.

3.3 DASH enzyme activities in quiescent cells

To study the modifying effect of cell growth rate on the specific DASH expression pattern, DPP-IV-like enzyme activities were evaluated in quiescent cells deprived of growth factors by withdrawal of serum from culture medium. The effect was monitored within 6 to 72 hours period. This treatment caused different results in the individual cell lines.

In glioma IV grade U87 and U138 cells, the enzyme activity attributable to Attractin increased three times, whereas the remaining two DPP-IV-like activity fractions remained unchanged. On the contrary, about double elevation of the hypothetical QPP, together with a similar increase in Attractin activity, was observed in glioma I grade Hs683 cells. Double rise of QPP was found also in Attractin negative T98G glioma IV grade cells. Slight elevation of QPP activity was also observed in melanoma cell lines SK-MEL-28 and Hs294T. All these changes were time dependent.

4. CONCLUDING REMARKS

Final matching of the particular DPP-IV-like enzyme activity with individual DASH molecule in the panel of cell lines used in this study still remains partially speculative. However, our data suggests the following conclusions:

(i) DASH expression pattern and its changes during transition of cells from the proliferation to the quiescent stage of growth probably depend on cell type. Its functional impact may finally depend on the availability of catalytic partners in the vicinity of the particular cell.

(ii) Change of DASH formula, seemingly hidden behind the whole DPP-IV-like hydrolytic activity, may shift particular substrate preference, resulting in modified response of the cell to the signals carried by biologically active DASH substrates. It may therefore be not only the inappropriate production of a regulatory molecule, but also the disordered regulation of its processing system which could be of pathogenetic relevance. Together, we assume that DASH molecules may represent an important phenotypic feature related to the actual cell growth properties. Such conclusion is indirectly supported also by our preliminary studies of DASH downregulation by RNAi[5]. Inhibition of cell proliferation and decreased survival was observed in cells exposed to Attractin-specific siRNAs, whereas downregulation directed to other DASH (DPP-IV, 8, 9, FAPα, QPP) have only slight effect on cell growth (unpublished results).

(iii) Inhibitors of DPP-IV-like enzymatic activity are expected to be of a significant therapeutic value in the treatment of HIV infection, diabetes mellitus and as immunosuppressants in the transplantation surgery and autoimmune diseases, including multiple sclerosis[6]. Due to an almost ubiquitous expression and multifunctional nature of the majority of DASH, it may be difficult to predict the final physiological consequences of inhibition of a particular DASH in different organ systems. Hence, the prerequisite for exploiting DASH as the pharmacological targets is understanding the subtle differences in their enzymatic activity parameters, "specifity" derived from their organ/tissue dependent expression patterns and their dynamics, which are at least in some cases determined by the functional status of the cell[1]. Despite of these objections, modulation of DASH enzyme activity may be an attractive therapeutic option in the future.

NOTES

Corresponding author: AS, Tel/Fax +42 2 2496 5826, E-mail: Aleksi@mbox.cesnet.cz, www.lf1.cuni.cz/lbnb

ACKNOWLEDGEMENTS

This work was supported by grants 301/02/0962 and 7/2002/C from Grant Agency of Czech Republic and Grant Agency of Charles University respectively and the research project "Oncology" of the 1st Faculty of Medicine of Charles University. Skilful technical help of Mrs Květoslava Vlašicová and Karin Roubíčková is greatly appreciated.

REFERENCES

1. Šedo, A., Malík, R., 2001, Dipeptidyl peptidase IV – like molecules: homologous proteins or homologous activities? *Biochim. Biophys.Acta,* **1550:** 107-116.
2. Malík, R., Mareš, V., Kleibl, Z., Pohlreich, P., Vlašicová, K., Šedo, A., 2001, Expression of attractin and its differential enzyme activity in glioma cells. *Biochem. Biophys. Res. Commun.* **284:** 289-94.
3. Tang, W., Gunn, T.M., McLaughlin, D.F., Barsh, G.S., Schlossman, S.F., Duke-Cohan, J.S., 2000, Secreted and membrane attractin result from alternative splicing of the human ATRN gene. *Proc. Natl. Acad. Sci. USA* **97:** 6025-6030.
4. Scanlan, M.J., Raj, B.K., Calvo, B., Garin-Chesa, P., Sanz-Moncasi, M.P., Healey, J.H., Old, L.J., Rettig, W.J., 1994, Molecular cloning of fibroblast activation protein alpha, a member of the serine protease family selectively expressed in stromal fibroblasts of epithelial cancers. *Proc. Natl. Acad. Sci. U S A* **9:** 5657-61.
5. Elbashir, S. M., Harborth, J., Weber, K., Tuschl, T., 2002, Analysis of gene function in somatic mammalian cells using small interfering RNAs. *Methods* **26:** 199-213.
6. Augustyns, K., Bal, G., Thonus, G., Belyaev, A., Zhang, X.M., Bollaert, W., Lambeir, A.M., Durinx, C., Goossens, F., Haemers, A., 1999, The unique properties of dipeptidyl-peptidase IV (DPP IV/CD26) and the therapeutic potential of DPP IV inhibitors. *Curr. Med. Chem.* **6:** 311-27.

Characterisation of Human DP IV Produced by a *Pichia pastoris* Expression System

JOACHIM W. BAER*, BERND GERHARTZ#, TORSTEN HOFFMANN*, FRED ROSCHE*, and HANS-ULRICH DEMUTH*
Probiodrug AG, Weinberweg 22, D-06120 Halle, Germany, # Novartis Pharma AG, Klybeckstr. 141, CH-4057 Basel, Switzerland

1. INTRODUCTION

Dipeptidyl peptidase IV (DP IV, EC 3.4.12.5) is a serine protease, which removes N-terminal dipeptides processing proline or alanine residues at the N-terminal penultimate position. DP IV occurs as a type II plasma membrane protein as well as in a soluble form[1,2]. DP IV has received interest due to its involvement in diabetes mellitus, AIDS or immunoregulatory processes[3]. Because of the growing interest in DP IV-function and structure we looked for an expression system to produce larger amounts of functional active DP IV.

We decided to use the methylotrophic yeast *Pichia pastoris* which has gained a widespread attention as an expression system because of its potential to produce large quantities of heterologous protein[4]. We favored *P. pastoris* over *Saccharomyces cerevisiae* as posttranslational modifications are reported to be more similar to those in mammalian cells[5].

For recombinant protein expression a commonly used approach is to express the gene of interest under the control of the AOX1 promotor. This promotor controls the expression of alcohol oxidase I which is responsible for the first step in the methanol metabolism[6]. In wild type *Pichia* the alcohol oxidase content can reach up to 30 % of the total protein amount of the cell. In a fermentation process the yeast has shown the ability to grow to high cell densities capable of giving high levels of expressed protein. After removing

Dipeptidyl Aminopeptidases in Health and Disease, Edited by Hildebrandt et al.
Kluwer Academic/Plenum Publishers, New York, 2003

the membrane anchor of (Δ1-36) DP IV secretion into the media of the functional active extracellular part proved to be efficient.

In this article we describe the cloning, expression and purification and stated the comparable properties of soluble recombinant DP IV produced in the yeast *P. pastoris.*

2. MATERIAL AND METHODS

Cloning and expression. The coding sequence of soluble human DP IV (Δ1-36) plus His_6-tag was cloned into pPICαC (Invitrogen). Transfection and expression within the *P. pastoris* system was carried out regarding the protocols from Invitrogen. Fermentation was performed according to Invitrogen *Pichia* fermentation process guidelines in a Biostat C-fermenter from B.Braun.

Purification. The fermentation media was centrifuged at 40,000*g to remove the cells. The supernatant was filtered over a 0.45 μm membrane filter and concentrated to 100 ml using a tangential flow system (Satorius; cut-off: 30 kDa). Hydrophobic interaction chromatography of the obtained concentrate was carried out using Phenyl Sepharose Fast Flow (Pharmacia). Fractions with the highest DP IV activity were dialysed overnight against 4 L 300 mM NaCl, 5 mM imidazole, 50 mM NaH_2PO_4-buffer pH 7.6. Affinity chromatography was carried out using a Ni-NTA Sepharose column attached to a FPLC system (Pharmacia). Purification was controlled by DP IV-activity measurements and SDS-PAGE.

Characterisation. Kinetic experiments were conducted at 30°C in 0.04 M HEPES-buffer, pH 7.6. Fluorometric measurements of the cleavage of AMC substrates were carried out in a SFM25 fluorimeter (Kontron), photometrical measurements of pNA-substrates were carried out using an UV/Vis spectrometer (Perkin Elmer). The isoelectric point of recombinant DP IV was determined using Servalyte pre nets gels (gradient pH 3 to 10) on a Multiphor II (Pharmacia). MALDI-TOF mass spectrometry was performed on a LD-TOF Hewlett Packard G2025 system.

3. RESULTS

The coding sequence of human DP IV was cloned in the secretional *P. pastoris* expression vector pPICαC. The transmembrane and the cytosolic part of DP IV was removed (Δ1-36) and substituted with a His_6-tag to express the enzyme in a soluble form (Figure 1).

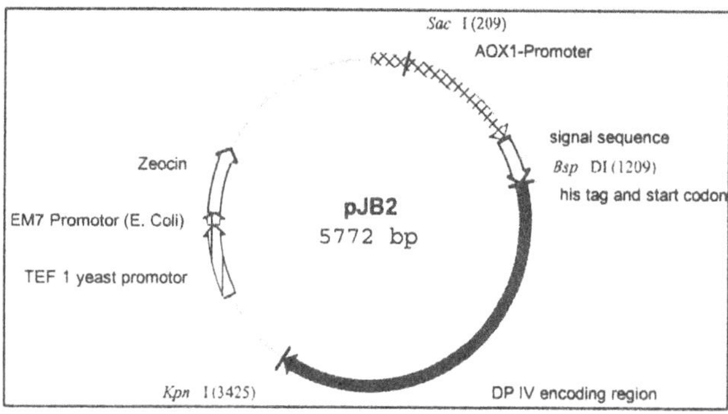

Figure 1. Construction of the expression vector pJB-2 containing signal sequence and AOX-promotor and inserted His_6-tag, and coding sequence for human (Δ 1-36) DP IV.

After transformation in the *P. pastoris* strain X-33 and expression in shaking flask cultures, fermentation in 5 liter scale has been established. Protein content and enzyme activity in the media elevated after induction of the AOX1 promotor with methanol (Figure 2). Secretion of DP IV into the media was found only partly complete, due to a significant amount of protein withhold in the periplasmatic space of *P. pastoris.*

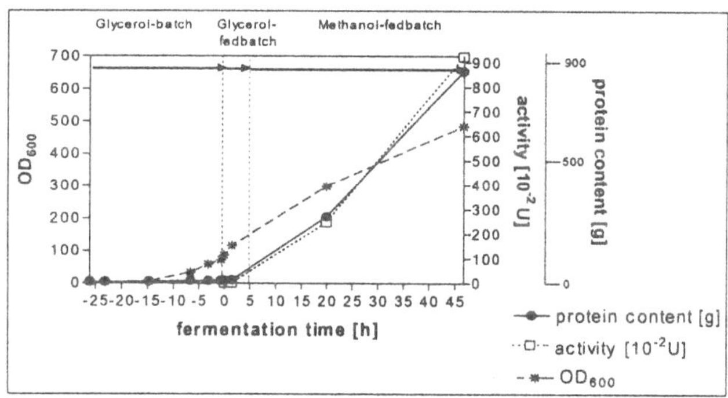

Figure 2. Fermentation protocol of DP IV expression. Accumulation of cell mass during the glycerol-batch and the glycerol-fed batch phase are represented by the dashed line (OD_{600}). Induction of the AOX promotor for DP IV-production in the methanol phase decelerated the growth rate while protein content (solid line) and DP IV activity (dotted line) arose.

DP IV accumulated in the media was purified by a two-step procedure using hydrophobic interaction chromatography and affinity chromatography on Ni-NTA. Purification of DP IV from the pellet and the supernatant yielded 2.7 mg DP IV in total with a specific activity of 32.3 U/mg (DP IV purified from porcine kidney (pDP IV): 27.2 U/mg). 29% of the DP IV activity was purified with a purification factor of 534.

The molecular weight of the monomeric recombinant DP IV was determined to be 103.6 kDa (porcine kidney DP IV: 100.8 kDa) by MALDI-TOF mass spectrometry. Gel filtration analysis confirmed that human recombinant DP IV exists as a dimer under native conditions (data not shown).

The isoelectric focusing revealed several isoforms exhibiting pI-values of 5.47 to 6.26, and indicating that the recombinant enzyme seems more heterologous glycosylated as the enzyme purified from porcine kidney (pI pDP IV: 5.57 and 6.16) (Figure 3).

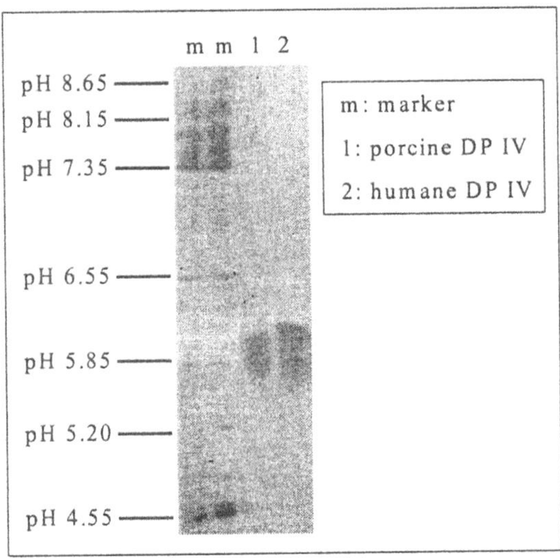

Figure 3. Determination of the isoelectric point on a pH 3-10 Servalyte pre nets gel (obtained by Serva, running on a Multiphore II, Pharmacia).

The kinetic characterization of human recombinant DP IV revealed comparable kinetic constants to those of pDP IV for the low molecular substrates and inhibitors tested (Table 1).

Table 1. Comparison of recombinant human and purified porcine DP IV regarding K_m- and K_i-values observed with different substrates and inhibitors.

	human DP IV	**porcine DP IV**
substrate	K_m [μM]	K_m [μM]
H-Gly-Pro-AMC	62.8±5.9	47.5±9.4
H-Gly-Pro-pNA	164.5±16.6	159.2±18.3
H-Ala-Ala-pNA	2741±1002	3571.0±291.4
inhibitor	K_i [nM]	K_i [nM]
Isoleucyl- thiazolidine	85.0±6.6	102.8±28.4
Isoleucyl-(2)- cyanopyrolidine	3.83±0.20	3.93±0.14

4. CONCLUSION

The expression and purification of soluble human DP IV was established in *P. pastoris* on a 5 liter fermentation scale (3 mg/fermentation, 32.27 U/mg). Thus it represents an alternate approach to previous published methods were DP IV was expressed in cell culture (Tanaka *et al.*[7]) with lower production rates and to the production in insect cells (Dobers *et al.*[8]) Biochemical and kinetic characterization demonstrated that the soluble recombinant DP IV displayed similar properties as DP IV purified from porcine kidney regarding size, activity, isoelectric point and glycosylation. Furthermore, the new expression method enables future structure-function related studies of DP IV.

ACKNOWLEDGEMENTS

For inhibitors, substrates and technical assistance and service we greatly acknowledge the support by Simon Hinke, Anja Weber, Susanne Manhart and Ingo Schulz. We thank Michael Wermann and Leona Wagner for providing purified porcine DP IV. This work was supported by the Federal Department of Science and Technology (BMBF-grant# 0312302 to HUD).

REFERENCES

1. Durinx, C., Lambeir, A.M., Bosmans, E., Falmagne, J.B., Berghmans, R., Haemers, A., Scharpe ,S., De Meester, I., 2000, Molecular characterization of dipeptidyl peptidase activity in serum: soluble CD26/dipeptidyl peptidase IV is responsible for the release of X-Pro dipeptides, *J Biochem.* **267(17):** 5608-13
2. Vanhoof, G., De Meester, I., van Sande, M., Scharpe, S., Yaron, A., 1992, Distribution of proline-specific aminopeptidases in human tissues and body fluids, *Eur J Clin Chem Clin Biochem.*, **30(6):** 333-8
3. Hildebrandt, M., Reutter, W., Arck, P., Rose, M., Klapp, B.F., 2000, A guardian angel: the involvement of dipeptidyl peptidase IV *in psychoneuroendocrine function, nutrition and immune defence. Clin Sci (Colch).*, **99(2):** 93-104.
4. Sreekrishna, K., Romanos, M.A., 1988, High-level expression of tetanus toxin fragment C in Pichia pastoris strains containing multiple tandem integration's of the gene, *Bio/Technology* **9:** 455-460
5. Trimble, R.B., Atkinson, P.H., Tschopp, J.F., Townsend, R., Maley, F., 1991, Structure of oligosaccharides on Saccharomyces SUC2 invertase secreted by the methylotrophic yeast Pichia pastoris, *J Biol Chem.* **266(34):** 22807-17.
6. Cregg, J.M., Madden, K.R., Barringer K.J., Thill G.P., Stillman C.A. 1989: Functional characterisation of the two alcohol oxidase genes from the yeast *Pichia pastoris*, *Molecular Cell Biology* **9:** 1316-1323
7. Tanaka, T., Camerini, D., Seed, B., Torimoto, Y., Dang, N.H., Kameoka, J., Dahlberg, H.N., Schlossman, S.F., Morimoto, C., 1992, Cloning and functional expression of the T cell activation antigen CD26, *J Immunol.*, 15;**149(2):** 481-6.
8. Dobers, J., Zimmermann-Kordmann, M., Leddermann, M., Schewe, T., Reutter, W., Fan, H., 2002, Expression, purification, and characterization of human dipeptidyl peptidase IV/CD26 in Sf9 insect cells, *Protein Expr Purif.*, **25(3):** 527.

Isolation and Characterization of Attractin-2

DANIEL FRIEDRICH, KERSTIN KÜHN-WACHE, TORSTEN HOFFMANN, and HANS-ULRICH DEMUTH
Probiodrug AG, Weinbergweg 22, 06120 Halle, Germany

1. INTRODUCTION

The mouse mahogany protein is involved in the control of pigmentation, myelination, immune functions and energy metabolism via modulation of the interaction of the agouti protein with melanocortin receptors and possibly other independent pathways[1]. Recently, attractin, the human homologue of the mahogany protein, has been identified as a soluble plasma protein with dipeptidyl peptidase IV-like activity[2]. Substrate, inhibitor and antibody specificities seem to be very similar to DP IV although the primary structures are completely different[2].

DP IV is a highly specific exopeptidase. Both in the membrane bound and the soluble form, DP IV is a glycosylated homodimeric enzyme with 110 kDa per subunit[3]. In contrast, attractin is a monomeric glycoprotein of 175 kDa. Attractin exists in soluble and transmembrane forms. So far, five alternative spliced isoforms of the human attractin have been described at the mRNA level (Swiss Prot), but only one form has been purified as yet. Attractin is a widely expressed protein. The mRNA of the protein has been found for example in placenta, colon, kidney, liver and especially in the central nervous system. The protein is expressed on activated T cells and a soluble form is released into the serum. As well as being a protease, attractin could also be a receptor due to its domain structure[4]. Here we present the first preparation of soluble attractin-2, one of the alternative spliced forms and a new reproducible preparation procedure of soluble attractin from human plasma based on characterisation results.

Dipeptidyl Aminopeptidases in Health and Disease, Edited by Hildebrandt et al.
Kluwer Academic/Plenum Publishers, New York, 2003

2. ISOLATION FROM HUMAN PLASMA

Recently, we have purified a soluble form of attractin from plasma by a two step procedure including ethanol precipitation and cationic exchange chromatography. Using these techniques, 100 μg of protein could be isolated from 100 ml plasma. The protein migrates as a single homogeneous band on SDS-PAGE under reducing conditions and clearly has a higher molecular weight than monomeric DP IV. The molecular weight of the protein was determined to be 175.5 kDa by MALDI-TOF mass spectrometry. Under native conditions the electrophoretic mobility of attractin was lower than that of the dimeric DP IV, indicating that attractin is a monomeric protein. Additionally isoelectric focusing of attractin revealed multiple active forms with isoelectric points ranging from 3.5 to 4.3. This has been described for DP IV-like activity[5] but not for DP IV. In contrast, the isoelectric points of DP IV from porcine kidney and from human placenta range from 5.3 to 6.1 and 4.9 to 5.5, respectively.

The N-terminus of purified attractin starts at Ala 84 of isoform 1, 2, 4, 5 and is located in an insertion that is not present in isoform 3[6]. Intriguingly, the N-terminus of the purified attractin was predicted by using the SignalP program (SwissProt) to be a signal sequence. However, the obtained N-terminal sequence of the isolated attractin questioned the predicted Ser 26 as an active site residue. This N-terminal sequence strongly indicates that the purified attractin is one of the soluble isoforms 2 or 4.

3. DEVELOPMENT OF A NEW OPTIMIZED PURIFICATION METHOD

3.1 Reduction of protein content

Based on the molecular weight and the pI-value of attractin, preparative two-dimensional electrophoresis was chosen as main the separation step. For a successful preparation, a reduction of plasma albumin was required as an initial step. This could be achieved with affinity chromatography (blue sepharose, Pharmacia) followed by cation exchange chromatography (sp-sepharose, Pharmacia). A reduction of 98% of the protein content under preservation of 52% of the total activity was possible.

3.2 Preparative two-dimensional electrophoresis

Preparative IEF: The Rotofor® System (BIO-RAD) was used for the preparative isoelectric focusing (IEF) of approximately 850 mg of protein. To build a wide-range of a pH-gradient, ampholytes (Serva) in the pH range 3-10 were employed. DPIV-like activity, protein content and the pH of the IEF fractions were measured after elution (**Fig. 1**). Both in the analytical and in the preparative IEF, the DP IV-like activity was focused around pH 3.5. The activity of the fractions correlated with the thickness of the protein band of attractin (178 kDa) in the SDS-PAGE (**Fig. 1**). Fractions 3 to 5 and 6 to 9 were collected and pooled. The latter was refractionated by IEF.

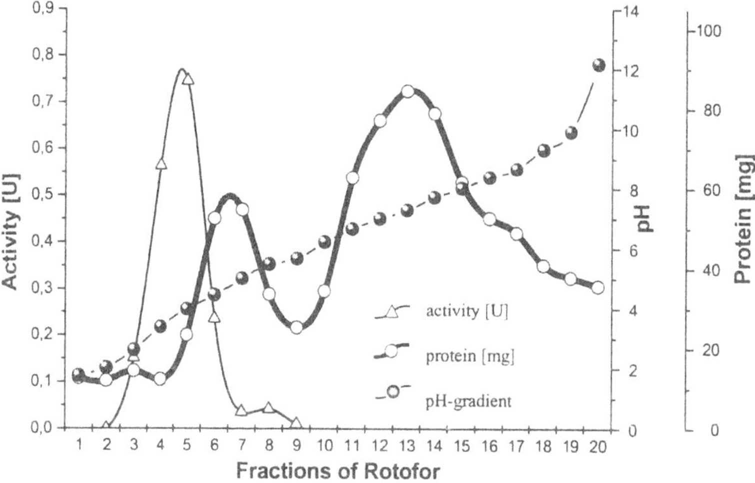

Figure 1. Isoelectric focusing on a Rotofor; Symbols: Δ: DP IV-like activity in U, ○: protein in mg, •: pH-gradient; Background: SDS-PAGE under reducing condition (7.5%) according to the fractions of Rotofor

Preparative PAGE: Preparative native PAGE was carried out using the Model 491 Prep Cell (BIO-RAD) and the discontinuous buffer system according to Laemmli. The total acrylamide concentrations were optimized at 4% and 6% for the stacking gel and the separating gel respectively. The sample contained approximately 70 mg of total protein.

After these procedures, approximately 7 mg attractin (178 kDa) with a specific activity of 0.144 U/mg contained a minor contamination.

3.3 Anion exchange chromatography

The remaining contamination could be eliminated by using high resolution anion exchange chromatography (Uno Q6, BIO-RAD). The obtained protein peak of attractin exactly matched the DP IV-like activity. Nearly 3 mg of the protein could be purified to homogenity. No other protein band was visible from serva blue and silver staining after SDS-PAGE. According to the theoretical prediction, the molecular weight was 132 kDa after PNGase treatment. The identity of the protein has been confirmed by N-terminal sequencing and could be detected by using a polyclonal anti-attractin antibody (C19, Santa Cruz).

Table 1. Purification of Attractin from human plasma (755 ml)

step		activity [U]	protein [mg]	specific activity [U/mg]	purification [-fold]	activity [%]	protein [%]
Plasma		8.305	39066	0.00021	1.0	100.0	100.0
Blue sepharose		5.026	3365.6	0.00149	7.0	60.5	8.6
sp-sepharose		4.354	850.5	0.00512	24.1	52.4	2.2
Rotofor		1.609	69.42	0.0232	109.0	19.4	0.18
Prep Cell		0.979	6.774	0.144	687.9	11.8	0.017
Uno Q6	Pool 1	0.664	2.101	0.316	1505.9	8.0	0.005
	Pool 2	0.158	0.865	0.182	867.3	1.9	0.002
	total	0.822	2.966	0.277	1319.6	9.9	0.008

4. CONCLUSION

Attractin 2/4 has been isolated to homogeneity from human plasma. Based on the native molecular weight of 178 kDa and pI-value around pH 3.5, a new reproducible purification procedure has been developed. N-terminal sequencing of attractin confirmed the predicted signal peptidase cleavage site in the insertion of the isoforms 1, 2, 4 and 5 and questioned the predicted Ser 26 as an active site residue.

We could not find differences between DP IV and attractin with respect to the specificity of inhibitors or substrates. Attractin is also capable to release dipeptides from higher molecular substrates such as neuropeptide Y.

ACKNOWLEDGEMENTS

The work was supported by a grant from the BMBF, no. 0312302.

REFERENCES

1. Dinulescu D. M., Fan W., Boston B. A., McCall K., Lamoreux M.L., Moore K.J., Montagno J., Cone R.D., 1998: Mahogany (mg) stimulates feeding and increases basal metabolic rate independent of its suppression of agouti, *Proc Natl Acad Sci USA.* **95**: 12707-12712
2. Duke-Cohan J.S., Tang W., Schlossman S.F., 2000: Attractin: A cub-family protease involved in T cell- monocyte/macrophage interactions, *Adv. Exp. Med. Biol.* **477**: 173-185
3. Demuth H.-U. & Heins J., (1995): On the catalytic mechanism of dipeptidyl peptidase IV., in *Dipeptidyl Peptidase IV (CD26) in Metabolism and the Immune Response* (Fleischer, B., ed.), pp. 1-37, R.G. Landes, Biomedical Publishers, Georgetown.
4. Gunn T.M., Miller K.A., He L., Hyman R.W., Davis R.W., Azarani A., Schlossman S.F., Duke-Cohan J.S., Barsh G.S., (1999): The mouse mahogany locus encodes a transmembrane form of human attractin, *Nature.* **398**: 152-156
5. Mentlein R., Dahms P., Grandt D., Kruger R., (1993): Proteolytic processing of neuropeptide Y and peptide YY by dipeptidyl peptidase IV, *Regul. Pept.* **49**: 133-44
6. Duke-Cohan J.S., Morimoto C., Rocker J.A., Schlossman S.F., (1996): Serum high molecular weight dipeptidyl peptidase IV (CD26) is similar to a novel antigen DPPT-L released from activated T cells, *The Journal of Immunology* **156**: 1714-21

Investigation of DP IV-dependent Protein-Protein Interactions using Surface Plasmon Resonance

JOERG STORK, TORSTEN HOFFMANN, and HANS-ULRICH DEMUTH
Probiodrug AG, Weinbergweg 22, D-06120 Halle, Germany

1. INTRODUCTION

Dipeptidyl peptidase IV (DP IV, EC 3.4.14.5) is a serine protease which removes N-terminal dipeptides with proline or alanine at the penultimate position[1]. DP IV is expressed on a variety of cells as a type II membrane protein. Additionally, in serum a soluble form of DP IV has been described[2]. Within the hematopoietic system DP IV was identified as CD26, a T-cell activation antigen which is predominately expressed on human T-lymphocytes[3]. DP IV/CD26 is reported to play a key role in T cell-mediated immune response by processing bioactive peptides such as cytokines, chemokines and neuronal and vasoactive peptides[4]. These processing results in activation or inactivation of the peptide or in a alteration of its receptor selectivity. In addition to its catalytic function DP IV is known to be a binding partner of adenosine deaminase (ADA)[5]. It has been proposed that DP IV is involved in binding to HIV[1] and in cell adhesion processes[6].

Dipeptidyl peptidase IV was described as a receptor for extracellular matrix proteins such as collagen and fibronectin which mediate the cell adhesion on extracellular matrix[7]. We investigated the interaction of DP IV with the extracellular matrix proteins fibronectin and collagen III applying surface plasmon resonance.

Dipeptidyl Aminopeptidases in Health and Disease, Edited by Hildebrandt et al.
Kluwer Academic/Plenum Publishers, New York, 2003

2. MATERIALS AND METHODS

Surface Plasmon Resonance (SPR) Analysis

The interaction of DP IV with fibronectin and collagen III was measured in real time utilising the BIAcore 3000 system (BIAcore AB, Uppsala, Sweden). Fibronectin, collagen III or DP IV were immobilised to a CM 5 sensor chip using the amino-coupling kit according to the manufacturer's recommendation. Experiments were performed at 25 °C at a flow rate of 5 µl/min. The surface was regenerated with 10 µl 2 M NH_4SCN-solution.

Cell Culture

Human glioma cell line U343 were cultured in RPMI 1640 supplemented with 10 % FCS and gentamicin 60 µg/ml. The cells were incubated at 37°C in a humidified atmosphere with 5% CO_2.

Membrane Preparation

Cells grown in 10 Petri dishes were detached with 10 mM EDTA in Hanks balanced salt solution and precipitated by centrifugation. The cell pellet was suspended in 5 ml of 20 mM Hepes buffer, pH 7.2, containing 0.25 M sucrose and an inhibitor mix (Sigma). Cells were lysed by sonication on ice. The homogenate was centrifuged at 800xg for 15 min to remove unbroken cells an nuclei, followed by centrifugation of the supernatant at 50,000xg for 1 h. The pellet containing cell membranes was resuspended in 20 mM Tris/HCl, pH 8.0, containing 1 % (v/v) Triton X-100 and centrifuged again at 50,000xg for 30 min to remove insoluble components. The buffer of the membrane protein solution was changed to PBS buffer utilising a HiPrep desalting column (Pharmacia, Uppsala, Sweden) in BioCAD 700E (Applied Biosystems, Weiterstadt, Germany).

3. RESULTS

Using the sensor chip technology, interaction of DP IV with fibronectin or collagen could be measured at salt conditions below physiological values. The recombinant human DP IV (80 µg/ml), expressed in *Pichia pastoris*[8], showed an ion strength dependent binding to sensor chip coupled fibronectin or collagen III. No interaction could be observed under physiological ion strength (Figure 1).

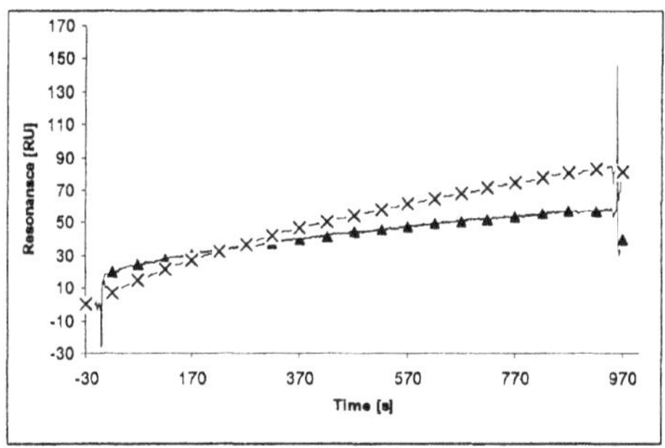

Figure 1: Time dependence SPR of binding of DP IV on fibronectin (σ) and collagen III (5) in PBS buffer with ion strength below physiological values. (25 °C, 70 mM NaCl, 9.5 mM phosphate, pH 7.6)

Furthermore, we used cell membrane preparations to investigate the role of DP IV in adhesion of cells to the extracellular matrix. Measurements using U343 membrane preparations showed, that these membrane interact with collagen III itself and that the addition of U343 membrane proteins enhance the binding of DP IV to collagen III (Figure 2). The height of the resonance curves after buffer change (dissociation phase) represents the amount of bound protein. The preparation of U343 membrane proteins adhere to collagen III and DP IV in a concentration depend manner (not shown). Similar results were achieved with membrane preparations of the human cell lines SY5Y and HepG2.

Gonzalez-Gronow *et al*[8] described that fibronectin and streptokinase have the sequence LTSRPA in common. This sequence binds to QLRCSGPGLPL, which serves as the binding region in DP IV. Incubation of increasing amounts of LTSRPA in the mixture of U343 membrane proteins caused a decreasing resonance signal. Whereas the peptide QLRCSGPGLPL interacted itself with sensor chip coupled DP IV.

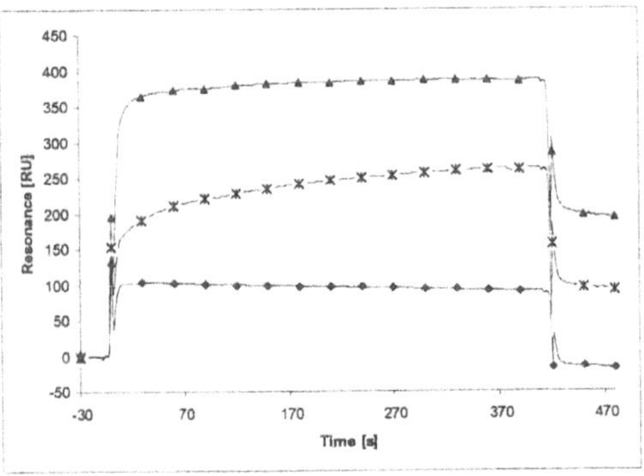

Figure 2: Time dependence SPR of binding of DP IV (υ), U343 membrane preparation (5) and a mixture of DP IV and membrane preparation (σ) on collagen III. (Analyte was replaced by running buffer at 400 second. (25 °C, 140 mM NaCl, 9.5 mM phosphate, pH 7.6)

4. CONCLUSION

Our data maintain the theory that DP IV is involved in adhesion of cells to the extracellular matrix.

The method developed here allows the testing of substances which can interfere with such interactions with extracellular matrix proteins.

Our data suggest the hypothesis that one ore more components localised at the membrane fraction are involved in DP IV mediated binding of cells on extracellular matrix.

ACKNOWLEDGEMENTS

For synthetic peptides we in greatly acknowledge the support of Susanne Manhart. We thank Joachim Baer for providing purified recombinant human DP IV. This work was supported by the Federal Department of Science and Technology (BMBF-grant#0312302 to HUD).

REFERENCES

1. De Meester, I., Vanhoof, G., Lambeir, A.-M., Scharpe, S., 1995, Use of immobilised adenosine deaminase (EC 3.5.4.4) for the rapid purification of native human CD26/dipeptidyl peptidase IV (EC3.4.14.5). *J. of Immun. Methods* **189:** 99-105.
2. Durinx, C., Lambeir, A.-M., Bosmans, E., Falmagne, J.-B., Berghamns, R., Haemers, A., Scharpe, S., De Meester, I., 2000, Molecular characterisation of dipeptidyl peptidase activity in serum. *Eur. J. Biochem.* **267:** 5608-5613.
3. Marguet, D., Bernard, A.-M., Vivier, J., Darmoul, D., Naquet, P., Pieres, M., 1992: C-DNA cloning for mouse thymocyte-activating molecule. *J. of Biol. Chem.* **267:** 2200-2208.
4. Tanaka, T., Kameoka, J., Yaron, A., Schlossman, S.F., Morimoto, C., 1993, The costimulatory activity of the CD26 antigen requires dipeptidyl peptidase IV enzymatic activity. *Proc. Nat. Acad. Sci.* **90:** 4586-4590.
5. Gutheil, W.G., Subramanyam, M., Flenthe, G.R., Sanford, D.G., Munoz, E., Huber, B.T., Bachovchin, W.W., 1994, Human Immunodeficiency Virus 1 Tat binds to Dipeptidyl Aminopeptidase IV (CD26): A possible mechanism for Tat's immunosuppressive activity. *Proc. Nat. Acad. Sci.* **91:** 6594-6598.
6. Hanski, C., Huhle, T. , Reuter, W., 1985, Involvement of Plasma Membrane Dipeptidyl Peptidase IV in Fibronectin-Mediated Adhesion of Cells and Collagen. *Biol. Chem.* **366:** 1169-1176.
7. Hanski, C., Huhle, T., Grossau, R. Reuter, W., 1988, Direct Evidence for Binding of Rat Liver DPP IV to Collagen in vitro. *Experimental Cell Research* **178:** 64-72.
8. Joachim Baer, T. Hoffmann, F. Rosche, H.-U. Demuth, B. Gerhartz, 2001, Poster presentation at 2nd General Meeting of the International Proteolysis Society (IPS), Freising
9. Gonzales-Gronow, M., Weber, M. R., Gawdi, G. and Pizzo, S. V., 1998, Dipeptidyl Peptidase IV(CD 26) is a Receptor for Streptokinase and Fibronectin in Rheumatoid Arthritis Human Synovial Fibroblasts. *Fibrinolysis and Protolysis* **12 (3):** 129-135.

3

IMMUNE MECHANISMS AND IMMUNE DISORDERS

Synergistic Action of DPIV and APN in the Regulation of T Cell Function

UWE LENDECKEL, MARCO ARNDT, ALICJA BUKOWSKA, JANINE TADJE, CARMEN WOLKE, THILO KÄHNE, KLAUS NEUBERT[3], JÜRGEN FAUST[3], ANNELORE ITTENSON[1], SIEGFRIED ANSORGE[2], and DIRK REINHOLD[1]

Institute of Experimental Internal Medicine, [1]Institute of Immunology, [2]Institute of Medicinal Technology Magdeburg, Otto-von-Guericke University Magdeburg, Leipziger Strasse 44, D-39120 Magdeburg, [3]Institute of Biochemistry, Martin-Luther-University, Kurt-Mothes-Str. 3, 06120 Halle, Germany

ABSTRACT

Inhibitors of the enzymatic activity of alanyl-aminopeptidases severely affect growth and typical functions of human peripheral T cells both in vitro and in vivo. The most prominent changes observed include the activation of cellular signal transduction pathways such as MAP kinases Erk1/2 or the Wnt-pathway, a decrease of production and release of „pro-inflammatory" cytokines (IL-2, IL-12) and, most importantly, an induction of expression and release of the immunosuppressive cytokine, TGF-β1. Similar effects on T cell proliferation and function have been observed in response to inhibition of DPIV, which is strongly suggestive of a functional synergism of APN and DPIV. In support of this hypothesis evidence is provided showing that the simultaneous application of inhibitors of DPIV and APN further enhances the anti-inflammatory and immunosuppressive effects provoked by the inhibition of APN or DPIV alone. Therefore, the simultaneous inhibition of these enzymes represents a promising strategy for the pharmacological therapy of T cell mediated diseases such as autoimmune disease, inflammation, allergy, and allograft rejection.

Dipeptidyl Aminopeptidases in Health and Disease, Edited by Hildebrandt et al.
Kluwer Academic/Plenum Publishers, New York, 2003

1. INTRODUCTION

Membrane alanyl aminopeptidase (EC 3.4.11.2, APN, CD13) is an 150 kDa metalloprotease of the M1 family of peptidases (clan MA, gluzincins[4]) which is - as far as the hematopoietic system is concerned – predominantly expressed on myelo-monocytic lineage cells. Apparently, various pathophysiological conditions such as chronic inflammation, malignant transformation and T cell activation induce APN gene and surface expression in human peripheral T cells via not yet well defined molecular mechanisms[7,9,15-17]. The human APN gene consists of 20 exons and is located on chromosome 15 (q25-q26)[11,12,22]. In monocytes and T cells transcription is initiated from the myeloid promoter, whereas in other tissues the epithelial promoter is used instead[18, 23].

In T cells, the inhibition of either APN gene expression or of APN enzymatic activity has profound effects on proliferation and function. Among the most prominent effects observed after pharmacological inhibition of APN are the induction of the immunosuppressive cytokine TGF-β1 and the decreased expression of IL-2 (for review see [9]).

Previous work revealed that two cellular pathways seem to participate in mediating "APN-derived" signals to the nucleus. In response to the aminopeptidase-inhibitors actinonin or probestin, respectively, there is an marked increase of expression and activity of the MAP-kinases Erk1/Erk2[8]. Furthermore, highly selective inhibitors of APN affect expression and activity of prominent members of the Wnt-pathway, namely Wnt-5a and glycogen synthase kinase-3β (GSK-3β)[10].

Notably, the pharmacological inhibition of the functionally related ectopeptidase dipeptidyl peptidase IV (DPIV, CD26) leads to changes of T cell function that are strikingly similar to those observed after APN inhibition (for review see [6]) and that include cell cycle arrest at G1/S transition, decreased DNA-synthesis, decrease of production and release of pro-inflammatory cytokines, and, most importantly, induction of TGF-β1 expression.

The aim of this study was to determine possible synergistic effects of simultaneously applied specific inhibitors of the enzymatic activities of APN and DPIV.

2. MATERIALS AND METHODS

2.1 Reagents

Actinonin and phebestin were from Sigma. Probestin was a kind gift of T. Aoyagi (Tokyo). I49 was synthesized by K. Neubert and J. Faust.

2.2 Cell culture

Mononuclear cells (MNC) were prepared from peripheral blood of healthy donors by Ficoll-Paque gradient centrifugation [1]. T cells were enriched from the MNC fraction by the nylon wool adherence technique [5]. T cells were kept overnight in IMDM-medium and then cultured at a density of 1×10^5 cells/ml with the additions and for periods of times indicated in the figures. $CD4^+CD25^+$ T cells (Treg cells) were obtained from MNC by positive selection using $CD25^+$ magnetic beads (Miltenyi Biotec).

2.3 RNA-isolation and RT-PCR

Total RNA was prepared by means of the RNeasy kit (Qiagen). 2 µg RNA were reverse-transcribed using AMV-RT (Stratagene) and 1/10th of the cDNA was used for RT-PCR.
Quantitative determination of IL-2 and TGF-β1 mRNA contents were performed using the iCycler (Bio-Rad) and the primer pairs available from Stratagene. 18S mRNA amounts were determined using the RT primer pair available from Ambion and used to normalise sample cDNA content. The fluorescence intensity of the double-strand specific SYBR-Green I, reflecting the amount of actually formed PCR-product, was read real-time at the end of each elongation step. Then specific initial template mRNA amounts were calculated by determining the time point at which the linear increase of sample PCR product started, relative to the corresponding points of a standard curve; these are given as artificial units.

2.4 Cytokine analyses

Amounts of IL-2 released into the culture medium were measured by means of the Quantikine Colorimetric Sandwich ELISA (R&D Systems) according to the recommended protocol. Amounts of TGF-β1 were determined by a non-commercial ELISA using a capture anti-TGF-β1 monoclonal antibody (mab) (Genzyme), chicken anti-TGF-β1 mab (clone BDA 19, R&D System), biotinylated rabbit anti-chicken polyclonal ab (IgG, Dianova), and streptavidin-horseradish peroxidase conjugate (Sigma).

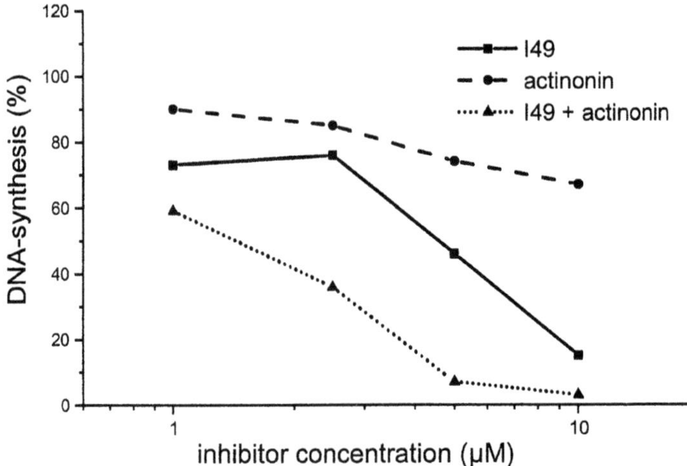

Figure 1: DNA synthesis of PHA-stimulated peripheral T cells is inhibited dose-dependently by I49 or actinonin. The combined administration of both inhibitors resulted in a stronger inhibition than could be achieved by each of the inhibitors alone. DNA synthesis was measured 72 hrs after T cell activation.

3. RESULTS

Synergistic action of APN and DPIV inhibitors on T cell proliferation

Both the aminopeptidase inhibitor actinonin and the inhibitor of DPIV, I49, showed a dose-dependent reduction in DNA synthesis of PHA-stimulated peripheral T cells. At 5µM concentrations actinonin reduced DNA synthesis to 75 %, whereas I49 caused a decrease by more than 50 %. When applied simultaneously at this concentration, these inhibitors reduced DNA synthesis to less than 10 % (Figure 1).

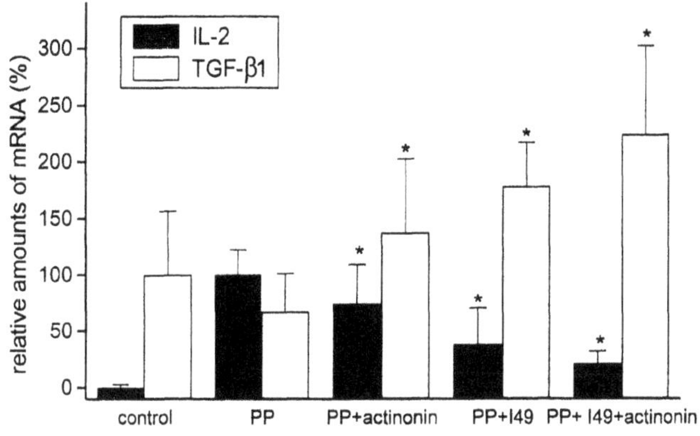

Figure 2: Inhibitors of APN (actinonin) and DPIV (I49) act synergistically in decreasing IL-2 and increasing TGF-β1 mRNA amounts in PHA/PMA (PP) -activated T cells. mRNA amounts were determined by quantitative RT-PCR using the iCycler (Bio-Rad) 24 hours (IL-2) or 3 hours after activation and inhibitor administration. (mean of three experiments ± sd) $^*p<0.05$

Synergistic Effects on IL-2 and TGF-β1 mRNA expression

Quantitative RT-PCR revealed that 24 hours after T cell activation there is an strong induction of IL-2 mRNA. Both actinonin and I49 caused a significant decrease of IL-2-mRNA amounts. The strongest reduction of IL-2 mRNA was observed after the simultaneous application of both inhibitors (21 ± 11 % compared to PHA/PMA) (Figure 2).

In contrast, there was an induction-dependent decrease of TGF-β1 mRNA amounts 3 hrs after Activation by PHA/PMA. When T cells were activated in the presence of either actinonin or I49, respectively, TGF-β1 mRNA amounts slightly increased. Again, the combination of both inhibitors resulted in maximum TGF-β1 mRNA expression (224 ± 78 % of control) (Figure 2).

Synergistic Effects on IL-2 and TGF-β1 protein

As observed with mRNA levels, there was an synergistic effect of I49 and actinonin on IL-2 and TGF-β1 protein expression. As shown in figure 3, the changes observed after administration of either I49 or actinonin alone were clearly surpassed by the combined inhibition of DPIV and APN: IL-2 concentrations were decreased to 195 ± 118 pg/ml and maximum amounts of TGF-β1 were 1330 ± 210 pg/ml (Figure 3).

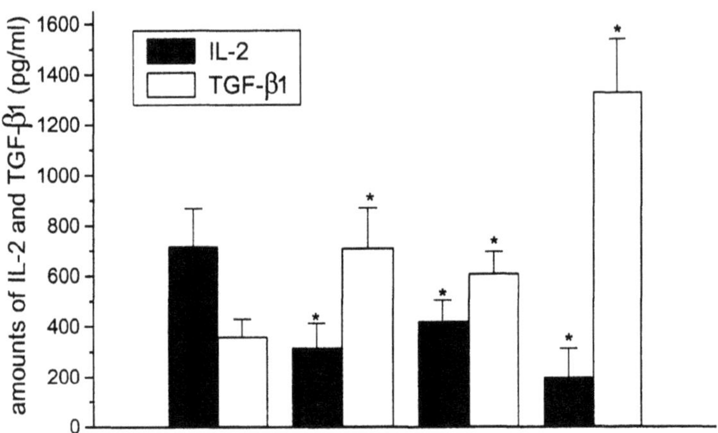

Figure 3: I49 and actinonin synergistically effect the production/release of IL-2 and TGF-β1 by pokeweed mitogen-activated T cells. *$p<0.05$.

Induction of TGF-β1 expression by aminopeptidase inhibition

$CD4^+CD25^+$ regulatory T cells isolated by two rounds of positive selection using CD25-magnetic beads (Miltenyi) showed a characteristic expression of surface markers, including surface-bound TGF-β1 (Table 1). 24 hrs after activation by PHA/PMA there was a slight decrease of TGF-β1 mRNA expression in Treg cells detectable. Activation in the presence of 10^{-6} M phebestin provoked a more than 25fold increase of TGF-β1 mRNA amounts compared to PHA/PMA ($p<0.05$) (Figure 4).

Under the same conditions, phebestin also slightly increased the surface expression (mean fluorescence intensity) of TGF-β1 by about 14 % (not shown).

Furthermore, the concentration of TGF-β1 (active + latent) in the culture supernatant was significantly increased by phebestin 31 % (565.9 vs. 432 ng/ml); $p<0.05$).

Table 1: Surface markers expressed on Treg cells

Surface marker	Positive cells (%)
CD25	97.6
CD4	93.7
TGF-β_1	94.7
CD45RB	98
CD152 (CTLA-4)	99

4. CONCLUSIONS

As it has been shown previously, inhibitors of APN and DPIV severly affect T cell proliferation and cytokine production. In summary, these effects are largely immunosuppressive and, to a significant extent, seem to depend on the increase of production and release of the immunosuppressive cytokine, TGF-β1.

The data presented here clearly show that in T cells the simultaneous inhibition of APN and DPIV leads to more drastic changes of proliferation and cytokine production than could be observed in response to each of the inhibitors alone.

Most importantly, APN and DPIV inhibitors also increased the production and release of TGF-β1 in an additive manner. In contrast, T cell growth and IL-2 production were synergistically decreased under these conditions.

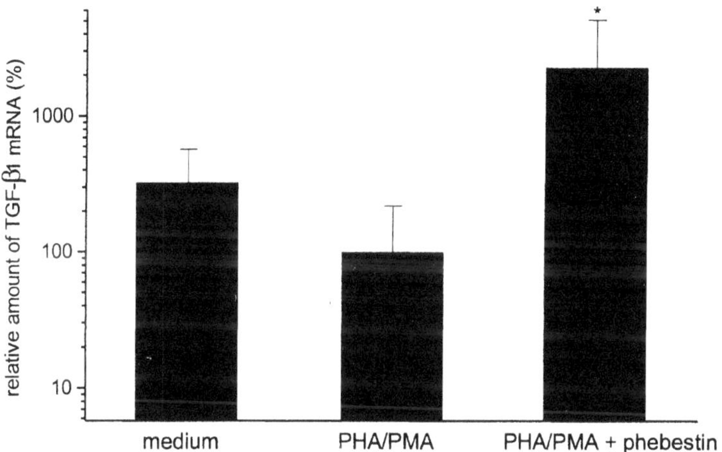

Figure 4: Induction of TGF-β1 mRNA expression in $CD4^+CD25^+$ human regulatory T cells by the inhibitor of alanyl-aminopeptidase, phebestin. TGF-I49 and actinonin synergistically effect the production/release of IL-2 and by pokeweed mitogen-activated T cells. *$p<0.05$

Due to their anti-inflammatory and anti-proliferative effects both APN and DPIV inhibitors have longly been regarded as potential therapeutics for treatment of T cell mediated diseases such as autoimmune disease or allograft rejections. Especially the use of DPIV inhibitors has recently been proven to effectively delay the onset and to diminish the score of disease in a mouse model of multiple sclerosis [20].

Our *in vitro* data imply that the combined inhibition of APN and DPIV might prove an even more effective strategy for the treatment of the above mentioned diseases. This has to be adressed in further *in vivo* experiments.

The crucial role of $CD4^{+}CD25^{+}$ regulatory T cells (Treg cells) in the regulation of autoimmunity and T cell homeostasis has been established during the last two years[3,14]. Treg cells via surface-bound TGF-β1 and direct cell-contact are capable of suppressing effector T cells and of bringing about antigen presenting cell (APC) anergy[13]. Lack of Treg cells or loss of their function leads to the development of autoimmune diseases such as colitis ulcerosa or multiple sclerosis[2,19,21].

We show here that the APN inhibitor phebestin increases TGF-β1 mRNA levels, surface expression and secretion into the medium of $CD4^{+}CD25^{+}$ Treg cells. To the best of our knowledge this is the first report to show a targeted induction of TGF-β1 in regulatory T cells. Of note, this induction could be achieved by the administration of a structurally relatively simple compound, which in addition seems to lack significant toxic effects, at least *in vitro*.

Summarizing our data we hypothesize that the anti-inflammatory and immunosuppressive effects provoked by inhibitors of the enzymatic activity of APN, and possibly also of DPIV, are largely due to an specific induction of TGF-β1 gene and surface expression in human peripheral regulatory T cells. Thus, a cellular immunosuppressive mechanism is provided that could be exploited as an alternative strategy for the prevention and treatment of autoimmune disease.

ACKNOWLEDGEMENTS

We thank Christine Wolf, Katja Mook, and Cornelia Müller for excellent technical assistance.

REFERENCES

1. Boyum, A., 1968, Isolation of mononuclear cells and granulocytes from human blood, *Scand. J. Clin. Lab. Invest. (suppl.)* **97:** 77-89
2. Furtado, G.C. et al., 2001, Regulatory T cells in spontaneous autoimmune encephalomyelitis, *Immunol. Rev.* **182:** 122-134
3. Gorelik, L. and Flavell, R.A., *2002, Nature Rev. Immunol.* **2:** 46-53
4. Hooper, N.M., 1994, Families of zinc metalloproteases, *FEBS Lett.* **354:** 1-6
5. Julius, M.H., et al., 1973, A rapid method for the isolation of functional thymus-derived murine lymphocytes, *Eur. J. Immunol.* **3:** 645

6. Kähne, T., et al., 1999, Dipeptidyl peptidase IV:A cell surface peptidase involved in regulating T cell growth, *Int. J. Mol. Med.* **4:** 3-15
7. Lendeckel, U., et al., 1996, Induction of the membrane alanyl aminopeptidase gene and surface expression in human T-cells by mitogenic activation, *Biochem. J.* **319:** 817-823
8. Lendeckel, U., et al., 1998, Inhibition of alanyl aminopeptidase induces MAP-kinase p42/ERK2 in the human T cell line KARPAS-299, *Biochem Biophys Res Commun* **252:** 5-9
9. Lendeckel, U., et al., 1999, Role of alanyl aminopeptidase in growth and function of human T cells (review), *Int. J. Mol. Med.* **4:** 17-27
10. Lendeckel, U., et al., 2000, Inhibition of alanyl-aminopeptidase suppresses the activation-dependent induction of glycogen synthase kinase-3β (GSK-3β) in human T cells, *Bioch. Biophys. Res. Commun.* **273:** 62-65
11. Lerche, C., et al., 1996, Human aminopeptidase N is encoded by 20 exons, *Mammalian Genome* **7:** 712-713
12. Look, A.T., et al., 1986, Molecular cloning, expression, chromosomal localization of the gene encoding a human myeloid membrane antigen (gp150), *J. Clin. Invest.* **78:** 914-921
13. Nakamura, K., et al., 2001, Cell contact-dependent immunosuppression by CD4(+)CD25(+) regulatory T cells is mediated by cell surface-bound transforming growth factor beta, *J. Exp. Med.* **194:** 629-644
14. Pontoux, C., et al., 2002, Natural CD4 CD25(+) regulatory T cells control the burst of superantigen-induced cytokine production: the role of IL-10, *Int. Immunol.* **14:** 233-239
15. Riemann, D., et al., 1995, Stimulation of the expression and the enzyme activity of aminopeptidase N/CD13 and dipeptidylpeptidase IV/CD26 on human renal cell carcinoma cells and renal tubular epithelial cells by T cell-derived cytokines, such as IL-4 and IL-13, *Clin. Exp. Immunol.* **100:** 277-283
16. Riemann, D., et al., 1993, Demonstration of CD13/aminopeptidase N on synovial fluid T cells from patients with different forms of joint effusions, *Immunobiol.* **187:** 24-35
17. Riemann, D., et al., 1994, Immunophenotype of lymphocytes in pericardial fluid from patients with different forms of heart disease, *Int. Arch. Allergy Immunol.* **104:** 48-56
18. Shapiro, L., 1991, Separate promoters control transcription of the human aminopeptidase N gene in myeoloid and intestinal epithelial cells, *J. Biol. Chem.* **266:** 11999-12007
19. Singh,B., et al, 2001, Control of intestinal inflammation by regulatory T cells, *Immunol. Rev.* **182:** 190-200
20. Steinbrecher, A., et al., 2001, Targeting dipeptidyl peptidase IV (CD26) suppresses autoimmune encephalomyelitis and up-regulates TGF-beta 1 secretion in vivo, *J. Immunol.* **166:** 2041-2048
21. Tung, K.S., et al., 2001, Regulatory T cell, endogenous antigen and neonatal environment in the prevention and induction of autoimmune disease, *Immunol. Rev.* **182:** 135-148
22. Watt, V.M., and Willard, H.F., 1990, The human aminopeptidase N gene: isolation, chromosome localization, and DNA polymorphism analysis, *Hum.Genet.* **85:** 651-654
23. Wex, T., et al., 1997, The activation-dependent induction of APN (CD13) in T cells is controlled at different levels of gene expression, *FEBS Lett.* **412:** 53-56

CD26/DPP IV in Experimental and Clinical Organ Transplantation

STEPHAN KOROM[*,#], INGRID DE MEESTER[¥], A. BELYAEV[¥], GEORG SCHMIDBAUER[*], and KONRAD SCHWEMMLE[*].
[*]*Dept. of General and Thoracic Surgery, Justus Liebig University, Giessen, Germany;* [#]*Division of Thoracic Surgery, University Hospital, Zurich, Switzerland;* [¥]*Dept. of Clinical Biochemistry, University of Antwerp, Antwerp, Belgium.*

ABSTRACT

The T-cell activation-Ag CD26 possesses dipeptidyl peptidase IV (DPP IV) enzymatic activity. Costimulatory efficacy and immunocompetence are associated with the enzymatic activity. *Goals:* In models of experimental cardiac allograft transplantation (HTx), we analyzed the role of CD26/DPP IV during organ rejection. Also, we investigated CD26 enzymatic and cellular expression in human recipients of kidney transplants (Tx). *Material and Methods:* Heterotopic HTx in rats, models of acute and accelerated rejection. Monitoring of DPP IV serum levels and humoral immunity. Pro-pro-diphenyl phosphonate was employed to inhibit DPP IV activity during rejection. In a prospective study, surface expression of CD26, 3, 4, 8, 45, 122 and ADA on PBL and DPP IV serum activity were measured in kidney recipients for 24 months post-transplantation. *Results:* Acute rejection was associated with increased serum DPP IV activity ($p<0.005$). Specific inhibition abrogated acute ($p<0.0001$) and accelerated ($p<0.01$) rejection, impairing cytotoxicity and allospecific Ig-synthesis. Kidney recipients displayed a significant drop in CD26 expression on PBL for up to 18 months postoperatively ($p<0.001$). CD4, 8, 45, 122 and ADA expression kinetics were only briefly affected. DPP IV enzymic activity stayed depressed for at least 12 months ($p<0.001$). *Conclusion:* CD26/DPP IV is pivotal in T-cell mediated immune responses toward allo-Ag. In clinical transplantation, engraftment/immunosuppression are reflected by CD26 cellular and enzymatic expression posttransplantation and may serve as an indicator for immunomodulation.

1. INTRODUCTION

The lymphocyte surface glycoprotein CD26 is a heterogenous molecule, characterized by an array of diverse functional properties. It belongs to an unique class of membrane-associated proteases, possessing dipeptidyl peptidase IV (DPP IV) enzymatic activity[1]. Simultaneously, CD26 acts as a T cell costimulator, and binds to CD45, adenosine deaminase (ADA), HIV Tat-1 protein, collagen and fibronectin[1,2]. It is involved in several immunological disorders, including severe combined immunodeficiency (SCID), acquired immunodeficiency syndrome (AIDS), multiple sclerosis and rheumatoid arthritis[1,2].

Accumulating evidence indicates a central role for CD26 expression in alloantigen-mediated immune pathways and during the memory T cell response[2]. DPP IV enzymatic activity, which has been correlated with immunological competence, is linked to the costimulatory efficacy of CD26[3,4]. Our interest was triggered in particular by the observation that DPP IV serum activity was decreased during clinical immunosuppression in transplant recipients[5,6]. The recent development of a novel low molecular weight inhibitor of DPP IV acctivity (pro-pro-diphenylphosphonate) [7,8] prompted us to further dissect the contribution of the catalytic moiety of CD26 toward the immune response triggered by in vivo alloantigen exposure. First, we analyzed the role of CD26 and its enzymatic activity during acute and accelerated cardiac allograft rejection in rats. Second, based on our findings in experimental organ grafting, we investigated the dynamics of CD26/DPP IV during the course of clinical transplantation in human recipients of kidney allografts.

2. MATERIAL AND METHODS

2.1 Animals and grafting technique

For the acute rejection model, LEW rat recipients were heterotopically engrafted with LBNF1 cardiac transplants onto the abdominal great vessels in standardized microsurgical technique[9]. Graft action could be monitred by palpation and were rated on a scale from 3 to 0 (3: 100% contractility; 2: 75% contractility; 1: 50% contractility; 0: no contractility, eg. rejection), For assessment of accelerated rejection, LEW recipients were pretreated with two BN skin grafts 7 days prior to HTx. In untreated recipients, cardiac transplants were acutely rejected within 7-8d, the accelerated rejection commenced within 24-36h.

2.2 Synthesis, treatment and measurement of DPP IV serum enzymatic activity

The synthesis and characterization of (S)-Pro-(S,R)-ProP(OPh)2 (pro-pro-diphenyl-phosphonate) has been described[8]. Pro-pro-diphenyl-phosphonate was solubilized in physiologic saline immediately before use. Treatment with pro-pro-diphenyl-phosphonate started at the day of transplantation (10mg/rat ip. and 20mg/rat sc.), followed by additional doses on day 3 and 6 (20mg/rat sc.). DPP IV activity was determined fluorometrically, as described[10] with modifications. One unit of enzymatic activity was defined as the amount of enzyme catalyzing the formation of 1 μmol of assay-product per minute under the conditions employed. The DPP IV activity determined prior to any invasive procedure in each rat was defined as 100%. The mean±SD of DPP IV activity in untreated animals was 29.7±4.8 U/l (n=50) [9].

2.3 Assessment of circulating allospecific antibodies

Circulating donor-specific IgM and IgG allo-Abs were determined in recipients serum as previously described[11]. The levels were expressed as mode channel fluorescence.

2.4 Analysis of PBL surface antigen expression

Flow cytometry staining (FACS) for PBL surfache Ag expression was performed in modification[11]. Levels were expressed as percentage of positive stained PBL.

2.5 Human recipients of kidney grafts

We report on the preliminary results of a prospective study with 56 kidney transplanted patients from 12/1997 to 12/1999 at the Dept. of General and Thoracic Surgery at the Justus-Liebig-University in Giessen. Directly pre-transplantation, and before induction therapy, PBL surface expression (FACS, as described[11]) of CD26, -3, -4, -8, -45, -122, ADA and serum DPP IV activity (fluorometrically, as described[10]) were assessed and monitored. At 1, 7, 14 and 21d, and 2, 4, 6, 12, 18 and 24 mo. after Tx, measurements were repeated.

Various induction and immunosuppressive regimens were employed.

3. RESULTS

3.1 Acute allograft rejection correlates with an increase of DPP IV serum activity

Since CD26/DPP IV has been shown to play a costimulatory role in T-cell activation in vitro[12-14] we investigated whether the exposure to allo-Ag in vivo affected host serum DPP IV levels. The rationale for this experiment was further based on earlier observations, indicating an association between immuncompetence in recipients of heart- and kidney allografts and the serum activity of DPP IV[5,6]. We employed a well defined rat transplantation model in which LBNF1 hearts are rejected acutely within 7±1d in untreated LEW recipients; LEW isografted recipients served as controls. In 80% of allografted animals (8 out of 10), after a three day lag, a consistent rise in

DPP IV activity could be detected, peaking at about 150% (45 U/l DPP IV; $p<0.005$) of its initial value by day 6, e.g. prior to the actual rejection, and returning to pre-operative levels by day 11 (data not shown). To rule out non-specific influence on shifts in DPP IV, we studied sham-operated rats. Indeed, DPP IV serum levels in these non-grafted animals did not display any significant changes after the surgery (data not shown). Thus, increase in DPP IV serum levels correlated with the host immune cascade leading to acute allograft rejection[9].

3.2 Specific DPP IV enzymatic activity inhibition abrogates acute rejection

Use of DPP IV inhibitors has been reported to suppress T cell proliferation in vitro[15,16] and to decrease antibody production in mice immunized with bovine serum[17]. Little if any is known about the putative role of CD26/DPP IV during the allo-Ag-mediated immune response. Therefore, we used pro-pro-diphenyl-phosphonate to inhibit DPP IV activity in vivo. The engrafted hosts displayed a marked decrease of serum activity, which dropped to <20% of pre-operative value within 24 h after administration of a single pro-pro-diphenyl-phosphonate dose (10mg/rat sc.). Moreover, treatment regimen started at the day of engraftment (30mg/animal sc.), followed by application on day 3 and 6 (20mg/rat sc.), resulted in a profound and sustained depression of DPP IV (day 3: 10.4± 6.7% and day 6: 9.4±2.8%). After cessation of treatment, DPP IV recovered slowly, reaching about 75% of its initial value by day 19 (data not shown). Having successfully inhibited circulating DPP IV levels, we then attempted to combat the host rejection response by using pro-pro-diphenyl-phosphonate treatment as a probe. Interestingly, pro-pro-diphenyl-phosphonate treatment regimen abrogated acute rejection at day 7, and prolonged the mean survival time of cardiac Tx to 14.0±0.9d (MST±SD; n=10; $p<0.0001$ as compared to untreated controls) (Fig. 1). As in allografted otherwise untreated recipients, the rise in DPP IV activity was paralleled by a decline in graft function, and preceded ultimate rejection. Thus, pro-pro-diphenyl-phosphonate-mediated depression of DPP IV serum activity abrogated acute rejection and significantly prolonged cardiac allograft survival[9].

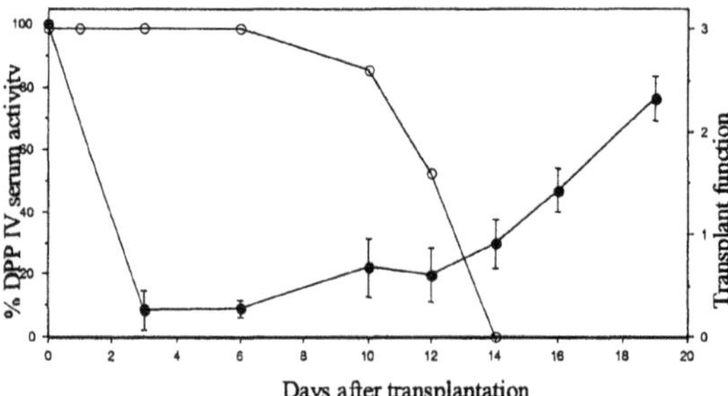

Figure 1: DPP IV serum activity (black circles) vs. graft function (white circles) in LEW rat recipients of LBNF1 heart transplants treated with pro-pro-diphenyl-phosphonate (day 0: 30mg sc., day 3: 20mg sc., and day 6: 20mg sc.). Heartaction is rated on a scale 0 to 3 (3: 100% contractility; 2: 75% contractility; 1: 50% contractility; 0: no contractility, e.g. rejection). Mean values and SD are shown (n=10).

3.3 Pro-pro-diphenyl phosphonate inhibiton impairs the synthesis of allospecific antibodies

To investigate the extent of specific DPP IV serum activity inhibition on the humoral immune response following exposure to allo-Ag, we analyzed allospecific Ab secretion in pro-pro-diphenyl-phosphonate-treated recipients. Although previous studies have linked CD26 enzymatic activity to Ab production after immunization[17,18] allospecific Ab kinetics in response to a perfused organ transplant have never been investigated in regard to their association with DPP IV activity. Animals treated with pro-pro-diphenyl-phosphonate displayed an absolute abrogation of allospecific IgM Ab secretion. In more than 15 treated animals, not one displayed any detectable IgM allo-Ab, in comparison to untreated recipients (Fig. 2).

Specific DPP IV activity inhibition delayed the appearance of allospecific IgG Ab in the serum for ca. 10 days, compared to rejecting controls. When IgG could finally be detected in the sera, absolute levels were lower, with greater interindividual variations (Fig. 3). In conclusion, inhibiting CD26 enzymatic activity following allo-Ag exposure profoundly impaired the quantitative and sequential humoral response[19].

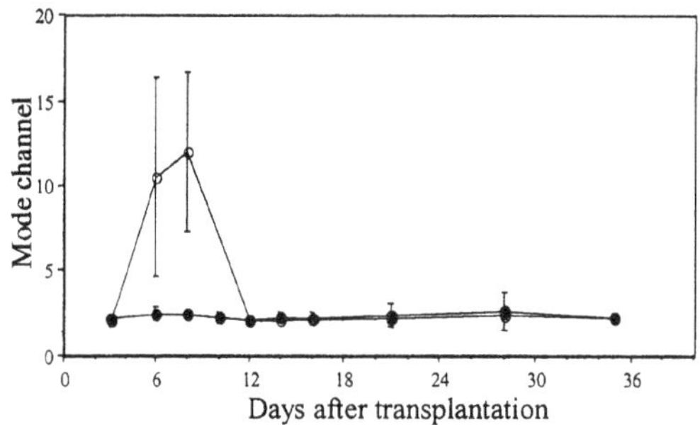

Figure 2: Circulating allospecific IgM Ab response following transplantation of LBNF1 cardiac allografts to LEW in pro-pro-diphenyl-phosphonate-treated (black circles) and untreated animals (open circles) by flow cytometric screening (mean values±SD are shown; n=5/10 animals/group). Note: Specific inhibition of DPP IV enzymatic activity completely abrogates allospecific IgM Ab secretion.

3.4 Pro-pro-diphenyl phosphonate inhibition abrogates accelerated allograft rejection

Pro-pro-diphenyl-phosphonate application diminished DPP IV serum levels ≤15% during the sensitization phase (data not shown). In 7 out of 13 animals blockade of DPP IV activity in vivo abrogated accelerated rejection and prolonged cardiac Tx survival (MST±SD=5.1±2.1d, $p<0.01$) (data not shown).

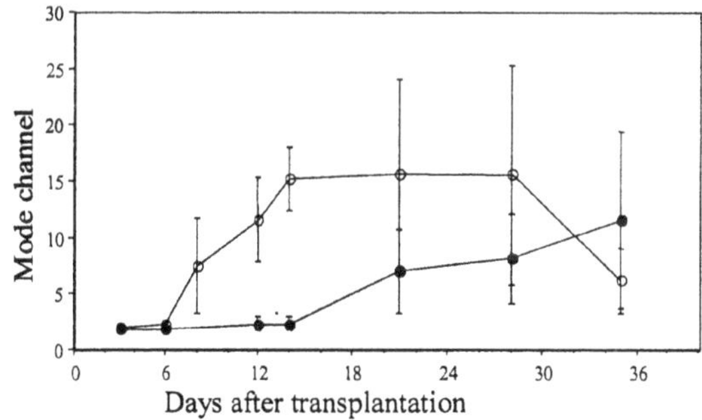

Figure 3. Circulating allospecific IgG Ab response following transplantation of LBNF1 cardiac allografts to LEW in pro-pro-diphenyl-phosphonate-treated (black circles) and untreated animals (open circles) by flow cytometric screening (mean values±SD are shown; n=5/7 animals/group). Note: Specific inhibition of DPP IV enzymatic activity leads to delayed and decreased allospecific IgG Ab secretion.

3.5 DPP IV enzymatic activity is decreased in human recipients of kidney grafts

To evaluate the impact of clinical solid organ transplantation and immunosuppressive therapy on DPP IV enzymatic activity over time, we monitored human recipients of kidney Tx. In 56 patients DPP IV enzymatic activity was measured for 24 months. Within 3 days following transplantation, DPP IV enzymatic activity significantly dropped (data not shown). This decline in activity could be monitored for the investigated population for at least 12 months (m=85±32%, $p<0.001$) (Fig. 4), with several patients displaying markedly lowered levels for up to 24 months (Fig. 4).

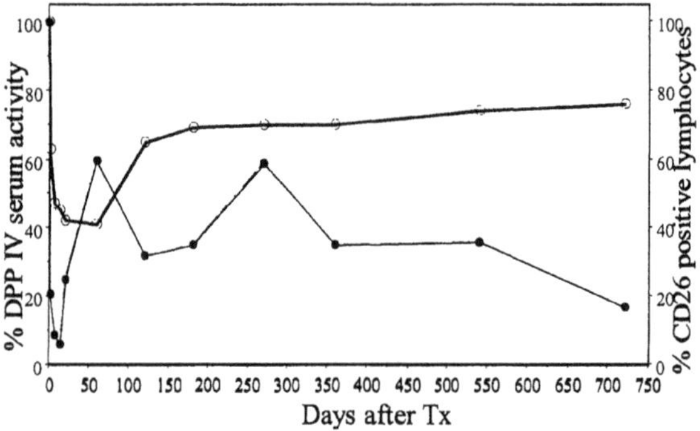

Figure 4: Dynamics of DPP IV serum activity (open circles) and CD26 positive PBL (black circles) plotted together over the period of 2 years following kidney transplantation of one patient (B. M.).

3.6 Kidney graft recipients show decreased CD26 expression on PBL

In parallel to measuring serum DPP IV enzymatic activity in human kidney graft recipients, we monitored the CD26 Ag expression kinetics PBL. Given the observation that lymhocytes are the main source for serum serum DPP IV, and based on the fact that CD26 cellular expression kinetics and circulating DPP IV enzymatic activity correlate during the early post-transplant period[9], we wanted to associate individual enzymatic serum levels with the cellular expression on PBL.

In addition, CD3, 4, 8, 45, 122 and ADA were measured as reference and key T cell surface Ags. Interestingly, CD26 Ag expression on PBL displayed a significant drop from baseline preop. (m=34±18%) for up to 18 months (m=15.2±11%, $p<0.001$) (Fig. 4). In contrast, CD4 (decreased expression for 6 months, $p<0.03$), CD8 (for 7d, $p<0.0001$), CD45 (for 2 months, $p=0.05$), CD3 (for 21d, $p<0.01$), CD122 (for 21d, $p<0.01$) and ADA (for 2 months, $p<0.01$) expression kinetics were only briefly affected[20].

4. CONCLUSION

Commencing experimental cardiac allograft rejection was reflected by increased DPP IV enzymatic serum activity. Employing a novel low molecular weight synthetic agent (pro-pro-diphenyl-phosphonate), we selectively inhibited CD26 catalytic activity, abrogating acute and accelerated rejection, thus significantly prolonging graft survival. Inhibition abolished circulating IgM titers and delayed IgG isotype switching.

Clinical kidney transplantation is associated with a significant decrease in serum DPP IV catalytic activity for 12 months. In addition, CD26 expression was significantly reduced on circulating PBL after renal engraftment for up to 18 months. Interestingly, this was in contrast with classical lymphocyte markers, which were only affected by transplantation for up to two months.

Our data extend earlier findings and provide further evidence that CD26/DPP IV plays a central role in T cell-mediated immune responses toward allo-Ag. In the further course of this study, we hope to clarify the role of the CD26 Ag and its enzymatic activity as possible clinical marker of immunomodulation in transplantation.

ACKNOWLEDGEMENTS

This study was supported by grants from the Deutsche Forschungsgemeinschaft (Ko 1637/1-1, 1637/2-1) and the Belgian National Fund for Scientific Research (NFWO).

REFERENCES

1. Morimoto, C., and Schlossman,S. F., 1994. CD26 a key costimulatory molecule on CD4 memory cells. *The Immunologist* **2/1**:4-7.
2. De Meester, I., Korom, S., Van Damme, J., and Scharpé, S., 1999. CD26, let it cut or cut it down. *Immunology Today* **20**: 367-375.
3. Tanaka, T., Kameoka, J., Yaron, A., Schlossman, S.F., Morimoto, C., 1993. The costimulatory activity of the CD26 antigen requires dipeptidyl peptidase IV enzymatic activity. *Proc Natl Acad Sci USA* **90**: 4586-4590.
4. Tanaka, T., Duke-Cohan, J. S., Kameoka, J., et al. 1994. Enhancement of antigen-induced T-cell proliferation by soluble CD26/dipeptidyl peptidase IV. *Proc Natl Acad Sci USA* **91**: 3082-3086.
5. Scharpé, S., De Meester, I., Vanhoof, G., et al. 1990. Serum dipeptidyl peptidase IV activity in transplant recipients. *Clin Chem* **36**:984.
6. Sanda, M.G., Pierson, R., Smith, C., and Rose, E., 1989. Serum dipeptidyl peptidase IV in cardiac transplant recipients. *Transplant Proc* **21**:2525-2526.
7. Belyaev, A., Borloo, M., Augustyns, K., et al. 1995. A new synthetic method for proline diphenyl phosphonates. *Tetrahedron Letters* **36**:3755-3758.
8. Lambeir, A.M., Borloo, M., De Meester, I., et. al. 1996. Dipeptidyl-derived diphenyl phosphonate esters: mechanism-based inhibitors of dipeptidyl-peptidase IV. *Biochimica Biophys Acta* **1290**:76-82.
9. Korom, S., De Meester, I., Stadlbauer, T.H.W., et al. 1997. Inhibition of CD26/dipeptidyl peptidase IV activity in vivo prolongs cardiac allograft survival in rat recipients. *Transplantation* **63**:1495-1500.
10. Scharpé, S., De Meester, I., Vanhoof, G., Hendriks, D., van Sande, M., Van Camp, K., and Yaron, A., 1988. Assay of dipeptidyl peptidase IV in serum by fluorometry of 4-methoxy-2-naphtylamine. *Clinical Chemistry* **34**:2299-2301.
11. Binder, J., Lehmann, M., Graser, E., Hancock, W.W., et al., 1996. The effects of nondepleting CD4 targeted therapy in presensitized rat recipients of cardiac allografts. *Transplantation* **61**:804-811.
12. Fleischer, B., 1987. A novel pathway of human T cell activation via a 103 kD T cell activationmolecule. *J Immunol* **138**:1346-1350.
13. Dang, N.H., Hafler, D., Schlossman, S.F., and Breitmeyer, J., 1990. FcR-mediated cross-linking of Tal (CDw26) induces human T lymphocyte activation. *Cell Immunol* **125**:42-57.
14. Dang, N.H., Torimoto, Y., Sugita, K., Daley, J.F., Schow, P., Prado, C., Schlossman, S.F., and Morimoro, C., 1990. Cell surface modulation of CD26 by anti-1F7 monoclonal antibody. Analysis of surface expression and human T cell activation. *J Immunol* **145**:3963-3971.
15. Schön, E., Jahn, S., Kiessig, S.T., et al. 1987. The role of dipeptidyl peptidase IV in human T lymphocyte activation. Inhibitors and antibodies against dipeptidyl peptidase IV suppress lymphocyte proliferation and immunoglobulin synthesis in vitro. *Eur J Immunol* **17**:1821-1826.
16. Flentke, G.R., Muñoz, E., Huber, B.T., Plaut, A.G., Kettner, C.A., and Bachovchin, W., 1991. Inhibition of dipeptidyl peptidase IV (DP-IV) by Xaa-boroPro dipeptides and use of these inhibitors to examine the role of DP-IV in T-cell function. *Proc Natl Acad Sci USA* **88**:1556-1559.

17. Kubota, T., Iizuka, H., Bachovchin, W.W., and Stollar, B.D., 1994. Dipeptidyl peptidase IV (DP IV) activity in serum and on lymphocytes of MRL/Mp-lpr/lpr mice correlates with disease onset. *Clin Exp Immunol* **96**:292-296.
18. Bednarczyk, J., Caroll, S.M., Marin, C., and McIntyre, B., 1991. Triggering of the proteinase dipeptidyl peptidase IV (CD26) amplifies human T lymphocyte proliferation. *J Cell Biochem* **46**:206-218.
19. Korom, S., De Meester, I., Schmidbauer, G., et al. 1999. Specific inhibition of CD26/DPP IV enzymatic activity in allograft recipients: effects on humoral immunity. *Transplant Proceedings*, **31**: 873.
20. Korom, S., De Meester, I., Maas, E., et al., 2002. CD26 expression and enzymatic activity in recipients of kidney allografts. *Transplant Proceedings*, **34**: 1753-1754.

CD26 is Involved in the Regulation of T-Cell Plasma Membrane Compartmentation

JUAN LOJO, FRANCISCO J. SALGADO, MONTSERRAT NOGUEIRA and OSCAR J. CORDERO
[*] *University of Santiago de Compostela. Department of Biochemistry and Molecular Biology. 15782 Santiago de Compostela, Galicia (Spain).*

1. INTRODUCTION

In recent years, major progress has been made in understanding proximal TCR signal-transduction events. Antigen presentation results in the clustering of protein tyrosine kinases (PTKs) that associate with the CD3 and TCR subunits and the co-receptors CD4 or CD8. The transmembrane tyrosine-phosphatase CD45 is essential in this process since dephosphorylates at least the inhibitory site of Src-family kinases, responsible for the phosphorylation of the immunoreceptor tyrosine-base activation motifs (ITAMs) [1,2].

An advance in this field came from the discovery of specialized membrane domains that serve as the sites of attachment of a variety of lipid-modified proteins and also integral membrane and cytoplasmic proteins (e.g. Src-family kinases Lck and Fyn, CD4, CD8 and LAT). These rafts, called GEMs (ganglioside-enriched membranes) or DRMs (detergent-resistant membranes) because contain a high density of sphingolipids and cholesterol [2,3], are particularly important for understanding CD45 function since this extremely active phosphatase does not require ligand binding for optimum catalytic activity and CD45-dependent dephosphorylation of key substrates such as Src or ZAP-70/Syk PTKs and ITAMs should be avoided. In the model of compartmentation, the immunological synapse initiates the TCR clustering and stabilization by the formation of a large lipid microdomain that accumulates (e.g. CD4 and CD8) and segregates (e.g. CD45 and LFA-1) several membrane proteins [4]. As transient tyrosine phosphorylation decreases within minutes after the initial response, it has been proposed that other phosphatases (not CD45) are recruited later to these rafts [2].

However, the role of the extracellular domain of CD45 remains elusive in spite that its structure strongly suggest ligand-receptor interactions [5]. The diversity of CD45 isoforms is cell type-dependent and regulated. Upon activation, naïve T cells switch from isoforms containing A, B or C epitopes to the lowest MW isoform CD45R0 that lacks sequences coded by A, B or C exons [6,7]. Several experiments reported distinct CD45 interactions on naïve and memory cells and T cell lines transfected with different isoform CD45 cDNAs, or cells from transgenic and knock-out mice, had differential responses to Ag. Moreover, CD45 has been found associated with surface molecules such as Thy-1, TCR, CD2, CD3, CD4, CD7, CD8, CD26, CD28, LFA-1, BCR, LPAP and CD45 itself. As the different CD45 isoforms have similar PTP activities, these data suggest that they may differentially interact with other surface molecules and alter PTP accessibility to substrates, which could in turn modify the signals received through Ag receptors, IL-Rs, and integrin-mediated adhesion to either augment or inhibit T cell activation [6-8].

T cells expressing high levels of CD26 constitute a subpopulation of $CD45R0^+$ cells with type 1 helper (T_H1) activities and transendothelial migration capacity [9]. We have reported a strong IL-12 (an inducing T_H1 response cytokine)-dependent CD26 up-regulation on activated T cells, including effector/memory CD45R0 cells, associated to a weaker staining of blasts with anti-CD45R0 UCHL-1 mAb. However, PTP activity, as well as DPPIV activity, was enhanced when IL-12 was present in the cultures, although IL-12 did not affect isoform switching to other CD45 isoforms (RA, RB, RC) and loss of CD45R0 staining was sialic acid-independent [10-13].

By confocal microscopy, we rejected a CD45 internalization as a possible explanation, and our anti-CD26 mAb coimmunoprecipitated R0, but anti-CD45R0 Abs did not coprecipitate CD26. In IL-12- cultured cells, anti-CD26 Ab coimmunoprecipitated more CD45 together to more CD26. An attractive model to explain these results all together is that IL-12 up-regulates a CD26 expression that interacts in *cis* with CD45R0 and/or other molecules, masking the UCHL-1 epitope [14].

This study describes the distribution of CD26 and CD45R0 molecules in plasma membrane microdomains of fresh human T cells. IL-12 also changed CD26 and CD45R0 membrane compartmentation on activated T cells. The significance of this finding is discussed.

2. RESULTS

2.1 CD26 and CD45 are present in PM microdomains and their distribution changes with T cell activation in the presence or absence of IL-12

The dual distribution of CD26, with approximately 28% of the material detected in the light-density fractions (LDF, 4 to 6) and 51% in the heavy-density fractions (HDF, 10-11) of sucrose gradient, is shown in human PBMCs (T cells since monocytes, B and NK cells are essentially CD26$^-$) (*Fig. 1*). Note the distribution observed for the activated T cells (37% and 30%, respectively) since, in addition to a higher intensity, CD26 was redistributed to the rafts (LDF, see *Fig. 1* legend). IL-12 enhanced CD26 intensity and the intermediate fractions were enriched for CD26.

Similarly, CD45 can be found associated with GEMs in human lymphocytes. The ratio LDF/HDF is higher for total CD45 and CD45R0 than that of CD26. Intensity differences between PBMC and PHA-activated lymphocyte fractions reflect total and R0 CD45 up-regulation. In comparison with CD26, PHA hardly affects CD45 redistribution. Moreover, in contrast to CD26 where the ratio LDF/HDF was constant, the presence of IL-12 in T-cell activation reduced the percentages of CD45 in the LDF (or fractions near LDF) of gradients. Very spectacular, this redistribution can be clearly attributed to the R0 isoform.

We confirmed these findings by other approaches that also demonstrated that hypothesized CD26-CD45R0 association is near (but on the outer edge of) GEMs.

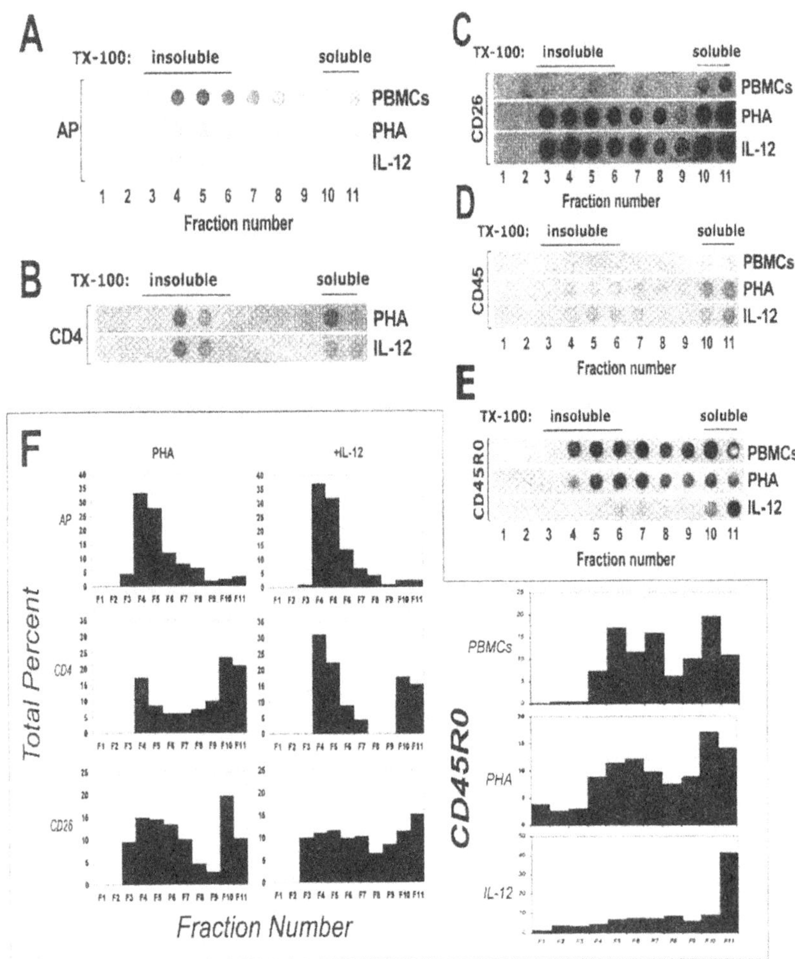

Figure 1. Detection of membrane associated proteins CD26, CD45, CD4 and AP in sucrose density gradient fractions. Costimulation with IL-12 excludes CD45R0 from rafts. Protein from PBMCs, PHA or PHA+IL-12 lymphoblasts was subjected to equilibrium density gradient centrifugation and 11 fractions were collected, dotted, and probed for the indicated Ag, except AP. CD26, total CD45 and CD45R0 are present, in addition to 9-10 and 10-11 (respectively) soluble HDF, in 4-6 and 3-5 insoluble LDF of PBMCs and PHA-blasts ($CD3^+$ cells)(*C-E*). Controls from the same gradients show the GPI-anchored AP and CD4 (with the well-known bipolar distribution)(*A,B*). The signal strength (arbitrary values) of each dot is expressed as percentage of the total for the respective Ag.

2.2 CD26 is involved in pulling out CD45R0 from rafts

The addition of antisense oligonucleotides for the CD26 mRNAs to the same culture conditions as above resulted in a drop of the cell surface CD26 expression, as expected, both in T cells activated without or with IL-12 (*Fig. 2, left*). This effect can be seen from day 2 of culture. In this way, we demonstrate that loss of CD45R0 staining observed in IL-12 cultures is specificly –oligonucleotide controls did not show this effect (data not shown)- associated (probably directly) to a certain kind of de novo CD26 protein expression that can be observed even with PHA stimulation alone (*Fig. 2, right*).

Experiments in progress (we could not finish them in time to show the data) are indicating that inhibition of CD26 biosynthesis in activated cells has only a mild effect on CD26 redistribution to membrane compartments, perhaps a slight diminution of soluble membrane CD26, and that IL-12-dependent redistribution of CD45R0 from GEMs to soluble membrane is avoided in cells treated with antisense –but not controls- oligonucleotides.

3. DISCUSSION

This study reveals that a fraction of the transmembrane proteins CD26 and CD45 is associated to GEMs in fresh and activated human lymphocytes. CD26, unlike CD4 presenting a bipolar pattern, is present in almost all sucrose-gradient fractions. CD26 has been proposed as a costimulatory molecule of TCR-dependent T cell activation [9,15]. Independent of its enzymatic activity [9], important in the T cell response through modulation of the activity of several biological factors (chemokines, etc.), co-crosslinking of anti-CD26 and CD3 mAbs enhanced phosphorylation of CD3 tyrosine residues and increased CD4-associated $p56^{lck}$ tyrosine kinase activity [15,17].

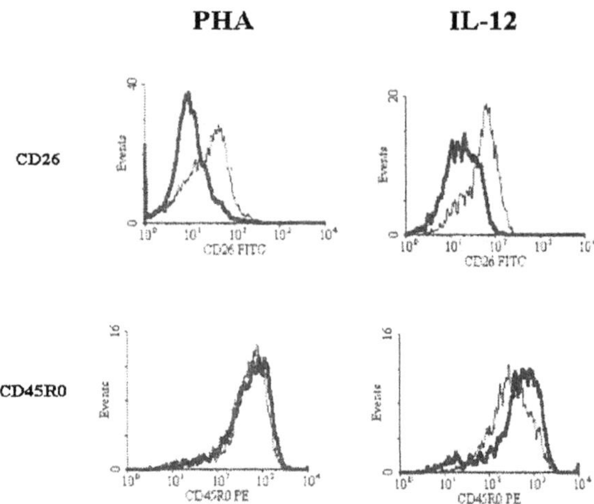

Figure 2. **Relationship of CD26-CD45R0 association with GEMs.** In *A*, flow cytometry analyses show that IL-12-dependent loss of CD45R0 staining is dependent of cell surface CD26 de novo expression. PBMCs -previously treated with CD26 antisense (*thick lines*), or control (not shown) oligos, or mock-treated (*thin lines*)- (10^6 cells/ml) were stimulated with PHA (1 μg/ml) in the absence (*left*) or presence (*right*) of purified rIL-12 (2 ng/ml). On day 5, cultured cells were subjected to one-color immunofluorescence with TP1/16-FITC mAb. Cells from the same donor were also stained with anti-CD45R0 FITC-UCHL-1 or anti-CD3 UCTH-1 (not shown) mAbs. Anti-CD3 staining was never affected.

This effect can be ascribed to the aggregation of lipid rafts that facilitates colocalization of kinases and TCR thereby triggering tyrosine-phosphorylation, as observed with other GPI-associated proteins [18].

However, a role in this process for a CD26-CD45 association (as it has been recently suggested) [19] cannot be discarded as a percentage of the PTP CD45 is associated to GEMs in human fresh T cells (B lymphocytes from PBMCs constitutively lack CD45 in GEMs [20]). This result is not unexpected since CD45 was found associated with the GPI-protein Thy-1 (CD90) or the raft-associated TCR in T-cell lines (cells probably resembling an effector state) [21], in spite that models of membrane compartmentation after B- or T-response to antigen are considering that CD45 is excluded from the signaling complex raft [1-3,22]. It is interesting to note for later discussion that these experiments have used B or Jurkat cells and both do not express CD26.

This CD45, present in rafts of effector/memory cells at similar proportions than CD26, can be ascribed to the R0 isoform. In fact, a fraction of CD26 and CD45R0 is very resistant to the treatment, suggesting that it is inside the raft core. A recent report confirms that CD26 and CD45R0 can be found in membrane rafts of CD26-transfected Jurkat cells under certain conditions [19]. These data do not invalidate the model of compartmentation, implying increased stability of phosphorylation [2,3,22], if we consider naïve T cells (that do not express the R0 isoform) but they point out to a role in effector/memory CD45R0$^+$ cells. As the effector/memory T cells are far more efficient in responding to stimulation, (low MW CD45 isoform) PTP activity could be necessary to maintain activation of Src kinases and to prime them to engaged TCR in these cells with the rafts bigger than naïve cells where CD45 can be exposed at the edges of smaller rafts [3].

We show that IL-12 does change dramatically the distribution of CD45R0 from rafts to soluble fraction. The finding that IL-12 drives CD26 and CD45R0 to particular PM regions is very important. First, IL-12 might pull out CD45R0 from rafts as IL-2 pulls out the α chain to bind the β and γ chain of IL-2R complex within soluble fractions[24]. As conclusion, interleukins are implied in the rules governing the inclusion and exclusion of proteins into rafts. Second, the impaired Ca^{2+} responses we have observed upon activation via TCR in IL-12 cultured cells (data not shown) can be explained by a redistribution of R0 from rafts, perhaps to control IL-12R-dependent signal transduction, as it was recently shown for other cytokine receptors through suppression of the JAK-STAT pathway[23].

Our results of FACS, confocal microscopy, PTP activity and immunoprecipitation, fitted with a model in which when cells are triggered by IL-12, more CD26 and CD45R0 proteins become associated (leading to a reduction of Ab binding for one of them). Together with our data from CD26 expression inhibition, the new CD26 molecules are directed to the soluble fraction near the lipid raft core and anchore CD45R0 molecules. If IL-12-dependent CD45R0 membrane compartment redistribution is avoided in the absence of CD26 de novo expression, CD26 could be part of a shuttling mechanism for CD45. Raft CD45 exchanges with the larger pool of freely diffusing CD45 after the formation of the immunological synapse[25].

Only a fraction of CD26 molecules associate to CD45R0. A recent report shows that CD26 (in spite of its very short cytoplasmic region) binds directly to the cytoplasmic region (PTP domain 2) but not to the extracellular domain of CD45[19]. Our results agree better with the fact that the extracellular domain of CD45 (in particular low MW) controls its association with the CD4-TCR complex[26].

As PTP activity is important in the regulation of adhesion-triggered tyrosine kinase cascades[2,27], and CD26 has been proposed as a functional adhesion receptor under certain circumstances [9,15-17,28,29] (note that several of those studies were done with CD26$^-$ Jurkat cells), a CD26/CD45R0 membrane redistribution could also regulate the presence of CD45 at integrin-dependent adhesion sites. Finally, both CD26 and CD45 molecules are involved in the pathophysiology of AIDS [30,31]. At least HIV-1 gp-120 [32], but perhaps also Tat[33] and Gag[34], can modulate CD4 lateral interactions with both. CD45 and CD26 are involved in the regulation of lymphocyte death at least under some circumstances[35]. We would like to speculate on the importance of the mechanism described in this article for the normal behaviour of the T cell and, further on, the possibility that this mechanism was blocked by particles or molecules from HIV-1 or other pathogens.

ACKNOWLEDGEMENTS

We wish to thank Dr. F. Sanchez-Madrid (Service of Immunology, Hospital de la Princesa, Universidad de Madrid) for providing us with anti-CD26 TP1/16 hybridoma. We thank the Centro de Transfusión de Galicia for buffy coats, Prof. S.F. Schlossman (Dana-Farber Cancer Institute, Harvard Medical School, Boston, MA) for his kind gift of 1F7 anti-CD26 Ab, and J. Trotter (Scripps Institute, LaJolla,CA) for the WinMDI software.

REFERENCES

1. van Leeuwen, J.E., and Samelson, L.E. (1999) *Curr. Opin. Immunol.* **11**, 242-248.
2. Thomas, M.L. (1999) *Curr. Opin. Immunol.* **11**, 270-276.
3. Xavier, R., and Seed, B. (1999) *Curr. Opin. Immunol.* **11**, 265-269.
4. Penninger, J.M., Irie-Sasaki, J., Sasaki, T., and Oliveira-dos-Santos, A.J. (2001) *Nat. Immunol.* **2**, 389-396.
5. Irie-Sasaki, J., Sasaki, T., Matsumoto, W., Opavsky, A., Cheng, M., Welstead, G., Griffiths, E., Krawczyk, C., Richardson, C.D., Aitken, K., Iscove, N., Koretzky, G., Johnson, P., Liu, P., Rothstein, D.M., and Penninger, J.M. (2001) *Nature* **409**, 349-354.
6. Johnson, P., Maiti, A., and Ng, D.H.W. (1996) *Weir's Handbook of Exp. Immunol.* Vol. 2, *Cell Surface and Messenger Molecules of the Immune System*, 62.1-62.16. L.A. Herzenberg and C. Blackwell, eds. Blackwell Science.
7. Trowbridge, I.S., and Thomas, M.L. (1994) *Annu. Rev. Immunol.* **12**, 85-116.
8. Shenoi, H., Seavitt, J., Zheleznyak, A., Thomas, M.L., Brown, E.J. (1999) *J. Immunol.* **162**, 7120-7127.
9. De Meester, I., Korom, S., Van Damme, J., and Scharpé, S. (1999) *Immunol. Today* **20**, 367-375.
10. Cordero, O.J., Salgado, F.J., Viñuela, J.E., and Nogueira, M. (1997) *Immunobiology* **197** 522-533.
11. Cordero, O.J., Salgado, F.J., Viñuela, J.E., and Nogueira, M. (1998) *Immunol. Lett.* **61**, 7-13.

12. Salgado, F.J., Vela, E., Martín, M., Franco, R., Nogueira, M., and Cordero, O.J. (2000) *Cytokine* **12**, 1136-1141.
13. Cordero, O.J., Salgado, F.J., Fernández-Alonso, C.M., Herrera, C., Lluis, C., Franco, R., and Nogueira, M. (2001) *J. Leukoc. Biol.* **70**, 920-30.
14. Salgado, F.J., Lojo, J., Viñuela, J., Nogueira, M., and Cordero, O.J. (2000) *18th International IUBMB Congress*, 242.
15. Tanaka, T., Camerini, D., Seed, B., Torimoto, Y., Dang, N.H., Kameoka, J., Dahlberg, H.N., Schlossman, S.F., and Morimoto, C. (1992) *J. Immunol.* **149**, 481-486.
16. von Bonin, A., Huhn, J., and Fleischer, B. (1998) *Immunol. Rev.* **161**, 43-53.
17. Morimoto, C., and Schlossman, S.F. (1998) *Immunol. Rev.* **161**, 55-70.
18. Ilangumaran, S., He, H.T., and Hoessli, D.C. (2000) *Immunol. Today* **21**, 2-7.
19. Ishii, T., Ohnuma, K., Murakami, A., Takasawa, N., Kobayashi, S., Dang, N.H., Schlossman, S.F., and Morimoto, C. (2001) *Proc. Natl. Acad. Sci. U.S.A.* **98**, 12138-12143.
20. Cheng, P.C., Dykstra, M.L., Mitchell, R.N., and Pierce, S.K. (1999) *J. Exp. Med.* **190**, 1549-1560.
21. Volarevic, S., Burns, C.M., Sussman, J.J., and Ashwell, J.D. (1990) *Proc. Natl. Acad. Sci. U.S.A.* **87**, 7085-7089.
22. Rodgers, W., and Rose, J.K. (1996) *J. Cell Biol.* **135**, 1515-1523.
23. Blank, N., Kriegel, M., Hieronymus, T., Geiler, T., Winkler, S., Kalden, J.R., Lorenz, H.M. (2001) *J. Immunol.* **166**, 6034-6040.
24. Marmor, M.D., and Julius, M. (2001) *Blood* **98**, 1489-1497.
25. Johnson, K.G., Bromley, S.K., Dustin, M.L., and Thomas, M.L. (2000) *Proc. Natl. Acad. Sci. U. S. A.* **97**, 10138-10143.
26. Leitenberg, D., Boutin, Y., Lu, D.D, and Bottomly, K. (1999) *Immunity* **10**, 701-711.
27. Krauss, K., and Altevogt, P. (1999) *J. Biol. Chem.* **274**, 36921-36927.
28. Cheng, H.C., Abdel-Ghany, M., Elble, R.C., and Pauli, B.U. (1998) *J. Biol. Chem.* **273**, 24207-24015.
29. Oravecz, T., Pall, M., Roderiquez, G., Gorrell, M.D., Ditto, M., Nguyen, N.Y., Boykins, R., Unsworth, E., and Norcross, M.A. (1997) *J. Exp. Med.* **186**, 1865-1872.
30. Struyf, S., Proost, P., Schols, D., De Clercq, E., Opdenakker, G., Lenaerts, J.P., Detheux, M., Parmentier, M., De Meester, I., Scharpé, S., and van Damme, J. (1999) *J. Immunol.* **162**, 4903-4909.
31. Blázquez, M.V., Madueno, J.A., González, R., Jurado, R., Bachovchin, W.W., Peña, J., Muñoz, E. (1992) *J. Immunol.* **149**, 3073-3077.
32. Feito, M.J., Bragardo, M., Buonfiglio, D., Bonissoni, S., Bottarel, F., Malavasi, F., and Dianzani, U. (1997) *Int. Immunol.* **9**, 1141-1147.
33. Wrenger, S., Hoffmann, T., Faust, J., Mrestani-Klaus, C., Brandt, W., Neubert, K., Kraft, M., Olek, S., Frank, R., Ansorge, S., and Reinhold, D. (1997) *J. Biol. Chem.* **272**, 30283-30288.
34. Nguyen, D.H., and Hildreth, J.E. (2000) *J. Virol.* **74**, 3264-3272.
35. Klaus, S.J., Sidorenko, S.P., and Clark, E.A. (1996) *J. Immunol.* **156**, 2743-2753.

Inhibition of Dipeptidylpeptidase IV (DPP IV, CD26) Activity Modulates Surface Expression of CTLA-4 in Stress-Induced Abortions

JENS RÜTER*, HANS-ULRICH DEMUTH#, PETRA C. ARCK*, TORSTEN HOFMANN#, BURGHARD F. KLAPP,* and MARTIN HILDEBRANDT*

*Department of Internal Medicine/Psychosomatics, Charité Campus Mitte, Luisenstr. 13A, Berlin, Germany; #probiodrug AG, Weinbergweg 22, Halle/Saale, Germany

1. INTRODUCTION

Dipeptidylpeptidase IV (DPP IV, CD26) is a serine-type protease which preferentially cleaves N-terminal dipeptides from polypeptides containing proline or alanine as the penultimate amino acid. DPP IV has been shown to induce and enhance T-cell activation by virtue of its enzymatic activity[1] and by crosslinking of CD26, i.e. membrane-bound DPP IV, with either CD2 or CD3[2].

Various *in-vitro* experiments have shown specific effects of synthetic DPP IV inhibitors on the function of immune cells (reviewed in[3]). Thus, it has been shown that DPP IV inhibitors are capable of suppressing the proliferation of human PBMC stimulated with mitogens[4]. Also, DPP IV inhibitors have been shown to exert suppressive effects on the production of IFN-γ and tumor necrosis factor (TNF)-α[5] and stimulating effects on the production of transforming growth factor (TGF)-β[6,7]. *In vivo* experiments in animal models of experimental autoimmune encephalomyelitis (EAE)[8] and cardiac allograft survival[9] suggest a potential therapeutic use for DPP-IV inhibitors in immunological reactions.

Dipeptidyl Aminopeptidases in Health and Disease, Edited by Hildebrandt et al.
Kluwer Academic/Plenum Publishers, New York, 2003

We were able to demonstrate earlier that inhibition of DPP IV activity abrogates the stress-related increase in abortions and a concomitant increase in γ-interferon (IFN-γ)[10] in a murine model of stress-induced abortions. These stress-induced abortions are known to be mediated by immunological mechanisms[11] started by an increased production of T-helper-1 (Th1) cytokines in the decidua of the placenta[12].

Beyond the observed modulatory effect on cytokines, it is conceivable that DPP IV/ CD26 may also interact with costimulatory molecules which form an essential part of T cell responses. CTLA-4 (CD152) is a co-stimulatory receptor that shares 31% homology with CD28. It is rapidly expressed on T-cells following activation and acts as a functional antagonist of CD28 at the two B7 ligands (B7-1 and B7-2)[13]. Signals transduced through CTLA-4 result in abortive T-cell activation, poor Interleukin-2 (IL-2) production, and anergy[14], a state in which T cells fail to progress through the cell cycle and subsequently fail to divide upon restimulation[15], while CD28 promotes IL-2 production, clonal expansion, and anergy avoidance[16]

The aim of the present study was to investigate the effect of DPP IV-inhibition on the expression of the surface antigen CTLA-4 on decidual T-lymphocytes in murine stress-induced abortions as a model for the study of immunotolerance.

2. MATERIAL AND METHODS

2.1 Animals

Female CBA/J and male DBA/2 mice were purchased from Charles River, Germany, and maintained in an animal facility with a 12 hour light/dark cycle. Animal care and experimental procedures followed institutional ethics guidelines and conformed to requirement of the state authority for animal research conduct (LAGetSi, Berlin). After overnight cohabitation of CBA/J females with a DBA/2 male, females with vaginal plugs (day 0.5 of pregnancy) were segregated. Subgroups of at least 20 mated mice received daily intraperitoneal injections of the DPP IV inhibitor Ile-Cyanopyrrolidide (0.5 μmol/kg/d in 0.2 ml PBS; hereafter denoted as DPP IV inhibitor) starting on day 5.5 of pregnancy. Injection of the inactive stereoisomeric form of Ile-Cyanopyrrolidide was used as a control. The groups were divided, and half of the animals receiving the DPP IV inhibitor or the control substance, respectively, were exposed to sonic stress for 24 hours beginning on day 5.5 to boost abortion rates as described previously[16,18]. All animals were sacrificed on day 13.5, the numbers of normal and resorbing sites were determined, and cells were taken for further study.

2.2 Preparation of cell suspensions and flow cytometry

Mononuclear cells were prepared from murine deciduae as described[16,18]. All monoclonal antibodies (mAb) were purchased from Pharmingen (Heidelberg, Germany) and included the following: R-Phycoerythrin-labeled (R-PE) hamster anti-mouse mAb against CD3 (clone G4.18), CD4 (clone GK1.5), CD28 (clone 37.51); Fluorescein Isothiocyanate (FITC)-conjugated rat anti-mouse mAb against CD4 (clone RM 4-5), CD26 (clone H194-112); and Allophycocyanin (APC)-conjugated rat anti-mouse against CD8a (clone 53-6.7). Unconjugated hamster anti-mouse CTLA-4/CD152 (clone UC10-4F10-11) was used. Each antibody was diluted 1:100 in cytoflow buffer (PBS, 1 % BSA, 5 mM EDTA, and 0.1 % sodium azide). All incubations were performed for 20 minutes at 20°C in the dark. The unconjugated CTLA-4 mAb was labelled with biotin-conjugated mouse anti-hamster IgG mAb cocktail (clones G70-204 and G94-56) and Streptavidin-Phycoerythrin (PE; Pharmingen, Heidelberg, Germany). A nonspecific hamster IgG antibody was used for control staining procedures instead of unconjugated CTLA-4 mAb.

Cell fluorescence was measured using a Becton Dickinson FACSCalibur Flow Cytometer. A gate for the lymphocyte population was defined by forward and side light scatter characteristics (size/complexity criteria). The results of the cytofluorometric analyses were recorded

1) as the percentage number of cells positive for the respective antibody or, in double-staining techniques, for two different antibodies, as compared to an isotype control, and
2) as the mean fluorescence intensity of the respective antibody-positive population.

2.3 Enzyme-linked-immunoassay (ELISA)

Pre-coated ELISA-kits (Quantikine®M, R&D Systems, Wiesbaden, Germany) were used for quantitative determination of IFN-γ and IL-10 in serum. Sample/reagent-preparation and assay procedure were performed as suggested by the manufacturer.

2.4 Synthesis of 2-(S)cyano-1-isoleucylpyrrolidine

1. *Boc-Ile-Pro-NH$_2$:* Di-isopropylamine was added to a solution of H-ProNH$_2$*HCl (225 mg, 1.50 mmol) in dry CH_2Cl_2 (15 ml) until the pH was adjusted to 9. BocIleOSu was added in one portion and the mixture stirred for 16 h under an argon atmosphere. The solvent was evaporated and the residue treated in a standard way, i.e. the residue was partitioned between ethylacetate (60 ml) and 0.3 N $KHSO_4$ solution (10 ml). The organic layer was further washed with saturated $NaCHO_3$ solution (10 ml), water (10 ml) and brine (5 ml). The solution was dried and evaporated at reduced pressure.

2. *Boc-Ile-Pyrr-CN:* Imidazole (84 mg, 1.24 mmol) was added to a solution of Boc-Ile-Pro-NH$_2$ in dry pyridine (10 ml) under an argon atmosphere. The solution was cooled to –35°C, before the drop by drop addition of $POCl_3$ (0.25 ml, 2.48 mmol). The reaction was stirred at –30°C to –20°C for 60 min. The solution was then evaporated and the crude residue subjected to column chromatography (silica gel) to yield 180 mg (94 %) of 2-(S)-cyano-1-[*tert*-(butoxycarbonyl)isoleucylpyrrolidine as a colorless oil.

3. *H-Ile-Pyrr-CN*TFA:* Deprotection was carried out by stirring with triflouro acetic acid for 60 min. Evaporation and lyophilization from water afforded 60 mg of 2-(S)cyano-1-isoleucylpyrrolidine as a white solid.

2.5 Statistical analysis

Statistical analyses were performed as a cross-sectional analysis of the resorption rates and flow cytometric data in all four groups. All calculations were performed using SPSS 9.0 computer software. Among the tests applied were the following: Student´s t-Test, one-factor variance analysis (Oneway) with post-hoc comparison (Scheffé), cross-table analyses and Fisher's exact test.

3. RESULTS

Stress-induced, cytokine-mediated abortions were abrogated by inhibition of DPP IV (30.1% in stressed animals without the inhibitor, 8.3% in stressed animals with inhibitor, $p<0.01$, table 1).

In animals exposed to stress, higher serum concentrations of γ-Interferon (IFN-γ) were detected than in animals exposed to stress that had been injected with the DPP IV inhibitor (mean +/- SD: 7.3+/-4.4 vs. 0.6+/-0.8; P=.001, table 1). Interleukin-10 (IL-10) concentrations showed only a tendency towards being lower in non-stressed animals receiving the DPP IV inhibitor than in non-stressed controls (mean +/- SD: 10.5+/-13.5 vs. 28.2+/-13.9, P=.087).

Table 1: Characterization of decidual lymphocyte subpopulations

	Control	**Control/+Inh**	**Stressor**	**Stressor/+Inh.**
Abortions (% of implants)	10.9 ± 10.7	9.3 ± 13.5	30.1 ± 12.3	8.3 ± 14.2
IFN-γ (pg/ml)	1.2 ± 1.8	2.5 ± 3.9	7.3 ± 4.4	.63 ± .82
CD3+ (%)	24	31	38	12
CD3+/CD4+ (%)	17	12	10	8
CD4+/CD26+ (%)	4	6	7	2
MFI CD26 on $CD4^{pos}$	1580	440**	1507	368**
CD3+/CD8+ (%)	2.8	3.2	5.4	7.3
CD8+/CD26+ (%)	1	1	3	4
CD3+/CD28+ (%)	12	20	22	2***
MFI CD28 on $CD3^{pos}$	253	163**	221	120**
MFI CTLA-4 on $CD26^{pos}$	91	117	$30^{###}$	94***
MFI CTLA-4 on $CD4^{pos}$	510	286**	543	227***

MFI=mean fluorescence intensity; Inh.=DPP-IV inhibitor; **= p<.01 (Inh. vs. respective non-Inh.); ***= p<.001 (Inh. vs. respective control); ###= p<.001 (Stressor vs. control)

Inhibition of DPP IV activity was associated with a lower expression of CD26/DPP IV on CD4-positive decidual lymphocytes, irrespective of the exposure to stress (1580 vs. 440 mean fluorescence units, p<0.01; Figure 2a). Only a low number of CD8-positive cells were detected (Table 1), rendering a flow cytometric assessment of changes in the surface expression of CD26 impossible.

In stressed animals, a lower surface densitiy of CTLA-4 on decidual CD26-positive lymphocytes was observed than in non-stressed animals; inhibition of DPP IV restored CTLA-4 surface density to normal (94,3 vs. 30,2 mean fluorescence units; p<0,01, Fig. 1b). No change in intracellular CTLA-4 expression could be detected. A representative assessment of CTLA-4 expression on CD26-positive cells is shown in Figure 1a.

Decidual lymphocytes showed a lower percentage of cells coexpressing the antigens CD3 and CD28 as well as a lower surface density of CD28 in animals who had received the DPP IV inhibitor irrespective of stress (Fig.2b).

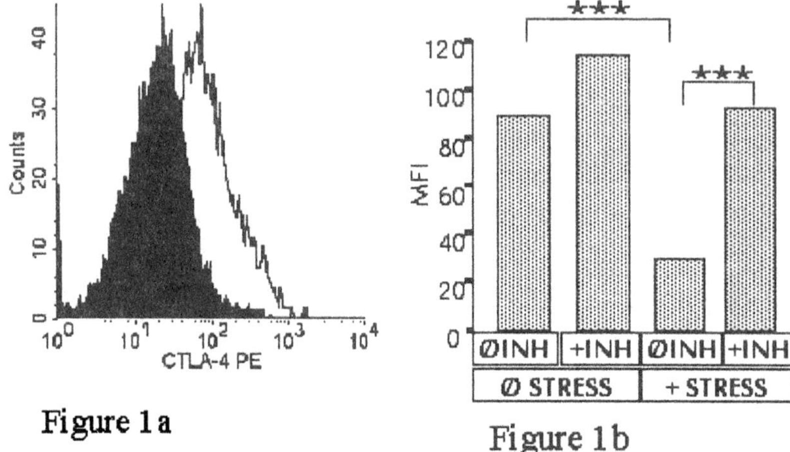

Figure 1: CTLA-4 expression on decidual lymphocytes
(a) representative histogram showing mean fluorescence intensity (MFI) of CTLA-4 on decidual C26-positive cells in a mouse that had received the DPP-IV inhibitor and stress (WHITE) versus one that had received inactive stereoisomer and stress (BLACK)
(b) bar diagram of all annimals, ØINH: treated with inactive stereoisome, +INH: treated with DPP-IV inihibitor, ØSTRESS: received no stress, +STRESS: received stress

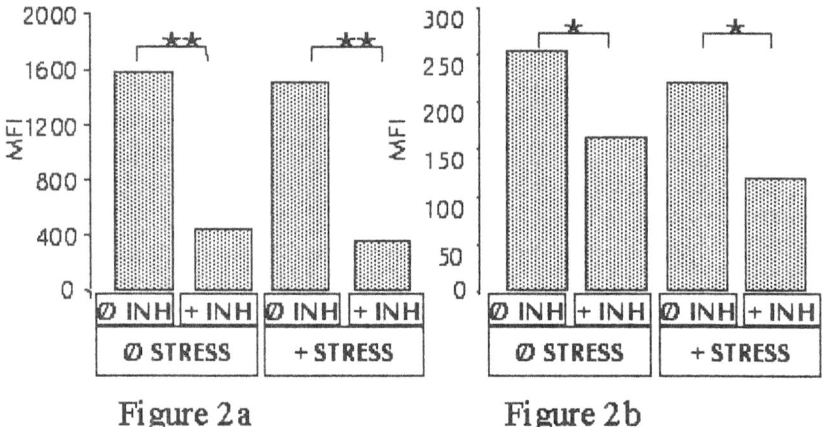

Figure 2:
(a) CD26-Expression on CD4+ cells and
(b) CD28-Expression on CD3+ cells expressed as mean fluorescence intensity (MFI)
same legend as in Fig. 1

4. DISCUSSION

In the present study, we investigated the effect of the DPP IV-inhibitor 2-(S)cyano-1-isoleucylpyrrolidine on the expression of CTLA-4, CD28 and CD26 on decidual T-lymphocytes in a murine model of abortion. As shown before[16], inhibition of DPP IV activity abrogated stress-induced abortions and an associated increase in IFN-γ-production. The present study shows that CTLA-4-expression on decidual T-lymphocytes is increased in mice receiving the DPP IV-inhibitor while stress-induced abortions are associated with a decrease of CTLA-4-expression on decidual CD26-positive lymphocytes. In addition, CD26- and CD28-expression on CD4-positive decidual T-lymphocytes are decreased in mice receiving the DPP IV-inhibitor.

CTLA-4 is expressed mainly intracellularly on resting T-cells, and its intracellular and surface expression are induced by IFN-γ and IL-2 *in vitro*[23] and by soluble peptides in transgenic mice with a restricted T-cell receptor repertoire *in vivo*[24]. Our data suggest that stress-induced abortions, which are mediated by cytokines produced by T-helper-1(Th1) cells such as TNF-α and IFN-γ[25,26], are associated with a decreased surface expression of CTLA-4 on decidual T-lymphocytes. Since CTLA-4 competes with CD28 for the same ligand (B7-1/2) on APC's[19], this downregulation of CTLA-4 in the context of stress-induced abortions may contribute to the immunological dysbalances observed in murine stress-induced abortions by allowing for an enhanced costimulation of T-cells through the CD28/B7-pathway. A decreased CTLA-4 expression as a cause of immune dysbalance is most strikingly demonstrated by the phenotype of CTLA-4-deficient (CTLA-4$^{-/-}$) mice that develop fatal lymphoproliferative disease with multiorgan tissue destruction[27,28].

Our data suggest that inhibition of DPP IV affects the expression of CTLA-4 and costimulatory molecules such as CD28 and CD26. When stress-induced abortions are abrogated by inhibition of DPP IV activity, CTLA-4 expression returns to the levels observed in non-stressed animals and CD26 and CD28 are downregulated. Thus, it seems possible that this changed profile of antigen expressionon T-lymphocytes upon inhibition of DPP IV may contribute to the modulation of immune mechanisms leading to the observed prevention of abortions.

The modulation of CTLA-4 expression by inhibition of DPP IV activity was observed on CD26-positive cells only: CTLA-4 expression on decidual CD4-positive cells was lower upon DPP IV inhibition than in controls, irrespective of the exposure to stress. It is conceivable that, in CD4-positive cells, the lower expression of CTLA-4 upon inhibition of DPP IV activity reflects a status of diminished T cell activation, as reflected by lower amounts of CD26 and of CD28 in the same groups of animals.

Possibly, the effects of DPP IV inhibition on the surface expression of CTLA-4 are secondary to a modulation of intracellular signal transduction. The inhibition of DPP IV activity interacts with early phosphorylation events in the process of T cell activation, as shown for the hyperphosphorylation of $p56^{lck}$, in a reversible and dose-dependent manner[29]. Interestingly, phosphorylation of CTLA-4 by $p56^{lck}$ and other tyrosine kinases has been shown to prevent the internalization of CTLA-4 into the cell[30]. This observation would, although not proven in this study, link DPP IV activity and, conversely, inhibition of DPP IV activity, to the regulation of CTLA-4 expression.

Given the fact that DPP IV/CD26 induces T cell proliferation through CD86 (B7-2) upregulation on APCs[31], a picture emerges in which DPP IV inhibition not only prevents CD86 upregulation, but also enhances CTLA-4 surface expression, thus interfering with T cell activation and costimulatory events in a potent and multi-faceted fashion.

It should be noted that a stress-related increase in surface expression of CD26 was observed on CD4-positive cells only. This observation suggests a distinct, yet undefined role for CD4-positive lymphocytes in stress-induced, DPP IV/CD26-mediated events, whereas CD8-positive cells may exert rather an effector function in stress-induced, cytokine-mediated abortions as discussed previously[32].

5. CONCLUSION

Our results point at a therapeutic use of DPP IV inhibitors in immune reactions, e.g. allograft rejection. A modulation of CTLA-4 expression due to inhibitors of DPP IV activity may well explain the potent anti-abortogenic effect observed here and lead to novel therapeutic applications of DPP IV inhibitors in immune-related disorders and autoimmune diseases. Further studies are needed to elucidate the intracellular and molecular mechanisms underlying this effect.

REFERENCES

1. Tanaka,T., Duke-Cohan,J.S., Kameoka,J., Yaron,A., Lee,I., Schlossman,S.F., and Morimoto,C., 1994, Enhancement of antigen-induced T-cell proliferation by soluble CD26/dipeptidyl peptidase IV, *Proc. Natl. Acad. Sci U. S. A*, **91:** 3082-3086.

2. Dang,, N., Torimoto,Y., Deusch,K., Schlossman,S., and Morimoto,C., 1990, Comitogenic effect of solid-phase immobilized anti-1F7 on human CD4 T cell activation via CD3 and CD2 Pathways, *J Immunol*, **144:** 4092-4100.
3. Kahne, T., Lendeckel, U., Wrenger, S., Neubert, K., Ansorge, S., and Reinhold, D., 1999, Dipeptidyl peptidase IV: a cell surface peptidase involved in regulating T cell growth (review), Int. *J Mol. Med*, **4:** 3-15.
4. Schön,E., Mansfeld,H., Demuth,H., Barth,A., and Ansorge,S., 1985, The dipeptidyl peptidase IV, a membrane enzyme involved in the proliferation of T lymphocytes, *Biomed Biochem Acta*, **44:** K9-K15.
5. Reinhold,D., Hemmer,B., Gran,B., Born,I., Faust,J., Neubert,K., McFarland,H.F., Martin,R., and Ansorge,S., 1998, Inhibitors of dipeptidyl peptidase IV/CD26 suppress activation of human MBP-specific CD4+ T cell clones, *J Neuroimmunol.*, **87:** 203-209.
6. Reinhold,D., Bank,U., Buhling,F., Lendeckel,U., Faust,J., Neubert,K., and Ansorge,S., 1997, Inhibitors of dipeptidyl peptidase IV induce secretion of transforming growth factor-beta 1 in PWM-stimulated PBMC and T cells, *Immunology*, **91:** 354-60.
7. Reinhold,D., Bank,U., Buhling,F., Tager,M., Born,I., Faust,J., Neubert,K., and Ansorge,S., 1997, Inhibitors of dipeptidyl peptidase IV (DP IV, CD26) induces secretion of transforming growth factor-beta 1 (TGF-beta 1) in stimulated mouse splenocytes and thymocytes, *Immunology Letters*, **58:** 29-35.
8. Steinbrecher,A., Reinhold,D., Quigley,L., Gado,A., Tresser,N., Izikson,L., Born,I., Faust,J., Neubert,K., Martin,R., Ansorge,S., and Brocke,S., 2001, Targeting dipeptidyl peptidase IV (CD26) suppresses autoimmune encephalomyelitis and up-regulates TGF-beta 1 secretion in vivo, *J Immunol*, **166:** 2041-2048.
9. Korom,S., De Meester,I., Stadlbauer,T., Chandraker,A., Schaub,M., Sayegh,M., Belyaev,A., Haemers,A., Scharpé,S., and Kupiec-Weglinski,J., 1997, Inhibition of CD26/Dipeptidylpeptidase IV Activity in vivo prolongs cardiac allograft survival in rat recipients, *Transplantation*, **63:** 1495-1500.
10. Hildebrandt,M., Arck,P.C., Kruber,S., Demuth,H.U., Reutter,W., and Klapp,B.F., 2001, Inhibition of dipeptidyl peptidase IV (DP IV, CD26) activity abrogates stress-induced, cytokine-mediated murine abortions, *Scand J Immunol*, **53:** 449-454.
11. Clark,D.A., Banwatt,D., and Chaouat,G., 1993, Stress-triggered abortion in mice prevented by alloimmunization, *Am J Reprod. Immunol*, **29:** 141-147.
12. Arck,P.C., Merali,F.S., Manuel,J., Chaouat,G., and Clark,D.A., 1995, Stress-triggered abortion: inhibition of protective suppression and promotion of tumor necrosis factor-alpha (TNF-alpha) release as a mechanism triggering resorptions in mice, *Am J Reprod. Immunol*, **33:** 74-80.
13. Oosterwegel,M.A., Greenwald,R.J., Mandelbrot,D.A., Lorsbach,R.B., and Sharpe,A.H., 1999, CTLA-4 and T cell activation, *Curr. Opin. Immunol*, **11:** 294-300.
14. Wells, A.D., Walsh,M.C., Bluestone,J.A., and Turka,L.A., 2001, Signaling through CD28 and CTLA-4 controls two distinct forms of T cell anergy, *J Clin Invest*, **108:** 895-903.
15. Guinan, E., Gribben,J., Boussiotis,V., Freeman,G., and Nadler,L., 1994, Pivotal Role of the B7:CD28 Pathway in Transplantation Tolerance and Tumor Immunity, *Blood*, **84:** 3261-3282.
16. Harding, F.A., McArthur,J.G., Gross,J.A., Raulet,D.H., and Allison,J.P., 1992, CD28-mediated signalling co-stimulates murine T cells and prevents induction of anergy in T-cell clones, *Nature*, **356:** 607-609.
17. Wang,X.B., Zheng,C.Y., Giscombe,R., and Lefvert,A.K., Regulation of surface and intracellular expression of CTLA-4 on human peripheral T cells, *Scand J Immunol*, **54:** 453-458.

Dipeptidyl Peptidase IV/CD26 in T Cell Activation, Cytokine Secretion and Immunoglobulin Production

HUA FAN, SHULING YAN, SABINE STEHLING, DIDIER MARGUET[#], DETLEF SCHUPPAN[+], and WERNER REUTTER
Institut für Molekularbiologie und Biochemie, UKBF, Freie Universität Berlin, Arnimallee 22, D-14195 Berlin, Germany; [#]Centre d'Immunologie INSERM-CNRS de Marseille-Luminy, Marseille, France; [+]Klinik für Innere Medizin I, Abt. Hepatologie, Universität Erlangen-Nürnberg, Ulmenweg 18, D-91054 Erlangen, Germany

1. INTRODUCTION

Dipeptidyl peptidase IV (DPPIV/CD26; EC. 3.4.14.5), a widely distributed multifunctional type II plasma membrane glycoprotein, is involved in different biological processes. It is a serine protease associated with uptake and transmembrane transport of proline-containing peptides as well as with processing of physiological active peptides[1]. As an exopeptidase it cleaves *N*-terminal dipeptides after proline or alanine residues. Furthermore, an endopeptidase activity of DPPIV has also been reported[2].

Due to its interaction with proteins of the extracellular matrix (ECM), such as collagen and fibronectin, DPPIV/CD26 can also be considered as a cell adhesion molecule[3]. The interaction of DPPIV/CD26 with proteins of the ECM has been determined by several binding assays[4-6]. Although the binding properties of DPPIV/CD26 to proteins of the ECM have been well characterized, the significance of this interaction for its biological function, especially for the immune regulation is still unknown.

It has been shown that DPPIV/CD26 plays a crucial role in T cell activation and immune regulation[7-10]. The expression level of DPPIV/CD26 is tightly regulated during the development of T lymphocytes and its density on the plasma membrane is markedly enhanced after lymphocyte activation.

Thus this protein is regarded as an activation marker for T, B and NK cells[11]. Some immunoregulative hormones and chemokines closely related to the immune function have been shown to be substrates of DPPIV/CD26. Among these substrates are substance P, neuropeptide Y, endomorphin-2, GLP-1, RANTES (regulated on activation normal T-cell expressed and secreted), eotaxin, MDC (monocyte-derived chemokine) and SDF-1α and SDF-1ß (stromal derived factor)[10, 12-14]. Also of great importance is that the T cell surface molecule DPPIV/CD26 serves as a co-stimulator in the antigen-stimulated activation of T lymphocytes[15,16] and mediates signaling by direct interaction with CD45[17,18]. On human T cells, DPPIV/CD26, as a receptor of adenosine deaminase (ADA), may play an important role for the regulation of the immune response[19].

Synthetic inhibitors of the enzymatic activity of DPPIV have been shown to suppress certain immune reactions *in vitro* and *in vivo*[20,21]. However, in the immune system DPPIV/CD26 is regarded not only as an enzyme, but also as a co-stimulator for T cell activation, a receptor for ADA and a receptor for collagen. So it can be postulated that DPPIV/CD26, depending on the physiological environment and/or pathological conditions, could function in different ways. In the present work we investigated the role of the diverse functions of DPPIV/CD26 in T cell activation. We have found that collagen inhibits the co-stimulating effect of DPPIV/CD26, and hence can modulate the activation of T cells. The enzymatic activity of this molecule is not for its co-stimulating effect on T cell activation. In consideration of the multi-functions of DPPIV/CD26, our experiments with DPPIV/CD26 knockout mice could provide useful information about the physiological role of this plasma membrane glycoprotein in the immune response, as well as additional insights into the underlying molecular mechanisms of signal transduction processes.

2. RESULTS AND DISCUSSION

2.1 Rat DPPIV/CD26 exhibits a co-stimulatory effect on human T cell activation

Rat DPPIV/CD26 exhibits a high (85%) homology to human DPPIV/CD26. Hence studies on the mechanisms of DPPIV/CD26 in T cell activation were performed with Jurkat stable transfectants containing wild type rat DPPIV/CD26 (rCD26/Jurkat). Cells were stimulated with specific antibodies adsorbed on the surface of wells of microtiter plates (solid-phase immobilization). As an examination of cell activation, the production of the cytokine IL-2 was measured[22]. After stimulation with anti-CD3 mAb, rCD26/Jurkat transfectants showed a two-fold higher secretion of IL-2 in comparison with Jurkat cells lacking DPPIV/CD26 (Fig. 1). When the cells were co-stimulated with anti-CD3 mAb and anti-rat-DPPIV/CD26 polyclonal antibodies, a four-fold raise in IL-2 secretion was measured in rCD26/Jurkat as compared to the controls (Fig. 1). This suggests that rat-DPPIV/CD26 could also act as co-stimulator and thus mediate the antigen-stimulated activation of human T cells. Since the human ADA does not bind to rat-DPPIV/CD26[23] we can conclude from our results that the interaction of DPPIV/CD26 with ADA is not necessary for the DPPIV/CD26-mediated T cell co-stimulation, which is in accordance with the suggestion of Dong et al.[19].

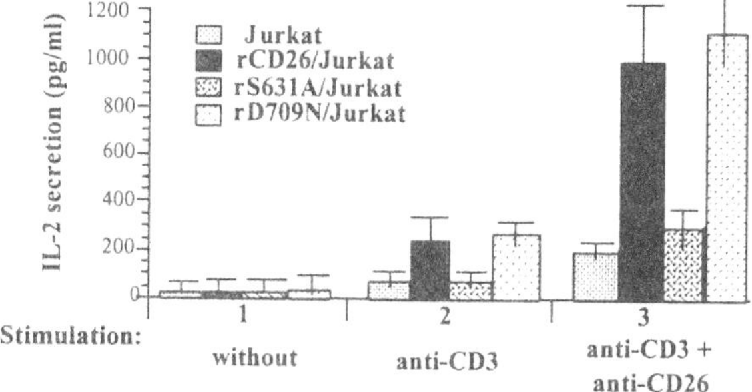

Figure 1. IL-2 secretion by Jurkat cells and Jurkat transfectants after stimulation. Jurkat cells and jurkat stable transfectants were stimulated with solid-phase immobilized anti-human CD3 mAb (5 µg/ml), or with co-immobilized anti-human CD3 mAb (5 µg/ml) and anti-rat DPPIV/CD26 polyclonal antibodies (6 µg/ml). After 24 hours, the concentration of IL-2 in cell culture supernatants was assayed by ELISA.

2.2 The enzymatic activity of DPPIV is not required for the DPPIV/CD26 co-stimulatory effect on T cell activation

The role of the enzymatic activity of DPPIV/CD26 in its co-stimulatory effect on T cell activation represents a controversial issue in the area of DPPIV/CD26[15,24]. In the present work two enzymatic inactive mutants (S631A and D709N) were tested for their co-stimulatory effect on IL-2 secretion of Jurkat cells, in order to clarify this debatable point. As expected these mutants did not exhibit any enzymatic activity since Ser^{631} and Asp^{709} are components of the amino acid triad of the catalytic site of DPPIV/CD26, but their expression on the Jurkat cell surface was comparable to that of wild type DPPIV/CD26. We observed, in agreement with Tanaka et al. that the

S631A mutation markedly inhibited IL-2 secretion (Fig. 1)[15]. But unexpectedly the D709N mutation did not affect the signaling pathway leading to raised secretion of IL-2, despite that this mutant did not exhibit any enzymatic activity (Fig. 1). Based on the results obtained with the mutant S631A, Tanaka et al., have postulated that the enzymatic activity of DPPIV/CD26 is required for its co-stimulatory effect on IL-2 secretion during T cell activation. Steeg et al. found no reduction in IL-2 secretion of DPPIV/CD26 Jurkat transfectants incubated with inhibitors of the enzymatic activity of this protein[24]. In the present work with the D709N mutant we demonstrate that the enzymatic activity of DPPIV/CD26 is not required for its co-stimulatory effect on IL-2 secretion of T cells. We postulate that the inability of the S631A mutant to mediate the signal transduction pathway of IL-2 secretion may be due to the requirement of the serine residue for signal transduction, or to alterations in the protein functional conformation, rather than to its lack of enzymatic activity (Fan et al., in preparation).

2.3 Collagen inhibits the CD26 co-stimulatory effect on T cell activation

The adhesive interactions of cells with the extracellular matrix play a central role in the functions of the immune system, which regulate the migration of lymphocytes and the interactions of activated cells during the immune response. It has been determined that native DPPIV/CD26 not only binds to collagens but also that its binding capacity depends markedly on the type of collagen[6]. We then asked whether the interaction of DPPIV with specific types of collagen differentially influences the DPPIV/CD26 co-stimulatory capacity for T cell activation. For this purpose IL-2 secretion of rCD26/Jurkat cells was measured after incubation with different types of collagen during stimulation with anti-CD3 mAb. We observed that the secretion of IL-2 was inhibited by collagen type I, type XIV, type II, type VI and type III in different levels. While collagen type I showed a stronger inhibitory capacity, collagen type IV had no significant influence on the IL-2 secretion (Fig was analyzed with its isolated single chains. The α1(I) chain showed the strongest inhibitory capacity, more than 95%, whereas the α2(I) chain caused an inhibition of 40%, only (Fig.2). This is also in accordance with the results of the *in vitro* binding assay, in which a strong binding of the α1(I) chain and a significantly weaker affinity of the α2(I) chain to DPPIV/CD26 were demonstrated[6].

The monoclonal anti-DDIV/CD26 antibody mAb 13.4 also completely inhibited the IL-2 secretion (Fig. 2). It has been reported that the binding of DPPIV/CD26 to collagen was abolished by this antibody[6]. We could show here that both mAb 13.4 and collagen type I could inhibit the DPPIV/CD26 mediated T cell activation. This suggests that the binding domain(s) of DPPIV/CD26 for collagen and the mAb 13.4 is (are) essential components of the signal transduction pathway of IL-2 secretion in T cells. We show that collagen can modulate the T cell activation via its interaction with DPPIV/CD26. Further experiments are necessary to clarify whether this interaction has additional significance for adhesion, migration and signal transduction of T cells and other leukocytes.. 2). These results are consistent with those of *in vitro* binding tests[6] in which collagen type I showed stronger affinity, while collagen type IV bound poorly to DPPIV/CD26. The inhibitory effect of collagen type I.

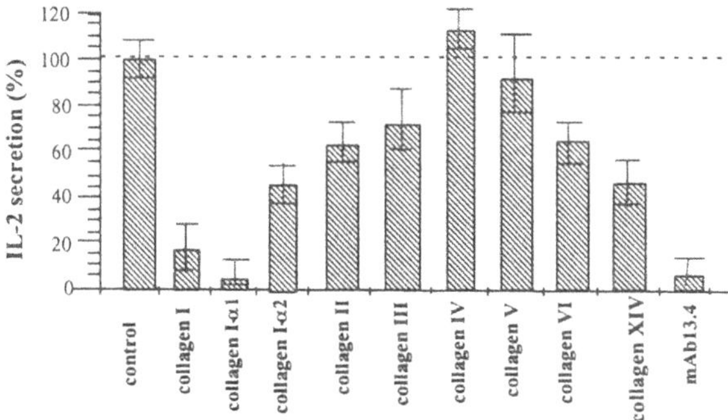

Figure 2. Influence of different collagen types on the IL-2 secretion of rCD26/Jurkat transfectants induced by anti-human CD3 mAb. The rCD26/Jurkat transfectants were stimulated with solid-phase immobilized anti-human CD3 mAb (5 μg/ml) alone (control) or in the presence of different types of co-immobilized collagens (1 μg/ml) for 24 hours. The IL-2 concentrations in cell culture supernatants were assayed by ELISA.

2.4 Influence of DPPIV/CD26 on T cell development and differentiation

Investigations of the last few years as well as the present work have provided evidence for the involvement of DPPIV/CD26 in processes like T cell activation and cell adhesion to the extracellular matrix[6,8-10,18]. However, the physiological function of this multifunctional molecule in the intact animal has not yet been clarified. Therefore, we studied the role of DPPIV/CD26 in activation and differentiation of lymphocytes after stimulation with different mitogens in DPPIV/CD26 gene knockout mice.

DPPIV/CD26 gene knockout mice show an apparently normal phenotype. However, the percentage of $CD4^+$ cells (helper/memory cell population in mice spleen lymphocytes (MSLs) is about 30% lower in $DPPIV/CD26^{-/-}$ mice than in $DPPIV/CD26^{+/+}$ mice (data not shown), suggesting an involvement of DPPIV/CD26 in lymphocyte development and maturation *in vivo*. MSLs were stimulated *in vitro* for 72 h with the following mitogens: phytohemagglutinin (PHA), pokeweed mitogen (PWM), concanavalin A (ConA) and lipopolysaccharide (LPS). The proliferation rates of DPPIV/CD26$^{-/-}$ MSLs were decreased to 70% and 80% after stimulation with ConA and PWM, respectively, whereas no significant changes after stimulation with PHA or LPS were observed. The secretion of several cytokines in MSLs after stimulation with PWM was also determined. A significant reduction in the secretion of the Th2-type cytokine IL-4 of DPPIV/CD26$^{-/-}$ MSLs, to 20-40% of that of MSLs from wild type mice, was measured. An increase in the production of Th1-type cytokine IFN-γ of DPPIV/CD26$^{-/-}$ MSLs was observed, while no changes of the secretion of IL-2 and IL-6 were detected (Fig. 3) (Yan et al., in preparation).

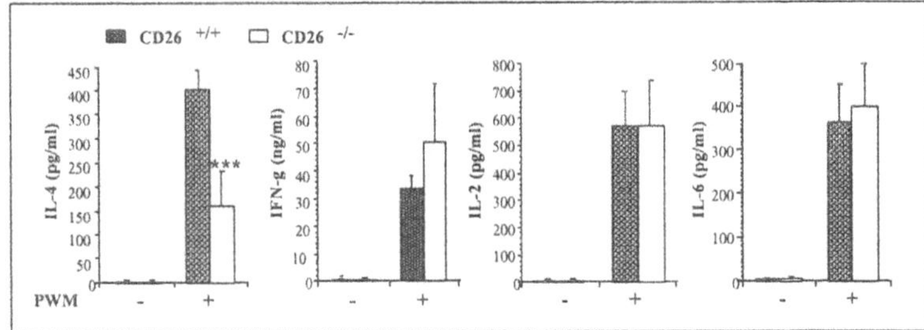

Figure 3. Cytokines secretion by mouse spleen lymphocytes after stimulation with PWM. Spleen lymphocytes from CD26$^{-/-}$ and CD26$^{+/+}$ mice were stimulated by PWM (0.5 μg/ml) for 72 h. Cytokine concentrations in culture supernatants were measured by ELISA. The noted values represent the mean ± SD of a minimum of seven mice. ***: $P<0.001$ as compared to CD26$^{+/+}$ mice.

The lower concentration of IL-4 in the supernatants of CD26$^{-/-}$ MSLs is probably due to the lower percentage of CD4^{+} lymphocytes in the knockout mice (s. above), since CD4^{+} cells are the major source of this cytokine. However, the lower percentage of CD4^{+} cells could not explain the increased IFN-γ production, and it did not result in a decreased production of IL-2 and IL-6 in CD26$^{-/-}$ MSLs. The reduction of IL-4 corresponded with an imperfect differentiation of the Th2 subset[25,26], indicating an involvement of DPPIV/CD26 expression in the regulation of cytokine secretion, as well as in the differentiation of T-lymphocytes.

It has been found that blocking of the enzymatic activity of DPPIV/CD26 with specific inhibitors suppressed the IL-2 and IFN-γ production of human T-cells, and reduced the IL-2 and IL-12 secretion of human peripheral blood mononuclear cells (PBMC) stimulated with PWM[11,27,28]. The apparent contrast between the results of the present work and those of Arndt et al. can be explained on the basis of the different leukocyte populations used. The activation and differentiation of T cells *in vivo* is influenced by the interaction with other subsets of lymphocytes and antigen presenting cells (APCs), so that the cytokine production from mixed lymphocyte populations, as in our study, may be different to that from purified T cells or PBMC. Moreover, as discussed earlier, DPPIV/CD26 has in the immune system a co-stimulatory effect on lymphocyte activation, which is independent of its enzymatic activity. Therefore, the use of enzymatic inhibitors of DPPIV/CD26 should only interfere with signal transduction events dependent on the proteolytic activity of this plasma membrane glycoprotein. Experiments using DPPIV specific inhibitors can help to elucidate the role of DPPIV/CD26 enzymatic activity in the immune response, whereas studies with DPPIV/CD26 knockout mice provide insights into the physiological role of this multifunctional molecule in the immune system.

2.5 The deficiency of DPPIV/CD26 results in reduced immunoglobulin concentrations after stimulation by PWM in vivo

To examine whether the differentiation and functions of B-lymphocytes were dependent upon the reduction of IL-4 secretion and ultimately on the expression of DPPIV/CD26, the concentrations of different immunoglobulins in sera were measured by ELISA after stimulation with PWM *in vivo*. As shown in Fig. 4, DPPIV/CD26$^{-/-}$ mice presented significantly reduced IgG concentrations six days after immunization with PWM. The IgE concentrations were also reduced after six days, but the most significant reduction was observed 19 days after immunization with PWM, while no significant differences in the IgM levels between both kinds of mice were detected (Fig 4) (Yan et al., in preparation). These results suggest that the deficiency of DPPIV/CD26 *in vivo* leads to a specific impairment of the production of immunoglobulins. Moreover, this impairment could be related to the lower IL-4 levels of the DPPIV/CD26$^{-/-}$ mice, since this cytokine plays a crucial role in the antibody forming process[25,26]. IL-4 induces activation and differentiation of B cells, as well as acts as a growth factor for T cells, thereby promoting differentiation of Th2 cells and reinforcing the antibody response and plasma cells isotype switching to IgG and IgE production[25,26]. Regarding the importance of DPPIV/CD26 on T cell function, Ohnuma et al. found that soluble DPPIV/CD26 up-regulated the expression of the co-stimulatory molecule CD86 on monocytes, and enhanced the T cell immune response[29]. In the present *in vivo* study, however, we cannot exclude the possibility of a direct effect of DPPIV/CD26 on B cells. Buhling et al. reported that stimulation of isolated

CD20-positive B cells with PWM led to a raise in the proportion of CD26-positive cells, from 5% to 51%, suggesting an involvement of CD26 in B cell activation[30].

In DPPIV/CD26 knockout mice, we have found above all for the first time a disturbed immune response to PWM. Further experiments with stimulation or immunization with other antigens and pathogens *in vivo*, and with purified subsets of lymphocyte *in vitro* are necessary, in order to understand the role of this protein in the pathomechanisms of several diseases with immunological implications.

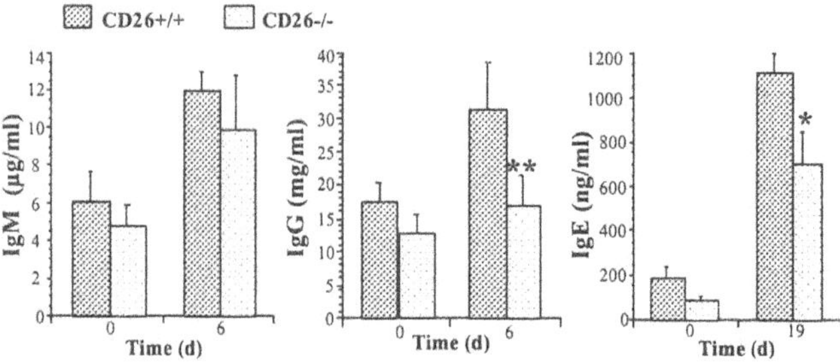

Figure 4. Immunoglobulins productions by CD26$^{+/+}$ and CD26$^{-/-}$ mice stimulated with PWM. CD26$^{+/+}$ and CD26$^{-/-}$ mice were injected (i. p.) with 200 µl PWM (40 mg/ml). After indicated days the concentration of immunoglobulins in sera was analyzed by ELISA. The noted values represent the mean ± SD of a minimum of five mice. *: $P<0.05$, **: $P<0.01$ as compared to CD26$^{+/+}$ mice.

3. CONCLUSION

Collagen type I inhibits the co-stimulatory activity of DPPIV/CD26, and hence modulates the T cell activation. The enzymatic activity of DPPIV is not necessary for the co-stimulatory activity of this molecule in T cell activation. DPPIV/CD26 plays an important role in the development, maturation, activation and differentiation of T cells as well as on their functions in the immune system. Deficiency of DPPIV/CD26 results in an impaired development and maturation of CD4 lymphocytes and a disturbed response to PWM stimulation. The IL-4 secretion was decreased, and hence the IgG production was reduced and isotype switching to IgE was affected.

ACKNOWLEDGEMENTS

This work was supported by a grant from the *Deutsche Forschungsgemeinschaft* Bonn (Sonderforschungsbereich 366 and

Graduiertenkolleg 276), the *Sonnenfeld-Stiftung* and the *Fonds der Chemischen Industrie*, Frankfurt/Main.

REFERENCES

1. Trugnan, G., 1995, Control of dipeptidyl peptidase IV/CD26 cell surface espression in intestinal cells. In *Dipeptidyl Peptidase IV (CD26) in Metabolism and the Immune Response* (B. Fleischer, eds), R. G. Landes Company, Texas, pp. 79-98.
2. Bermpohl, F., Löster, K., Reutter, W. & Baum, O., 1998, Rat dipeptidyl peptidase IV (DPP IV) exhibits endopeptidase activity with specificity for denatured fibrillar collagens. *FEBS Lett*, **428:** 152-6.
3. Reutter, W., Baum, O., Löster, K., Fan, H., Bork, J.P., Bernt, K., Hanski, C. & Tauber, R., 1995, Functional aspects of the three extracellular domains of dipeptidyl peptidase IV: Characterization of glycosylation events, of the collagen-binding site and endopeptidase activity. In *Dipeptidyl Peptidase IV (CD26) in Metabolism and the Immune Response* (B. Fleischer, eds), R. G. Landes Company, Texas, 55-79.
4. Hanski, C., Huhle, T. & Reutter, W., 1985, Involvement of plasma membrane dipeptidyl peptidase IV in fibronectin-mediated adhesion of cells on collagen. *Biol Chem Hoppe Seyler,* **366:** 1169-76.
5. Hanski, C., Huhle, T., Gossrau, R. & Reutter, W., 1988, Direct evidence for the binding of rat liver DPP IV to collagen in vitro. *Exp Cell Res,* **178:** 64-72.
6. Löster, K., Zeilinger, K., Schuppan, D. & Reutter, W., 1995, The cysteine-rich region of dipeptidyl peptidase IV (CD 26) is the collagen-binding site. *Biochem Biophys Res Commun,* **217:** 341-8.
7. Fleischer, B., 1994, CD26: a surface protease involved in T-cell activation. *Immunol Today,* **15:** 180-4.
8. Morimoto, C. & Schlossman, S.F., 1998, The structure and function of CD26 in the T-cell immune response. *Immunol Rev,* **161:** 55-70.
9. Franco, R., Valenzuela, A., Lluis, C. & Blanco, J., 1998, Enzymatic and extraenzymatic role of ecto-adenosine deaminase in lymphocytes. *Immunol Rev,* **161:** 27-42.
10. De Meester, I., Korom, S., Van Damme, J. & Scharpe, S., 1999, CD26, let it cut or cut it down. *Immunol Today*, **20**: 367-75.
11. Ansorge, S., Kähne, T., Lendeckel, U., Reinhold, D., Neubert, K., Steinbrecher, A. & Brocke, S., 2001, Dipeptidyl peptidase IV/CD26 and aminopeptidase N/CD13 in regulation of the immune responce. In *Cell-surface aminopeptidase: Basic and clinical aspects* (S. Mizutani, eds), Elsevier, Amsterdam, 85-94.
12. Hildebrandt, M., Reutter, W., Arck, P., Rose, M. & Klapp, B.F., 2000, A guardian angel: the involvement of dipeptidyl peptidase IV in psychoneuroendocrine function, nutrition and immune defence. *Clin Sci (Lond),* **99:** 93-104.
13. Marguet, D., Baggio, L., Kobayashi, T., Bernard, A.M., Pierres, M., Nielsen, P.F., Ribel, U., Watanabe, T., Drucker, D.J. & Wagtmann, N., 2000, Enhanced insulin secretion and improved glucose tolerance in mice lacking CD26. *Proc Natl Acad Sci U S A*, **97:** 6874-9.
14. Proost, P., De Meester, I., Schols, D., Struyf, S., Lambeir, A.M., Wuyts, A., Opdenakker, G., De Clercq, E., Scharpe, S. & Van Damme, J., 1998, Amino-terminal truncation of chemokines by CD26/dipeptidyl-peptidase IV. Conversion of RANTES into a potent inhibitor of monocyte chemotaxis and HIV-1-infection. *J Biol Chem*, **273:** 7222-7.

15. Tanaka, T., Kameoka, J., Yaron, A., Schlossman, S.F. & Morimoto, C., 1993, The costimulatory activity of the CD26 antigen requires dipeptidyl peptidase IV enzymatic activity. *Proc Natl Acad Sci U S A*, **90:** 4586-90.
16. Dang, N.H., Torimoto, Y., Deusch, K., Schlossman, S.F. & Morimoto, C., 1990, Comitogenic effect of solid-phase immobilized anti-1F7 on human CD4 T cell activation via CD3 and CD2 pathways. *J Immunol*, **144:** 4092-100.
17. Ishii, T., Ohnuma, K., Murakami, A., Takasawa, N., Kobayashi, S., Dang, N.H., Schlossman, S.F. & Morimoto, C., 2001, CD26-mediated signaling for T cell activation occurs in lipid rafts through its association with CD45RO. *Proc Natl Acad Sci U S A*, **98:** 12138-43.
18. von Bonin, A., Huhn, J. & Fleischer, B., 1998, Dipeptidyl-peptidase IV/CD26 on T cells: analysis of an alternative T-cell activation pathway. *Immunol Rev*, **161:** 43-53.
19. Dong, R.P., Tachibana, K., Hegen, M., Munakata, Y., Cho, D., Schlossman, S.F. & Morimoto, C., 1997, Determination of adenosine deaminase binding domain on CD26 and its immunoregulatory effect on T cell activation. *J Immunol*, **159:** 6070-6.
20. Augustyns, K., Bal, G., Thonus, G., Belyaev, A., Zhang, X.M., Bollaert, W., Lambeir, A.M., Durinx, C., Goossens, F. & Haemers, A., 1999, The unique properties of dipeptidyl-peptidase IV (DPP IV / CD26) and the therapeutic potential of DPP IV inhibitors. *Curr Med Chem*, **6:** 311-27.
21. Kubota, T., Flentke, G.R., Bachovchin, W.W. & Stollar, B.D., 1992, Involvement of dipeptidyl peptidase IV in an in vivo immune response. *Clin Exp Immunol*, **89:** 192-7.
22. Tanaka, T., Camerini, D., Seed, B., Torimoto, Y., Dang, N.H., Kameoka, J., Dahlberg, H.N., Schlossman, S.F. & Morimoto, C., 1992, Cloning and functional expression of the T cell activation antigen CD26. *J Immunol*, **149:** 481-6.
23. Dinjens, W.N., ten Kate, J., Wijnen, J.T., van der Linden, E.P., Beek, C.J., Lenders, M.H., Khan, P.M. & Bosman, F.T., 1989, Distribution of adenosine deaminase-complexing protein in murine tissues. *J Biol Chem*, **264:** 19215-20.
24. Steeg, C., Hartwig, U. & Fleischer, B., 1995, Unchanged signaling capacity of mutant CD26/dipeptidylpeptidase IV molecules devoid of enzymatic activity. *Cell Immunol*, **164:** 311-5.
25. Abbas, A.K., Murphy, K.M. & Sher, A., 1996, Functional diversity of helper T lymphocytes. *Nature*, **383:** 787-93.
26. Sallusto, F., Lanzavecchia, A. & Mackay, C.R., 1998, Chemokines and chemokine receptors in T-cell priming and Th1/Th2-mediated responses. *Immunol Today*, **19:** 568-74.
27. Arndt, M., Reinhold, D., Lendeckel, U., Spiess, A., Faust, J., Neubert, K. & Ansorge, S., 2000, Specific inhibitors of dipeptidyl peptidase IV suppress mRNA expression of DP IV/CD26 and cytokines. *Adv Exp Med Biol*, **477:** 139-43.
28. Arndt, M., Lendeckel, U., Spiess, A., Faust, J., Neubert, K., Reinhold, D. & Ansorge, S., 2000, Dipeptidyl peptidase IV (DP IV/CD26) mRNA expression in PWM-stimulated T-cells is suppressed by specific DP IV inhibition, an effect mediated by TGF-beta(1). *Biochem Biophys Res Commun*, **274:** 410-4.
29. Ohnuma, K., Munakata, Y., Ishii, T., Iwata, S., Kobayashi, S., Hosono, O., Kawasaki, H., Dang, N.H. & Morimoto, C., 2001, Soluble CD26/dipeptidyl peptidase IV induces T cell proliferation through CD86 up-regulation on APCs. *J Immunol*, **167:** 6745-55.
30. Bühling, F., Junker, U., Reinhold, D., Neubert, K., Jager, L. & Ansorge, S., 1995, Functional role of CD26 on human B lymphocytes. *Immunol Lett*, **45:** 47-51.

Dipeptidyl Peptidase IV Inhibitors with the N-terminal MXP Sequence: Structure-Activity-Relationships

JÜRGEN FAUST[*], PETRA FUCHS[*], SABINE WRENGER[#], DIRK REINHOLD[#], ANGELA STÖCKEL-MASCHEK[*], THILO KÄHNE[§], SIEGFRIED ANSORGE[§], and KLAUS NEUBERT[*]
[*]*Institute of Biochemistry, Department of Biochemistry/Biotechnology, Martin-Luther-University Halle-Wittenberg, Germany;* [#]*Institute of Immunology, Otto-von-Guericke-University Magdeburg, Germany;* [§]*Institute of Experimental Internal Medicine, Otto-von-Guericke-University Magdeburg, Germany*

1. INTRODUCTION

Dipeptidyl peptidase IV (DP IV, EC 3.4.14.5) is a transmembrane type II glycoprotein, which is present on most of the mammalian cells. As an exopeptidase DP IV catalyzes the release of N-terminal dipeptides from oligo- and polypeptides with protonated N-terminus if the penultimate amino acid is proline or alanine.

DP IV is identical to the activation antigen CD26 expressed on T and B lymphocytes and natural killer cells and plays a key role in the regulation of differentiation and growth of lymphocytes. Specific synthetic inhibitors of DP IV, such as Lys[Z(NO_2)]-thiazolidide, suppress mitogen- and alloantigen-induced T cell proliferation, B cell differentiation, immunoglobulin secretion and modulate cytokine production[1].

We could show, that peptides with the N-terminal MXP motif inhibit DP IV and exhibit similar suppressive effects on the activation of immune cells as observed by using synthetic inhibitors[2].

The human immunodeficiency virus-1 (HIV-1) transactivator Tat (86 amino acids) containing this N-terminal MXP motif is described as the first known natural inhibitor of DP IV[3] and it suppresses antigen-, anti-CD-3- and mitogen-induced activation of human T cells.

Recently we have demonstrated, that the N-terminal part of Tat proteine is important for suppression of CD26 dependent T cell growth. To prove the specifity of the Tat-DP IV interaction and to identify the amino acids necessary for this interaction, we synthesized N-terminal nonapeptides derived from the Tat(1-9)-sequence as basic structure.

2. RESULTS AND DISCUSSION

The peptides were synthesized by solid phase peptide synthesis using Fmoc technique and purified by HPLC.

The inhibition of DP IV-catalyzed substrate hydrolysis by Tat(1-9) (MDPVDPNIE) was not significantly improved by amino acid exchanges at positions 1, 3, 4, 5 and 6. On the contrary amino acid exchanges at positions 5 and 6 partially strong weaken the inhibitory effect[4]. However, the exchange of D in position 2 by proteinogenic amino acids resulted in a peptide with enhanced DP IV inhibitory potentials (Fig. 1).

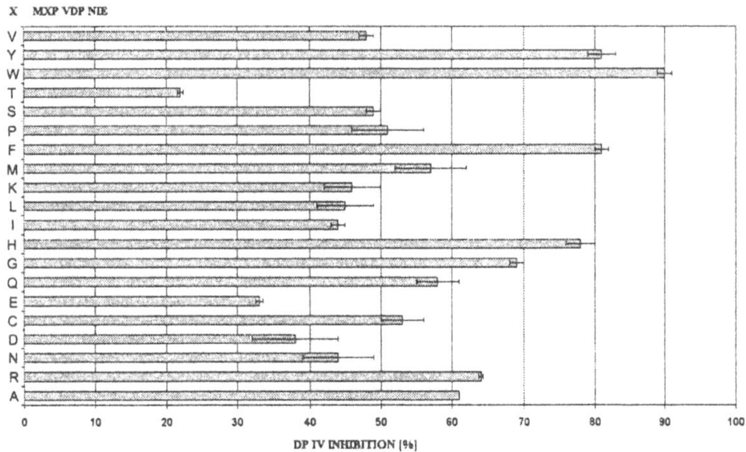

Figure 1. Influence of amino acid substitutions in Tat(1-9) on inhibition of DP IV (3μM $(Ala\text{-}Pro)_2$-rhodamine-110 substrate, 120μM of putative inhibitory peptide, pig kidney DP IV)

Thus, Tat(1-9) peptide analogues with aromatic amino acids W, Y, F in second position inhibit the DP IV activity more than Tat(1-9). W^2-Tat(1-9), carrying the N-terminal MWP, is the most potent DP IV inhibitor (K_i = 2.2 10^{-6}M) of this series.

Moreover, we investigated the exchange of amino acid in position 2 by tryptophan at further MXP peptides. Thus, N-terminal partial sequence of recombinant cytokine G-CSF (MTPLGPASS) and interleukin-2 (MAPTSSSTK) are significantly more potent DP IV inhibitors with tryptophan in position 2 (Fig. 2).

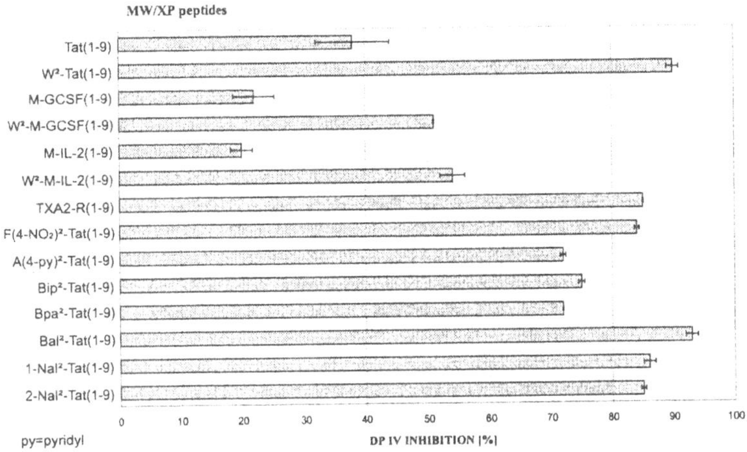

Figure 2. Influence of amino acid substitutions in MW/XP peptides on inhibition of DP IV (3μM $(Ala\text{-}Pro)_2$-rhodamine-110 substrate, 120 μM of putative inhibitory peptide, pig kidney DP IV)

These data suggest the importance of the N-terminal MWP motif for the inhibition of DP IV. It was presumed, that the aromatic indole ring of tryptophan is favoured to exhibit attractive interactions with DP IV. Therefore we synthesized a series of Tat(1-9)-derived peptides with unusual aromatic amino acids ($F(4\text{-}NO_2)$ = 4-nitrophenylalanine, A(4-py) = ß-(4-pyridyl)-alanine, Bip = ß-(3-biphenyl)-alanine, Bpa = 4-benzoyl-phenyl-alanine, Bal = ß-(3-benzothienyl)-alanine, 1-Nal = ß-(1-naphthyl)-alanine, 2-Nal = ß-(2-naphthyl)-alanine) in position 2. Most of these Tat(1-9) derivatives exhibit strong inhibition of DP IV comparable to W^2-Tat(1-9) (Fig.2).

The search for proteins extracellularly exposing the N-terminal MWP sequence (Swiss-Prot protein database) revealed the thromboxane A2 receptor (TXA2-R) sequence. TXA2-R(1-9) (MWPNGSSLG) inhibited DP IV-catalyzed substrate hydrolysis as potent as W^2-Tat(1-9) (Fig.2). Further investigations showed suppression of DNA synthesis of tetanus toxoid-stimulated peripheral blood mononuclear cells and modulation of cytokine production like other DP IV inhibitors[5]. This indicates, that TXA2-

R could be endogenous ligand of DP IV modulating T cell activation via inhibition of DP IV.

The potent DP IV inhibitors W^2-Tat(1-9), TXA2-R(1-9), W^2-M-G-CSF(1-9), W^2-M-IL-2(1-9) and some Tat(1-9) peptides derived by unnatural amino acids with aromatic rings are very different in their sequence (positions 4-9), illustrating the importance of the MW/XP motif for DP IV inhibition/CD26-mediated suppression of immune cell activation.

3. CONCLUSION

By amino acid exchanges based on the sequence of the moderate DP IV inhibitor Tat(1-9) with the N-terminal MDP sequence, we identified W^2-Tat(1-9) inhibiting DP IV clearly more efficiently.

Further potent DP IV inhibitors are some Tat(1-9) peptides derived by unusual amino acids which have an aromatic ring system.

The enhanced inhibitory potential of MW/XP-peptides may be explained with strong hydrophobic interactions between aromatic rings of inhibitor and DP IV.

ACKNOWLEDGEMENTS

This work was supported by Deutsche Forschungsgemeinschaft, SFB 387 and NE 501/2-1.

REFERENCES

1. Kähne, T., Lendeckel, U., Wrenger, S., Neubert, K., Ansorge, S., and Reinhold, D., 1999, Dipeptidyl peptidase IV: a cell surface peptidase involved in regulating T cell growth. *Int. J. Mol. Med.* **4**: 3-15.
2. Hoffmann, T., Reinhold, D., Kähne, T., Faust, J., Neubert, K., Frank, R., and Ansorge, S., 1995, Inhibition of dipeptidylpeptidase IV (DP IV) by anti – DP IV antibodies and non-substrate X-X-Pro-oligopeptides ascertained by capillary electrophoresis. *J. Chromatogr. A* **716**: 355-362.
3. Gutheil, W.G., Subramanyam, M., Flentke, G.R., Sanford, D.G., Munoz, E., Huber, B.T., and Bachovchin, W.W., 1994, Human immunodeficiency virus 1 Tat binds to dipeptidyl peptidase IV (CD26): A possible mechanism for Tat's immunosuppressive activity. *Proc. Natl. Acad. Sci. USA* **91**: 6594-6598.
4. Wrenger, S., Hoffmann, T., Faust, J., Mrestani-Klaus, C., Brandt, W., Neubert, K., Kraft, M., Frank, R., Ansorge, S., and Reinhold, D., 1997, The N-terminal structure of HIV-1 Tat is required for suppression of CD26-dependent T cell growth. *J. Biol. Chem* **272**: 232-234.

5. Wrenger, S., Faust, J., Mrestani-Klaus, C., Fengler, A., Stöckel-Maschek, A., Lorey, S., Kähne, T., Brandt, W., Neubert, K., and Ansorge, S., 2000, Down-regulation of T cell activation following inhibition of dipeptidyl peptidase IV / CD26 by the N-terminal part of the thromboxane A2 receptor. *J. Biol. Chem.* **275**: 22180-22186.

On the Role of Dipeptidyl Peptidase IV in the Digestion of an Immunodominant Epitope in Celiac Disease

SINA KOCH, DORIT ANTHONSEN, HANNE SKOVBJERG, and HANS SJÖSTRÖM
Biochemistry Laboratory C, Department of Medical Biochemistry and Genetics, The Panum Institute, University of Copenhagen, Copenhagen, Denmark

Best Poster Award, International Conference on Dipeptidyl Aminopeptidases, Berlin, September 2002

1. INTRODUCTION

Celiac disease is a chronic intestinal disease with intestinal villus atrophy and crypt hypertrophy, resulting in malabsorption. The disease often starts in childhood with symptoms like chronic diarrhoea, abdominal distension and failure to thrive. It may, however also start later in lifer, when the symptoms often are more unspecific as e.g. fatigue, anemia and neurological symptoms. The disease is quite common with a prevalence of about 1:300 in many countries. It is provoked by ingesting gluten of wheat and similar proteins of rye and barley. Gluten consists of two fractions, defined by their solubility in 70 % ethanol. The soluble part constitutes the gliadins and the insoluble part the glutenins. Most of the patients are human leukocyte antigen (HLA) DQ2 positive, and the disease is usually mentioned as an autoimmune disease, even if the provoking epitopes are exogeneously supplied. The only known treatment is to avoid eating foods with these components. For reviews see [1,2].

Lundin and collaborators [3] isolated gliadin-specific T lymphocytes from the intestine of celiac disease patients. This allowed the establishment of a

Dipeptidyl Aminopeptidases in Health and Disease, Edited by Hildebrandt et al.
Kluwer Academic/Plenum Publishers, New York, 2003

lymphocyte proliferation assay making a systematic search for disease provoking epitopes possible. Several such epitopes have been characterised and most of them exist in gliadins. In particular a peptide (QLQPFPQP**Q**LPYPQPQS) occurring in alpha-gliadins, seems to be an immunodominant epitope[4,5]. Most of the characterised epitopes contain a particular glutamine (bold Q) that is deamidated to glutamic acid to enhance the recognition by T cells[6]. This reaction is likely to be executed by the enzyme tissue transglutaminase, which among several glutamines seems to preferentially deamidate this particular glutamine[7]. The serum of celiac patients contains antibodies to transglutaminase and this can be used for diagnostic purposes[8].

The pathogenetic mechanism implies intestinal leakage of gliadin peptides to the lamina propria, where they are deamidated by tissue transglutaminase. Antigen-presenting cells present these gliadin peptides to T helper cells, which become activated and initiate a destruction of the mucosa. They also stimulate B-cells to produce anti-tissue transglutaminase and anti-gliadin[9].

A main question is why this type of reaction is initiated only by proteins of wheat and related cereals and not by other major food proteins like milk or muscle proteins. It has been suggested that the gliadins or part of the gliadins are more difficult to digest than other major food proteins. These proteins are rich in glutamine and proline and furthermore the immunodominant epitope is located in a very proline rich sequence. This provides a structural basis for a digestion with only a few types of proteolytic enzymes. It may finally be suggested, that there are also other pathogenetically important peptides, that may be isolated on the basis of their relative resistance to digestion. This paper gives some aspects of the mechanism for digestion of gliadin and speculates about connections to the pathogenesis of celiac disease.

2. DIGESTION WITH PEPSIN AND PANCREATIC ENDOPEPTIDASES

Digestions with pepsin and pancreatic endopeptidases were performed by either of two approaches: In one approach separate *in vitro* digestions of a synthetic peptide containing a sequence from the immunodominant epitope with protected N- and C-terminal ends (acyl-QPFPQPQLPYPQPQ-amide) were analysed by absorbance measurements after gel filtration chromatography. In the other approach *in vitro* digestion of a gliadin suspension with pepsin was followed by digestion with a mixture of trypsin, chymotrypsin and elastase. The generated products were analysed by absorbance measurements at 276 nm after gel filtration chromatography. The major changes occurred within the first hour of incubation, but some further digestion was noticed after repeated digestion with a new portion of enzyme(s).

The fractions were analysed by an inhibition immunoassay using monoclonal antibodies (prepared by C. Koch, Statens Serum Institut, Copenhagen, Denmark) raised against a peptide (KLQPFPQPELPYPQPQ-amide) containing the immunodominant epitope.

In accordance with earlier knowledge, gliadin was solubilised during pepsin treatment. The present procedure (1 hour digestion) mainly resulted in large fragments appearing early in the chromatographic elution, whereas a repeated pancreas endopeptidase digestion (3 times 1 hour) resulted in a broad range of mainly smaller-sized peptides. The immunological analysis, using an antibody preferring peptides containing the particular glutamine-deamidated epitope, showed distinct - however broad - peaks, containing a range of peptides of different sizes.

No digestion could be registered after *in vitro* digestion of the synthetic peptide acyl-QPFPQPQLPYPQPQ-amide with pepsin, trypsin or chymotrypsin under the used conditions. Altogether the data suggest that at least some peptides containing the immunodominant epitope are relatively resistant to hydrolysis by several endopeptidases. However, more experiments including gliadin digestions with higher amounts of enzyme are needed to verify this suggestion.

3. DIGESTION WITH SMALL INTESTINAL MICROVILLUS ENZYMES

The results of the initial experiments on gastric and pancreatic digestion suggest that the intestinal capacity to hydrolyse these resistant peptides is of importance in the degradation of the epitope. Intestinal digestion is mainly performed by enzymes bound to the microvillus membrane of the enterocytes. Besides several types of exopeptidases, also endopeptidases exist in the intestinal microvillus membrane. Aminopeptidase N and dipeptidyl peptidase IV are two important players in the digestion of peptides from the N-terminal end of intestinal peptides. These two enzymes are complementary: Whereas aminopeptidase N has a broad specificity to remove N-terminal amino acids, it cannot release amino acids bound to a proline. In these cases dipeptidyl peptidase IV instead liberates an aminoacyl-proline containing dipeptide.

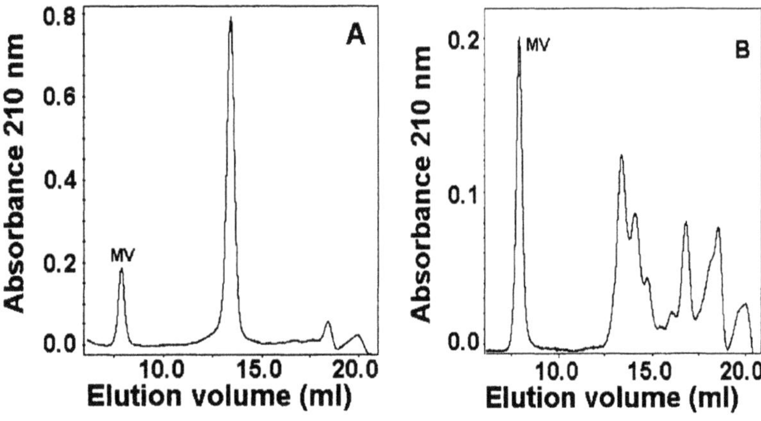

Figure 1. Digestion of 0,125 mM acyl-QPFPQPQLPYPQPQ-amide (A) and 0,125 mM QPFPQPELPYPQPQ-amide (B) with a Triton X-100 solubilised pig microvillus membrane preparation. After incubation for 60 min at 37 °C, the reaction was stopped by boiling for 5 min and a sample analysed by gel filtration chromatography on a Superdex Peptide HR 10/30 column. MV: microvillus material.

To investigate a possible intestinal endopeptidase activity hydrolysing the epitope, the peptide acyl-QPFPQPQLPYPQPQ-amide was incubated with a pig intestinal microvillus membrane preparation, solubilised with Triton X-100. Microvillus membranes were prepared essentially as described by[10]. As shown in fig. 1A no hydrolysis could be demonstrated. In a parallel experiment under identical conditions, it was demonstrated that the N-terminally non-protected peptide QPFPQPELPYPQPQ-amide was hydrolysed (fig. 1B). We have earlier suggested[11] that a corresponding protected peptide (acyl-QPFPQPELPYPQPQ-amide) might be sensitive to microvillus hydrolysis due to an endopeptidase activity specific for the glutamic acid. We have, however not been able to verify this in further experiments. The demonstrated hydrolysis of QPFPQPELPYPQPQ-amide is therefore suggested to be due to N-terminally acting exopeptidases. In conclusion, microvillus membrane endopeptidase activity does not seem to be of major importance in the hydrolysis of the epitope, at least in the pig.

Taking the high amount of prolines of the immunodominant antigenic peptide into consideration, dipeptidyl peptidase IV is a good candidate enzyme for the N-terminal digestion of the peptide by the intestinal microvillus membrane. To study the importance of this enzyme in relation to other possible microvillus membrane enzymes, the digestion was performed by incubation of the peptide QPFPQPQLPYPQPQ-amide with a pig microvillus membrane preparation in the presence of the dipeptidyl peptidase IV inhibitor valine-pyrrolidide (kind gift of Dr. J.J. Holst, University of Copenhagen, Denmark). The results are shown in fig. 2.

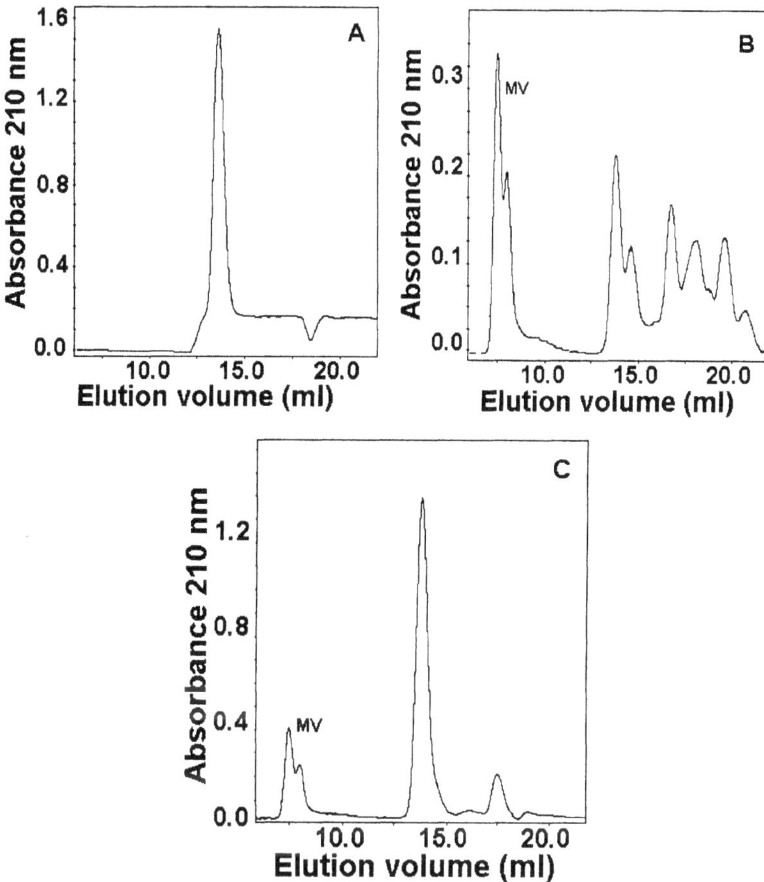

Figure 2. Effect of the dipeptidyl peptidase IV inhibitor valine-pyrrolidide (0,1 mM) on the pig microvillus membrane digestion of QPFPQPQLPYPQPQ-amide (0,125 mM). The samples were incubated for 60 min at 37 °C, the reaction stopped by boiling for 5 min and a sample analysed by gel filtration chromatography on a Superdex Peptide HR 10/30 column. A: Peptide only B: Peptide and microvillus membrane preparation C: Peptide, microvillus membrane preparation and valine-pyrrolidide. MV: microvillus material.

It can be seen that this peptide having an non-protected N-terminal is digested, and that the hydrolysis is efficiently inhibited by the dipeptidyl peptidase IV inhibitor. This demonstrates that dipeptidyl peptidase IV is the dominating, if not the only, small intestinal microvillus enzyme responsible for the release of aminoacyl-proline in the N-terminal digestion of the epitope.

4. CONCLUSION

An immunodominant gliadin epitope provoking celiac disease seems to be notably resistant to digestion with pepsin and pancreatic endopeptidases. This means that the capacity of the intestinal mucosa to digest it will be critical to the disease provoking capability of the peptide. We have shown that a peptide containing the immunodominant epitope is efficiently digested from the N-terminal end by dipeptidyl peptidase IV. Similar results have been obtained in a parallel work by Hausch et al.[12], who in addition demonstrated that dipeptidyl carboxypeptidase I is of importance for digestion of the epitope from the C-terminal end. We have earlier demonstrated that dipeptidyl peptidase IV is low in celiac disease and also that it shows a substantial depression in patients in remission[13]. Even if a variation of the enzyme level is not of primary importance in the pathogenesis of the disease, it may that the relative capacity for the digestion of the peptide can be exceeded after a meal rich in wheat proteins, and thereby contribute to the development of the disease.

ACKNOWLEDGEMENTS

This work was supported by grants from the Danish Medical Research Council and the Novo Nordisk Foundation.

REFERENCES

1. Marsh, M. N.,1992, Gluten, Major Histocompatibility Complex, and the Small Intestine. A Molecular and Immunobiologic Approach to the Spectrum of Gluten Sensitivity ('Celiac Sprue'). *Gastroenterology* , **102:** 330-354.
2. Sollid, L. M.,2000, Molecular Basis of Celiac Disease. *Annu. Rev. Immunol.* **18:** 53-81.
3. Lundin, K. E.; Scott, H.; Hansen, T.; Paulsen, G.; Halstensen, T. S.; Fausa, O.; Thorsby, E.; Sollid, L. M.,1993, Gliadin-Specific, HLA-DQ(Alpha 1*0501,Beta 1*0201) Restricted T Cells Isolated From the Small Intestinal Mucosa of Celiac Disease Patients. *J Exp. Med* , 178: 187-196.
4. Anderson, R. P.; Degano, P.; Godkin, A. J.; Jewell, D. P.; Hill, A. V.,2000, In Vivo Antigen Challenge in Celiac Disease Identifies a Single Transglutaminase-Modified Peptide As the Dominant A-Gliadin T-Cell Epitope. *Nat. Med* , **6:** 337-342.

5. Arentz-Hansen, H.; Korner, R.; Molberg, O.; Quarsten, H.; Vader, W.; Kooy, Y. M.; Lundin, K. E.; Koning, F.; Roepstorff, P.; Sollid, L. M.; McAdam, S. N.,2000, The Intestinal T Cell Response to Alpha-Gliadin in Adult Celiac Disease Is Focused on a Single Deamidated Glutamine Targeted by Tissue Transglutaminase. *J Exp. Med*, **191:** 603-612.
6. Sjöström, H.; Lundin, K. E.; Molberg, O.; Korner, R.; McAdam, S. N.; Anthonsen, D.; Quarsten, H.; Norén, O.; Roepstorff, P.; Thorsby, E.; Sollid, L. M.,1998, Identification of a Gliadin T-Cell Epitope in Coeliac Disease: General Importance of Gliadin Deamidation for Intestinal T-Cell Recognition. *Scand J Immunol,* **48:** 111-115.
7. Piper, J. L.; Gray, G. M.; Khosla, C.,2002, High Selectivity of Human Tissue Transglutaminase for Immunoactive Gliadin Peptides: Implications for Celiac Sprue. *Biochemistry*, **41:** 386-393.
8. Dieterich, W.; Ehnis, T.; Bauer, M.; Donner, P.; Volta, U.; Riecken, E. O.; Schuppan, D.,1997, Identification of Tissue Transglutaminase As the Autoantigen of Celiac Disease. *Nat. Med*, **3:** 797-801.
9. Schuppan, D.; Dieterich, W.; Riecken, E. O.,1998, Exposing Gliadin As a Tasty Food for Lymphocytes. *Nat. Med*, **4:** 666-667.
10. Kessler, M.; Acuto, O.; Storelli, C.; Murer, H.; Muller, M.; Semenza, G.,1978, A Modified Procedure for the Rapid Preparation of Efficiently Transporting Vesicles From Small Intestinal Brush Border Membranes. Their Use in Investigating Some Properties of D-Glucose and Choline Transport Systems. *Biochim. Biophys. Acta*, **506:** 136-154.
11. Sjöström, H. ,2001, Deamidation of gliadin peptides in the pathogenesis of celiac disease. In: Proocedings of the Workshop on Transglutaminases, Protein Cross-Linking and Coeliac Disease. (Mäki M, Tossavainen M. eds.) University Press, Tampere, Finland. pp. 117-122.
12. Hausch, F.; Shan, L.; Santiago, N. A.; Gray, G. M.; Khosla, C.,2002, Intestinal Digestive Resistance of Immunodominant Gliadin Peptides. *Am. J Physiol Gastrointest. Liver Physiol*, **283:** G996-G1003.
13. Sjöström, H.; Norén, O.; Krasilnikoff, P. A.; Gudmand-Hoyer, E.,1981, Intestinal Peptidases and Sucrase in Coeliac Disease. *Clin. Chim. Acta*, **109:** 53-58.

The Properties of Human and Bovine CD8+CD26+ T Cells Induced by a Microbial Superantigen

SANG-UN LEE[1], YONG-HO PARK[2], WILLIAM C. DAVIS[3], LAWRENCE K. FOX[3], and GREGORY A. BOHACH[1]
[1]University of Idaho, Moscow, ID 83844, USA, [2]Seoul National University, Seoul, Korea, and [3]Washington State University, Pullman, WA 99163, USA

1. INTRODUCTION

Until recently, staphylococcal enterotoxins (SE) produced by *Staphylococcus aureus* were only known as agents of food poisoning. However, it is now known that SEs are superantigens, binding to MHC class II molecules outside of the peptide-binding groove and being presented as unprocessed proteins to T cells expressing certain T-cell receptor Vβ genes. T cell stimulation in this manner contributes to their ability to induce toxic shock syndrome, immunosuppression, and probably other diseases[1].

Bovine mastitis, a very costly disease in dairy industry, is caused by many pathogens. Among them, *S. aureus* is the most difficult to control. We suspect that the immunomodulation induced by SEs contributes to persistence. Previously, we reported that the expression of an uncharacterized molecule, ACT3, was aberrantly increased on $CD8^+$ T cells after bovine peripheral blood mononuclear cells (PBMC) were incubated with the type C SE (SEC)[2]. This finding suggested that these cells may have important roles in immunomodulation of the bovine immune system and may contribute to the persistence of *S. aureus*. Using a variety of techniques, we showed that the ACT3 molecule is the bovine orthologue of CD26[3,4]. However, the function of CD26 in the aberrant $CD8^+$ T cells is still uncertain. In this study, we assessed whether superantigen induction of CD26 is species-specific. We also analyzed the $CD26^+$ T cell phenotypes

Dipeptidyl Aminopeptidases in Health and Disease, Edited by Hildebrandt et al.
Kluwer Academic/Plenum Publishers, New York, 2003

and their cytokine profiles by RT-PCR to characterize the cell populations induced.

2. CD26 EXPRESSION ON SUPERANTIGEN-STIMULATED HUMAN PBMC

Human PBMC were incubated with SEC to determine if the increase of $CD8^+CD26^+$ T cells is a general phenomenon of superantigen or is species-specific. Human PBMC were isolated from venous blood of healthy donors by gradient centrifugation. The cells were incubated with SEC type 1 (SEC1; 0.1 μg/ml) or Concanavalin A (Con A; 5 μg/ml) for various periods of time. Cells were harvested and stained for flow cytometry (FC) analysis.

FC results for human cell cultures showed that SEC1 induced a much larger increase in the percentage of human $CD8^+CD26^+$ T cells than Con A (Fig. 1). This result showed that SEs activate human PBMC cultures in a manner very similar to that which we reported previously for bovine PBMC cultures.

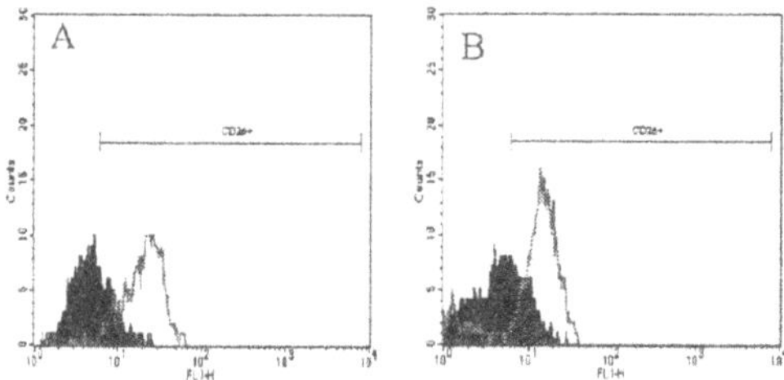

Figure 1. Representative histograms demonstrating the increased CD26 expression on human CD8+ T cells after 4 days (A) and 7 days (B) with SEC1. CD8+ T cells treated with Con A (closed); CD8+ T cells treated with SEC1 (open).

3. PHENOTYPES OF $CD26^+$ T CELLS INDUCED IN CULTURE WITH SEC1

To characterize $CD8^+CD26^+$ T cells proliferating in cultures treated with SEC1, a panel of monoclonal antibodies was used to stain representative cell surface molecules. The results showed that both $CD4^+CD26^+$ T cells and $CD8^+CD26^+$ T cells coexpress CD25, CD44, CD62L, and CD45R0, but not the γδ TCR. In addition, the increased expression of WC1 and uncharacterized molecule, ACT2 were observed on $CD8^+$ T cells but not

CD4⁺ T cells (Fig. 2). These results suggest that the CD26⁺ T cells are highly activated and confirm that expression of CD26 is indicative of high level activation as we proposed previously.

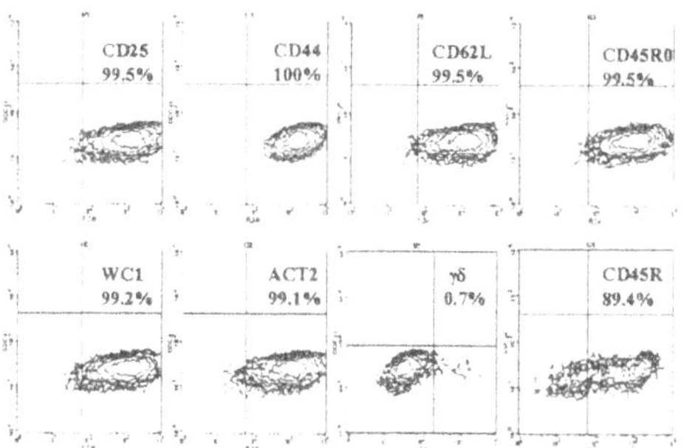

Figure 2. Representative plots demonstrating the subtypes of CD8+CD26+ T cells 7 days after treatment with SEC1.

4. THE CYTOKINE MESSENGER RNA EXPRESSION IN CULTURE WITH SEC1

Real time RT-PCR was applied to analyze IL-4, IL-10, IL-12, and TGF-β mRNA expression. Bovine PBMC were cultured as described above. The cells were stained and applied to a FACSVantage for sorting. The results showed that the proliferated $CD8^+CD26^+$ T cells, as well as $CD4^+CD26^+$ T cells, express IL-10 and TGF-β mRNA (Fig. 3). However, the expression of IL-4 and IL-12 was very low or undetectable (results not shown). These results indicate that superantigens induce the proliferation of bovine T cells having immunoregulatory roles.

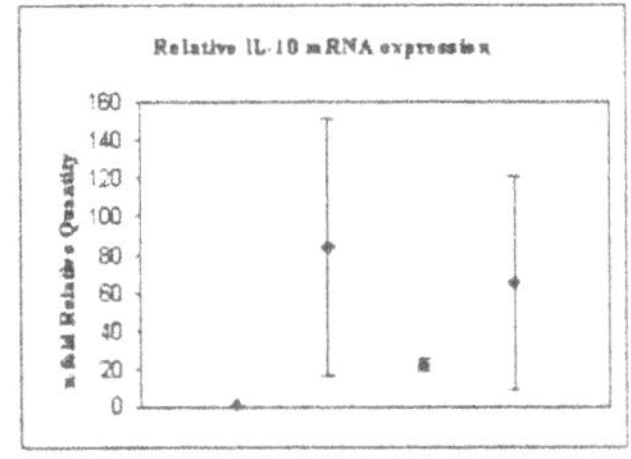

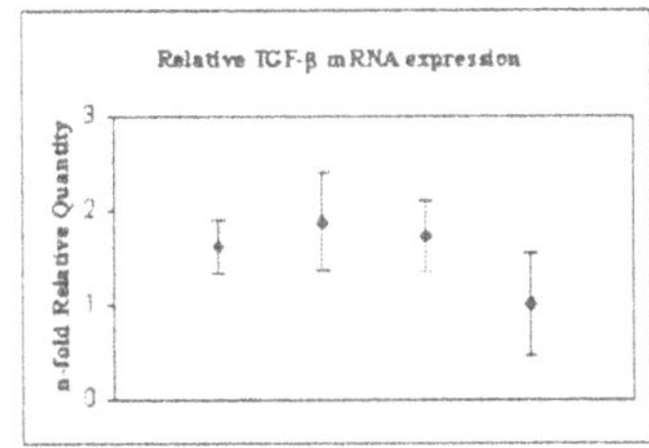

Figure 3. Relative cytokine mRNA expression in sorted T cell subtypes. The T cells were treated with SEC1 for 7 days and sorted for CD4+CD26- T cells, CD4+CD26+ T cells, CD8+CD26- T cells, and CD8+CD26+ T cells (from left to right). The sorted cells were acquired with a purity greater than 90%.

5. CONCLUSION

In this study, we demonstrated that the increased CD26 expression in $CD8^+$ T cells is a universal characteristic of superantigen action and the $CD8^+CD26^+$ T cells expressed IL-10 and TGF-β mRNA. Because superantigens are associated with immunosuppression such as hyporesponsiveness and anergy, our results suggest that superantigens induce T cells, which downregulate the immune function. In doing so, these cells likely protect the body from over responding to pathogens such as *S. aureus*. At the same time this response could contribute to the organism's persistence.

The Thl/Th2 paradigm has been used to explain the shift between humoral and cellular immunity for a long time. Recently, there has been increased interest in regulatory T (Tr) cells, which have distinct characteristics compared to the Thl/Th2 paradigm. For example, $CD8^+$ Tr cells secrete IL-10 and possibly TGF-β and express CD45R0[7]. We are now exploring the possibility that superantigen immunomodulation results from this more recently described pathway of activation.

ACKNOWLEDGMENTS

This work was supported by USDA NRICGP grants 99-35201-8581 (G.A.B) and 99-3504-8556 (W.C.D), USDA WNV grants 00144-0182085 (W.C.D.) and 9902050-0183734 (W.C.D.), PHS grants AI28401 (G.A.B.) and P20-RR15587 (G.A.B), the United Dairymen of Idaho (G.A.B), and the Idaho Agricultural Experiment Station (G.A.B).

REFERENCES

1. Foster, T.J. and Bohach, G.A. 2000, Staphylococcus aureus exotoxins. In *Gram positive pathogens* (Fischetti, V. et al., eds, ASM Press, Washington D.C., .), pp. 367-378.
2. Ferens, W.A., Davis, W.C., Hamilton, M.J., Park, Y.H., Deobald, C.F., Fox, L. and Bohach, G., 1998, Activation of bovine lymphocyte subpopulations by staphylococcal enterotoxin C. *Infect. Immun.* **66:** 573-580.
3. Lee, S.U., Ferens, W., Davis, W.C., Hamilton, M.J., Park, Y.H., Fox, L.K., Naessens, J. and Bohach, G.A., 2001, Identity of activation molecule 3 on superantigen-stimulated bovine cells is CD26. *Infect. Immun.* **69:** 7190-7193.
4. Lee, S.U., Park, Y.H., Davis, W.C., Hamilton, J., Naessens, J. and Bohach, G.A., 2002, Molecular characterization of bovine CD26 upregulated by a staphylococcal superantigen. *Immunogenetics* **54**: 216-220.

5. McGuirk, P. and Mills, K., 2002, Pathogen-specific regulatory T cells provoke a shift in the Th1/Th2 paradigm in immunity to infectious diseases. *Trends Immunol.* **23**: 450-455.

4

ANGIOGENESIS AND CANCER

DPPIV and Seprase in Cancer Invasion and Angiogenesis

W.-T. CHEN
Department of Medicine/Medical Oncology, State University of New York, Stony Brook New York 11794-8154, USA

1. INTRODUCTION

Several families of membrane proteases are distinguishable on the basis of their proteolytic activities, biologic functions, and structural organization. There are the membrane-type matrix metalloproteinases (MT-MMPs), the ADAM (a disintegrin and metalloprotease) family, the meprins, the secretases (also termed sheddases or convertases), and the metallo- and serine- peptidases. Localization of proteases is critical for their function in cellular activities. Increasing evidence indicates that the serine- peptidases and MT-MMPs accumulate at cell surface protrusions, termed invadopodia that may have a prominent role in processing soluble factors (including growth factors, chemokines, and other bioactive peptides) in addition to the well-established role of invadopodia in degrading the components of the extracellular matrix (ECM). Moreover, these membrane proteases may direct activation of either themselves or other workhorse soluble enzymes such as the 72-kDa matrix metalloprotease and plasmin[1]. It is generally agreed that a given membrane protease may have several functions (diversity) and that more than one membrane protease or one protease family may mediate the same function (redundancy) [2]. Different membrane proteases form complexes at invadopodia or other specialized locations that could provide distinct and overlapping actions. This may be necessary for complex regulatory processes where modulation is achieved by proteolysis

Dipeptidyl Aminopeptidases in Health and Disease, Edited by Hildebrandt et al.
Kluwer Academic/Plenum Publishers, New York, 2003

of several different molecules. For example, processing of various chemokines, activation of associated proteases, and ECM degradation.

Serine-type, integral membrane peptidases (SIMP) [3], including dipeptidyl peptidase IV (DPPIV/CD26), seprase/fibroblast activation protein alpha (FAPα) and related prolyl serine peptidases, exert their mechanisms of action on the cell surface. This presentation addresses potential roles that DPPIV and seprase may play in the localization and activation of cell surface proteases during the invasion of different cell types in tumor through the ECM.

2. EVIDENCES SUGGEST THE ROLE OF SEPRASE COMPLEXES IN CELL INVASION

It is still not understood about mechanisms of activation of membrane peptidases. Some membrane-bound proteases (meprin, DPPIV, seprase, QPP and PCP) must form an oligomeric structure for expression of proteolytic activity rather than proteolytic activation of a zymogen form, as in major soluble matrix metalloproteases. It is possible that membrane proteases have a lot in common with other well-characterized integral membrane glycoproteins such as integrins where heterodimeric interactions among subunits govern their functions.

An unique function of seprase was found to be exerted through its homo-dimeric conformation that can further complex with other cell surface proteins such as α3β1 integrin to engage in cell surface localization, focal ECM degradation and cancer cell invasion[4]. In addition, seprase complexed with homo-dimeric DPPIV at invadopodia, which elicit both gelatinase and prolyl peptidase activities, is activated on human connective tissue cells in response to wounding. The protease complex contributes to cell migration and invasion into collagenous matrix that is necessary for repair of connective tissue[5].

2.1 Localized ECM degradation

DPPIV contains a 110 kDa, type II transmembrane protein, that is expressed constitutively on brush border membranes of intestine and kidney epithelial cells[6] and transiently expressed in activated T-cells[7,8]. The active enzyme is a 200-220 kDa homodimer[9] that exhibits the dipeptidyl peptidase activity[10,11]. In addition, a recent report showed that DPPIV possesses a seprase-like gelatinolytic activity and therefore endopeptidase activity[12], suggesting its involvement in collagen degradation. Recent immunohistochemical studies suggested that DPPIV may also play important roles in ECM degradation; hence tissue remodeling such as in fibrosis and tumor invasion. The pattern of DPPIV expression is altered in cirrhotic human liver, with normal liver showing DPPIV expression in the bile canalicular domain of hepatocytes, whereas cirrhotic liver shows a loss of zonal expression and DPPIV is re-organized on proliferating bile ductules, leukocytes, and the basolateral domain of hepatocytes[13]. A retrospective study on follicular thyroid carcinoma supports this proposal that DPPIV became redistributed to sites interfacing the basement membrane and could be associated with cellular invasion[14].

Seprase, a 170 kDa, serine-type gelatinase and a homodimer of 97 kDa subunits, is proteolytically active on invadopodia of highly aggressive melanoma LOX cells[9,15]. Analysis of the deduced amino acid sequence from a cDNA that encodes the 97-kDa subunit[3] revealed that it is homologous to DPPIV, and is essentially identical to fibroblast activation alpha (FAPα)[16]. An alternatively spliced human seprase messenger was identified that encoded a novel truncated 27 kDa isoform, that precisely overlapped the carboxyl-terminal catalytic region of 97 kDa seprase subunit[17]. In contrast to the use of mAb F19 for immunohistochemistry staining of cancer tissues[18] that showed restricted distribution of FAPα in stroma cells activated by tumor[19], a detailed analysis using polyclonal antibodies directed against seprase was performed on malignant, premalignant, benign, and normal breast tissues[21]. Both 170-kDa gelatinase activity and immuno-reactivity of seprase were identified in tumor cells but not the stromal cells or morphologically normal epithelium of infiltrating ductal carcinomas. Kelly and colleagues, thus, concluded that the over-expression of seprase by carcinoma cells is consistent with seprase having a role in facilitating invasion and metastasis of infiltrating ductal carcinomas of the breast[21]. This cell localization result has been recently confirmed by two other studies from our laboratories using a panel of mAbs directed against seprase on breast cancer and gastric cancer[5]. Thus, the apparent difference in cellular localization of FAPα and seprase depicted by immunohistochemistry could be partially due to different isoforms, active sites or epitopes exhibited by seprase/FAPα in fibroblasts, endothelial and carcinoma cells.

These works point to the possibility that FAPα might be a key cell surface protease involved in promoting ECM degradation, tissue remodeling and fibrosis. Accordingly, FAPα was shown to be expressed at sites of liver tissue remodeling, e.g., stellate cells in cirrhotic human liver[13]. FAPα immuno-reactivity was most intense on perisinusoidal cells of the periseptal regions within regenerative nodules (15 of 15 cases); this pattern coincides with the tissue remodeling interface. Furthermore, the enzyme may not be essential or involving in housekeeping function, as a recent report shows that FAPα (-/-) mice are fertile, show no overt developmental defects, and have no general change in cancer susceptibility[22]. Further studies with Fap-/- lacZ showed that mice express β-Galactosidase at regions of active tissue remodeling during embryogenesis including somites and perichondrial mesenchyme from cartilage primordial[23].

Altered gene expression and inappropriate regulation of proteolytic activity occur often in diseases. As described above, there are many studies of the dys-regulated expression of DPPIV in leukocyte malignancies (leukemias, lymphomas, autoimmune diseases, HIV) as well as in solid tumor malignancies. Different expression patterns observed in FAPα (reactive stromal cells) and seprase (invasive cells) may reflect the dys-function of heterogeneous cells in cancer. However, the role of seprase/FAPα in cancer still awaits experimental proof using function-knockdown approach (i.e., by specific inhibitors or RNAi) and results of quantitative expression profiling studies (using real-time PCR).

2.2 Seprase-integrin complexes in cancer invasion

The integrin family of transmembrane adhesion proteins has been shown to exhibit multiple functions, including adhesion to ECM, cytoskeleton organization, and signal transduction[24,25]. Since integrin and integrin-associated molecules are enriched at invadopodia[26,27], integrins may also be involved in recruiting proteases to these sites of cell invasion. However, sequencing data on the 97-kDa protein subunit of seprase indicates only a short (six) amino acid sequence at the cytoplasmic amino terminus[3], suggesting that seprase localization at invadopodia may be dependent upon other membrane proteins such as integrins. Immunoprecipitation, immunofluorescence, and cell surface crosslinking experiments demonstrated that seprase and α3β1 integrin associate at invadopodia in a collagen-dependent manner to engage in cancer invasion[4]. Recently, monoclonal antibodies (mAbs) directed against β1 integrins were shown to block the localization of seprase to invadopodia, the local ECM degradation and invasion into collagenous gels by melanoma cells and fibroblasts[5].

2.3 Seprase-DPPIV complexes in wound healing

DPPIV and seprase are coordinated in the local degradation of denatured collagens by migratory cells[5]. Seprase and DPPIV form a complex on the cell surface that elicits both gelatin binding and gelatinase activities localized at invadopodia of cells migrating on collagenous fibers. The protease complex participates in the binding to gelatin and localized gelatin degradation, cellular migration, and monolayer wound closure. Serine-protease inhibitors can block the gelatinase activity and the localized gelatin degradation by cells. Antibodies to the gelatin-binding domain of DPPIV complexed with seprase reduce the proteases' and cellular abilities to degrade gelatin but do not affect cellular adhesion or spreading on type I collagen. Furthermore, expression of the seprase-DPPIV complex is restricted to migratory cells involved in wound closure *in vitro*, and in connective tissue cells during closure of gingival wounds, but not in differentiated tissue cells. Thus, cell surface proteolytic activities, which are nonmetallo-proteases, seprase and DPPIV, are responsible for the tissue invasive phenotype.

3. CONCLUSION

DPPIV and seprase are Pro-Xaa cleaving enzymes that form complexes on invadopodia of activated cells, becoming potent ECM-degrading proteases[5]. The protease complex exerts its mechanism of action on the surface of tissue and cancerous cells that are activated for matrix invasion. Not discussed in this presentation, these peptidases have potential to process bioactive peptides, chemokines and pro-angiogenic peptides, which in turn regulate chemotaxis of leukocytes, endothelial cell sprouting, and the activation of specific cellular function. Here, seprase complexes are shown to be involved in the adhesion to and proteolysis of components of ECM that activate signal transduction for specific cellular activities, including the migration and invasion of stromal and tumor cells.

The main functions of SIMPs reside in their proteolytic and adhesive capacities, thus influencing cellular activities, migration and invasion. These membrane proteases may form physically and functionally linked complexes with other proteases (5) and with integrins (4) at invadopodia, surface protrusions formed during activation of the cell to migrate and invade into the ECM. The capability of DPPIV and seprase to bind multiple molecules allows not only activation of themselves but also association with other membrane proteases and integrins to participate in cooperative ECM protein degradation at invadopodia during cancer invasion and angiogenesis.

REFERENCES

1. Sato, H., Takino, T.,Okada, Y., Cao, J., Shinagawa, A., Yamamoto, E., and Seiki, M., 1994, A matrix metalloproteinase expressed on the surface of invasive tumour cells [see comments], *Nature* **370:** 61.
2. Bauvois, B., 2001, Transmembrane proteases in focus: diversity and redundancy? *J Leukoc. Biol.* **70:** 11.
3. Goldstein, L.A., Ghersi, G., Piñeiro-Sánchez, M.L., Salamone, M., Yeh, Y.Y., Flessate, D., and Chen, W.-T., 1997, Molecular cloning of seprase: A serine integral membrane protease from human melanoma, *Biochimica et Biophysica Acta* **1361:** 11.
4. Mueller, S.C., Ghersi, G., Akiyama, S.K., Sang, Q.X., Howard, L., Pineiro-Sanchez, M., Nakahara, H., Yeh, Y., and Chen, W.-T., 1999, A novel protease-docking function of integrin at invadopodia., *J.Biol.Chem.* **274:** 24947.
5. Ghersi, G., Dong, H., Goldstein, L.A., Yeh, Y., Hakkinen, L., Larjava, H.S., and Chen, W.T., 2002, Regulation of fibroblast migration on collagenous matrix by a cell surface peptidase complex, *J.Biol.Chem.* **277:** 29231.
6. Yaron, A., and Naider, F., 1993, Proline-dependent structural and biological properties of peptides and proteins, *Crit.Rev.Biochem.Mol.Biol.* **28:**31
7. Morimoto, C., and Schlossman, S.F., 1994, CD26: A key costimulatory molecule on CD4 memory T cells, *The Immunologist* **2:**4.
8. Vivier, I., Marguet, D., Naquet, P., Bonicel, J., Black, D., Li, C.X., Bernard, A.M., Gorvel, J.P., and Pierres, M., 1991, Evidence that thymocyte-activating molecule is mouse CD26 (dipeptidyl peptidase IV), *J. Immunol.* **147:**447.
9. Pineiro-Sanchez, M.L., Goldstein, L.A., Dodt, J., Howard, L., Yeh, Y., Tran, H., Argraves, W.S., and Chen, W.-T., 1997, Identification of the 170-kDa melanoma membrane-bound gelatinase (seprase) as a serine integral membrane protease, *J.Biol.Chem.* **272:**7595.
10. Johnson, R.C., Zhu, D., Augustin-Voss, H.G., and Pauli, B.U., 1993, Lung endothelial dipeptidyl peptidase IV is an adhesion molecule for lung-metastatic rat breast and prostate carcinoma cells, *J.Cell Biol* **121:**1423.
11. Piazza, G.A., Callanan, H.M., Mowery, J., and Hixson, D.C., 1989, Evidence for a role of dipeptidyl peptidase IV in fibronectin- mediated interactions of hepatocytes with extracellular matrix, Biochemical Journal **262:**327.
12. Bermpohl, F., Löster, K., Reutter, W., and Baum, O., 1998, Rat dipeptidyl peptidase IV (DPP IV) exhibits endopeptidase activity with specificity for denatured fibrillar collagens, *FEBS Lett* **428:**152.
13. Levy, M.T., McCaughan, G.W., Abbott, C.A., Park, J.E., Cunningham, A.M., Muller, E., Rettig, W.J., and Gorrell, M.D., 1999, Fibroblast activation protein: a cell surface dipeptidyl peptidase and gelatinase expressed by stellate cells at the tissue remodelling interface in human cirrhosis, *Hepatology* **29:**1768.
14. Hirai, K., Kotani, T., Aratake, Y., Ohtaki, S., and Kuma, K., 1999, Dipeptidyl peptidase IV (DPP IV/CD26) staining predicts distant metastasis of 'benign' thyroid tumor [letter], *Pathology International* **49:**264.
15. Aoyama, A., and Chen, W.-T., 1990, A 170-kDa membrane-bound protease is associated with the expression of invasiveness by human malignant melanoma cells, *Proc.Natl.Acad.Sci.U.S.A.* **87:**8296.
16. Scanlan, M.J., Raj, B.K., Calvo, B., Garin-Chesa, P., Sanz-Moncasi, M.P., Healey, J.H., Old, L.J., and Rettig, W.J., 1994, Molecular cloning of fibroblast activation protein

alpha, a member of the serine protease family selectively expressed in stromal fibroblasts of epithelial cancers, *Proc.Natl.Acad.Sci.U.S.A.* **91**:5657.
17. Goldstein, L.A., and Chen, W.-T., 2000, Identification of an alternatively spliced seprase mRNA that encodes a novel intracellular isoform, *J Biol.Chem* **275**:2554.
18. Rettig, W.J., Garin-Chesa, P., Beresford, H.R., Oettgen, H.F., Melamed, M.R., and Old, L.J., 1988, Cell-surface glycoproteins of human sarcomas: differential expression in normal and malignant tissues and cultured cells, *Proc.Natl.Acad.Sci.USA* **85**:3110.
19. Garin-Chesa, P., Old, L.J., and Rettig, W.J., 1990, Cell surface glycoprotein of reactive stromal fibroblasts as a potential antibody target in human epithelial cancers, *Proc.Natl.Acad.Sci.USA* **87**:7235.
20. Rettig, W.J., Garin-Chesa, P., Healey, J.H., Su, S.L., Ozer, H.L., Schwab, M., Albino, A.P., and Old, L.J., 1993, Regulation and heteromeric structure of the fibroblast activation protein in normal and transformed cells of mesenchymal and neuroectodermal origin, *Cancer Res.* **53**:3327.
21. Kelly, T., Kechelava, S., Rozypal, T.L., West, K.W., and Korourian, S., 1998, Seprase, a membrane-bound protease, is overexpressed by invasive ductal carcinoma cells of human breast cancers, *Mod.Pathol.* **11**:855.
22. Niedermeyer, J., Kriz, M., Hilberg, F., Garin-Chesa, P., Bamberger, U., Lenter, M.C., Park, J., Viertel, B., Puschner, H., Mauz, M., Rettig, W.J., and Schnapp, A., 2000, Targeted disruption of mouse fibroblast activation protein, *Molecular & Cellular Biology* **20**:1089.
23. Niedermeyer, J., Garin-Chesa, P., Kriz, M., Hilberg, F., Mueller, E., Bamberger, U., Rettig, W.J., and Schnapp, A., 2001, Expression of the fibroblast activation protein during mouse embryo development, *Int. J. Dev. Biol.* **45**:445.
24. Lafrenie, R.M., and Yamada, K.M., 1996, Integrin-dependent signal transduction, *J. Cell. Biochem.* **61(4)**:543-53
25. Parsons, J.T., Integrin-mediated signalling: regulation by protein tyrosine kinases and small GTP-binding proteins, 1996, *Curr Opin Cell Biol* 8:146 (1996).
26. Nakahara, H., Mueller, S.C., Nomizu, M., Yamada, Y., Yeh, Y., and Chen, W.-T., 1998, Activation of beta1 integrin signaling stimulates tyrosine phosphorylation of p190RhoGAP and membrane-protrusive activities at invadopodia., *J. Biol. Chem.* **273**:9.
27. Coopman, P.J., Thomas, D.M., Gehlsen, K.R., and Mueller, S.C., 1996, Integrin α3β1 participates in the phagocytosis of extracellular matrix molecules by human breast cancer cells, *Mol. Biol. Cell* **7**: 1789.

Glutamate Carboxypeptidase II Inhibition as a Novel Therapeutic Target

ROJAS C, THOMAS AG, MAJER P, TSUKAMOTO T, LU XM, VORNOV JJ, WOZNIAK KM, SLUSHER BS.
Guilford Pharmaceuticals Inc., 6611 Tributary Street, Baltimore, Maryland 21224 USA

1. INTRODUCTION

Glutamate carboxypeptidase (GCP) II is a zinc peptidase that hydrolyzes the neuropeptide N-acetyl-aspartyl-glutamate (NAAG) to glutamate (G) and N-acetyl aspartate (NAA) (Fig. 1).

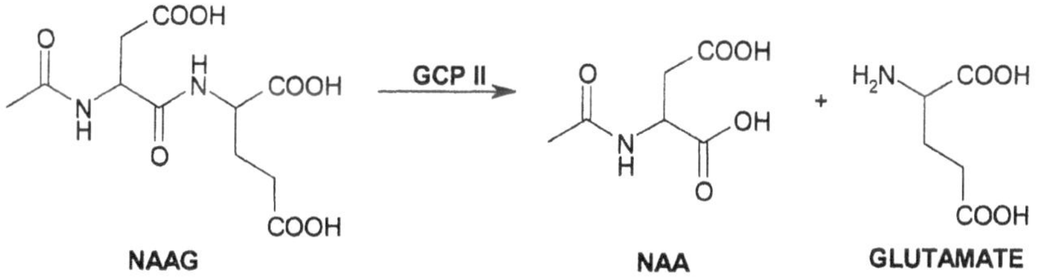

Fig. 1. Hydrolysis of NAAG catalysed by GCPII.

The enzyme was first identified, characterized and purified from the brain and kidneys of rodents[1]. In 1996, rat GCP II was cloned and found to be homologous to human prostate-specific membrane antigen (PSMA)[2] which is strongly expressed in prostate cancer[3]. GCP II has also been found in new vasculature of several solid tumors[4]. Additionally, GCP II catalyzes the hydrolysis of folate polyglutamate to folate and several molecules of glutamate in the membrane brush border of the small intestine[5]. GCP II has been referred to as NAALADase (**N**-**a**cetylated-α-**l**inked **a**cidic **d**ipeptid**ase**) when studying NAAG hydrolysis in the brain[6], as PSMA when studying the role of the enzyme in prostate cancer[7] or as folate hydrolase when focusing on the potential function of this enzyme in human nutrition[5]. However, the preferred official name for the enzyme is GCP II (EC 3.4.17.21).

1.1 Does GCP II exhibit dipeptidyl peptidase IV activity?

Pangalos and coworkers reported the cloning and expression of human NAALADase II[8], a type II integral membrane protein, similar to but distinct from NAALADase I (human PSMA or rat NAALADase). Also, there exists a NAALADase–"like" peptidase, NAALADase L, that exhibits sequence similarity but lacks NAALADase activity[8, 9]. Homogenates from *COS* cells transiently transfected with NAALADase I, II and L all exhibited DPP IV activity as measured by hydrolysis of Gly-Pro-AMC[8]. However, in a separate study, we found that recombinant human NAALADase I containing the extracellular portion of the enzyme (amino acids 44 - 750) did not exhibit DPP IV activity[10]. If confirmed, DPP IV activity by NAALADase would in all likelihood involve a separate active site from that involved in NAAG hydrolysis. This is because the two activities are known to proceed through distinct mechanisms. NAAG hydrolysis requires two zinc ions at the active site of the enzyme[11]; the zinc ions in these proteases are thought to act as electrophiles that polarize the carbonyl group of the peptide during hydrolysis[12]. DPP IV activity on the other hand, involves the catalytic triad Asp-His-Ser of serine proteases[13].

1.2 GCP II research at Guilford Pharmaceuticals

Research at Guilford has focused on the potential utility of GCP II inhibitors to treat central nervous system disorders where excess glutamate neurotransmission has been implicated. 2-(phosphonomethyl)pentanedioic acid (2-PMPA), a potent and specific GCP II inhibitor, provides neuroprotection in *in vitro* and *in vivo* models of cerebral ischemia[14] and attenuates neuropathic pain in a chronic constrictive injury (CCI) model.

2. 2-PMPA IS A POTENT AND SPECIFIC GCP II INHIBITOR IN VITRO

2-PMPA is a phosphonate analog of NAAG containing a glutamate analog moiety attached to phosphonic acid (Fig. 2).

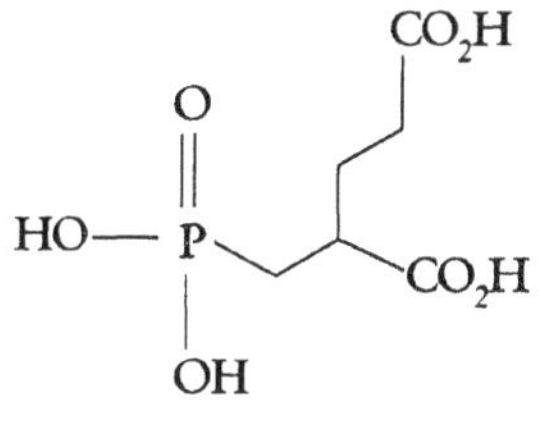

2-PMPA

Fig. 2. 2-(phosphonomethyl)pentanedioic acid (2-PMPA).

The glutamate portion of the molecule is thought to be necessary for recognition by GCP II and the phosphinic acid portion of 2-PMPA is thought to chelate the active site zinc ions[15]. 2-PMPA is a potent, competitive inhibitor of GCP II with a Ki of 0.2 nM[16]. It is a slow binding inhibitor with a fast association rate constant ($k_{on} = 3 \times 10^7\ M^{-1}\ s^{-1}$) and a slow dissociation rate constant ($k_{off} = 0.01\ s^{-1}$)[17]. The inhibitor seems to be quite specific for GCP II, i.e., no significant activities were observed at 10 μM (more than 10,000-fold higher than the K_i for NAALADase inhibition) in over 100 different receptor and enzyme assays, including glutamate receptors and transporters[14].

3. NEUROPROTECTION IN *IN VITRO* AND *IN VIVO* MODELS OF CEREBRAL ISCHEMIA

3.1 2-PMPA protects in an in vitro model of ischemia

The neuroprotective effects of 2-PMPA were first assessed in a cell culture model of ischemia. Ischemia was simulated, in rat cortical cultures, by exposing them to potassium cyanide and 2-deoxyglucose, inhibitors of oxidative metabolism and glycolysis, respectively. 2-PMPA or vehicle was added during the insult and throughout the recovery period. Injury was quantified by measuring the cytosolic lactate dehydrogenase (LDH) released into the medium over the 24-hour recovery period. As shown in Fig. 3, 2-PMPA afforded dose-dependent neuroprotection with a median effective concentration (EC_{50}) of 600 pM[14].

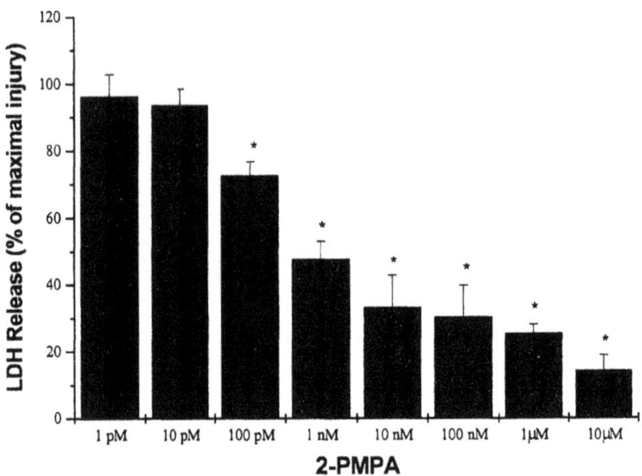

Fig. 3. 2-PMPA is neuroprotective in a cell culture model of ischemia. (* $p < 0.05$)

3.2 2-PMPA protects in an in vivo model of ischemia

Given the neuroprotective effects of GCP II inhibitors *in vitro*, the efficacy of GCP II inhibition was also assessed in an *in vivo* model of cerebral ischemia. Focal ischemia was induced in rats by middle cerebral artery occlusion (MCAO) for 2 hours, followed by 22 hours of reperfusion[18]. In the experiment, rats were randomly assigned to two treatment groups: a 2-PMPA group (10 mg/kg intraperitoneal bolus 30 min before occlusion followed by 2 mg/kg per hour intravenous infusion for 22 hours) and a control group (equivalent volumes of bolus and infused physiological saline). 22 hours after ischemia, the rats were sacrificed and their brains removed and evaluated by TTC (2,3,5-triphenyltetrazolium chloride) staining to determine brain injury volumes. As shown in Fig. 4, 2-PMPA significantly reduced the total brain injury volume, with predominant and significant effects in the cortical hemisphere[14].

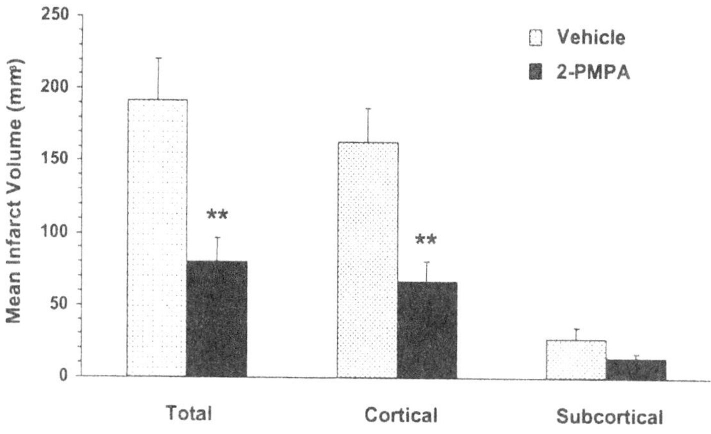

Figure 4. 2-PMPA reduces brain injury following MCAO. (**$p < 0.01$)

3.3 2-PMPA reduces excitotoxic glutamate

Since GCP II is known to hydrolyze NAAG to glutamate and NAA, changes in extracellular glutamate levels were examined during MCAO, to validate the mechanism of action of 2-PMPA. Rats implanted with microdialysis probes were treated with either 2-PMPA or vehicle and subjected to MCAO for 2 hours followed by 22 hours of reperfusion. Dialysates were collected up to 4 hours after occlusion and analysed for glutamate. While 2-PMPA had no significant effect on extracellular glutamate in normal, non-ischemic rats (left panel, Fig. 5), 2-PMPA significantly attenuated the ischemia-induced increase in extracellular glutamate seen in vehicle-treated rats (right panel, Fig. 5: 6,000% rise reduced to 1,200% ($p < 0.05$). The fact that 2-PMPA selectively attenuated the ischemia-induced rise in glutamate suggests a potential role for GCP II inhibitors in excitotoxic mechanisms[14].

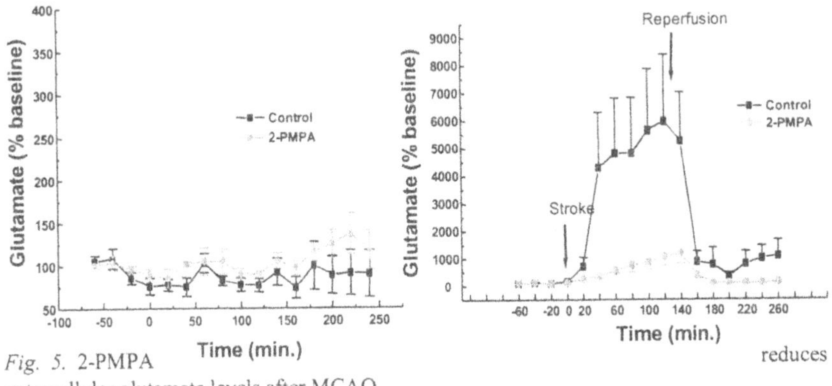

Fig. 5. 2-PMPA reduces extracellular glutamate levels after MCAO.

3.4 2-PMPA ATTENUATES HYPERALGESIA IN CHRONIC CONSTRICTIVE INJURY (CCI)

Animals were subjected to CCI, a surgical procedure described previously[19]. Briefly, one sciatic nerve was exposed by blunt dissection proximal to nerve trifurcation and four ligatures loosely tied at 1mm intervals. The other side was sham operated. Twelve days were allowed to elapse, after which an independent observer assessed thermal pain threshold (withdrawal latency) by means of the plantar test[20]. The ligated and non-ligated hind limbs of the CCI rats were tested and a difference score for each animal was determined by subtracting the mean withdrawal latency of the non-ligated (sham-operated) leg from the mean withdrawal latency of the ligated leg[19]. Therefore, negative values indicate a relative hyperalgesia on the operated side as compared to the sham side.

The results of 2-PMPA treatments on pain resulting from sciatic nerve ligation are presented in Fig. 6. The unoperated animals (untreated control) showed no difference between right and left leg withdrawal latencies, i.e., a difference score of about zero throughout the study. Vehicle-treated animals remained hyperalgesic over the period of testing as indicated by the negative withdrawal latency difference over the entire course of the study. 2-PMPA significantly attenuated the CCI-induced hyperalgesia beginning at 11 days of treatment and continuing through the end of the study on day 21.

These findings suggest that GCP II may be involved in the mediation of neuropathic pain and its inhibition may serve as a novel therapeutic arena for painful neuropathy.

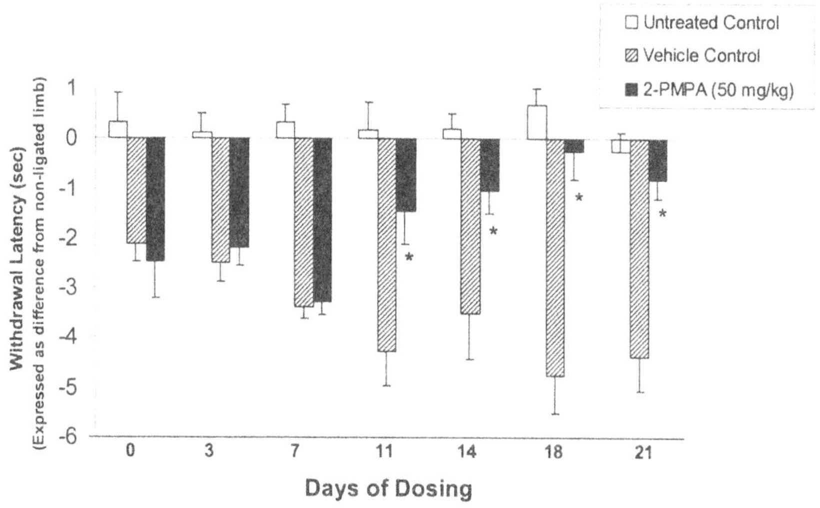

Fig. 6. 2-PMPA reduces neuropathic pain in CCI.

4. CONCLUSION

We have shown that 2-PMPA, an inhibitor of GCP II, was neuroprotective in *in vitro* and *in vivo* models of stroke and effectively attenuated neuropathic pain in a chronic constrictive injury model. These biological effects are likely mediated by GCP II inhibition-induced decreases in extracellular glutamate and/or increases in extracellular NAAG. Therefore, GCP II inhibition may be useful in neurological disorders in which excessive glutamate transmission is pathogenic. In addition, GCP II inhibition may represent a novel glutamate regulating strategy devoid of the side effects that have hampered the development of postsynaptic glutamate receptor antagonists. A lead GCP II inhibitor is currently completing Phase I clinical testing.

REFERENCES

1. Robinson, M.B., Blakely, R.D., Couto, R. and Coyle, J.T., 1987, Hydrolysis of the brain dipeptide N-acetyl-L-aspartyl-L-glutamate. Identification and characterization of a novel N-acetylated alpha- linked acidic dipeptidase activity from rat brain. *J Biol Chem* **262:**14498-14506.

2. Carter, R.E., Feldman, A.R. and Coyle, J.T., 1996, Prostate-specific membrane antigen is a hydrolase with substrate and pharmacologic characteristics of a neuropeptidase. *Proc Natl Acad Sci U S A* **93:**749-753.
3. Israeli, R.S., Powell, C.T., Fair, W.R. and Heston, W.D.,1993, Molecular cloning of a complementary DNA encoding a prostate-specific membrane antigen. *Cancer Res* **53:**227-230.
4. Chang, S.S., O'Keefe, D.S., Bacich, D.J., Reuter, V.E., Heston, W.D. and Gaudin, P.B., 1999, Prostate-specific membrane antigen is produced in tumor-associated neovasculature. *Clin Cancer Res* **5:**2674-2681.
5. Heston, W.D.,1997, Characterization and glutamyl preferring carboxypeptidase function of prostate specific membrane antigen: a novel folate hydrolase. *Urology* **49:**104-112.
6. Slusher, B.S., Robinson, M.B., Tsai, G., Simmons, M.L., Richards, S.S. and Coyle, J.T., 1990, Rat brain N-acetylated alpha-linked acidic dipeptidase activity. Purification and immunologic characterization. *J Biol Chem* **265:**21297-21301.
7. Pinto, J.T., Suffoletto, B.P., Berzin, T.M., Qiao, C.H., Lin, S., Tong, W.P., May, F., Mukherjee, B. and Heston, W.D., 1996, Prostate-specific membrane antigen: a novel folate hydrolase in human prostatic carcinoma cells. *Clin Cancer Res* **2:**1445-1451.
8. Pangalos, M.N., Neefs, J.M., Somers, M., Verhasselt, P., Bekkers, M., van der Helm, L., Fraiponts, E., Ashton, D. and Gordon, R.D., 1999, Isolation and expression of novel human glutamate carboxypeptidases with N-acetylated alpha-linked acidic dipeptidase and dipeptidyl peptidase IV activity. *J Biol Chem* **274:**8470-8483.
9. Shneider, B.L., Thevananther, S., Moyer, M.S., Walters, H.C., Rinaldo, P., Devarajan, P., Sun, A.Q., Dawson, P.A. and Ananthanarayanan, M., 1997, Cloning and characterization of a novel peptidase from rat and human ileum. *J Biol Chem* **272:**31006-31015.
10. Barinka, C., Rinnova, M., Sacha, P., Rojas, C., Majer, P., Slusher, B.S. and Konvalinka, J., 2002, Substrate specificity, inhibition and enzymological analysis of recombinant human glutamate carboxypeptidase II. *J Neurochem* **80:**477-487.
11. Rong, S.B., Zhang, J., Neale, J.H., Wroblewski, J.T., Wang, S. and Kozikowski, A.P., 2002, Molecular modeling of the interactions of glutamate carboxypeptidase II with its potent NAAG-based inhibitors. *J Med Chem* **45:**4140-4152.
12. Fersht, A. 1999, *Structure and Mechanism in Protein Science.* W. H. Freeman and Company, New York.
13. Hughes, T.E., Mone, M.D., Russell, M.E., Weldon, S.C. and Villhauer, E.B., 1999, NVP-DPP728 (1-[[[2-[(5-cyanopyridin-2-yl)amino]ethyl]amino]acetyl]-2-cyano-(S)-pyrrolidine), a slow-binding inhibitor of dipeptidyl peptidase IV. *Biochemistry* **38:**11597-11603.
14. Slusher, B.S., Vornov, J.J., Thomas, A.G., Hurn, P.D., Harukuni, I., Bhardwaj, A., Traystman, R.J., Robinson, M.B., Britton, P., Lu, X., Tortella, F.C., Wozniak, K.M., Yudkoff, M., Potter, B.M. and Jackson, P.F., 1999, Selective inhibition of NAALADase, which converts NAAG to glutamate, reduces ischemic brain injury. *Nat Med* **5:**1396-1402.
15. Jackson, P.F. and Slusher, B.S., 2001, Design of NAALADase inhibitors: a novel neuroprotective strategy. *Curr Med Chem* **8:**949-957.
16. Jackson, P.F., Cole, D.C., Slusher, B.S., Stetz, S.L., Ross, L.E., Donzanti, B.A. and Trainor, D.A., 1996, Design, synthesis, and biological activity of a potent inhibitor of the neuropeptidase N-acetylated alpha-linked acidic dipeptidase. *J Med Chem* **39:**619-622.

17. Tiffany, C.W., Cai, N.S., Rojas, C. and Slusher, B.S., 2001, Binding of the glutamate carboxypeptidase II (NAALADase) inhibitor 2-PMPA to rat brain membranes. *Eur J Pharmacol* **427**:91-96.
18. Longa, E.Z., Weinstein, P.R., Carlson, S. and Cummins, R., 1989, Reversible middle cerebral artery occlusion without craniectomy in rats. *Stroke* **20**:84-91.
19. Bennett, G.J. and Xie, Y.K., 1988, A peripheral mononeuropathy in rat that produces disorders of pain sensation like those seen in man. *Pain* **33**:87-107.
20. Hargreaves, K., Dubner, R., Brown, F., Flores, C. and Joris, J., 1988, A new and sensitive method for measuring thermal nociception in cutaneous hyperalgesia. *Pain* **32**:77-88

Dual Role of Dipeptidyl Peptidase IV (DPP IV) in Angiogenesis and Vascular Remodeling.

JOANNA KITLINSKA, EDWARD W. LEE, LIJUN LI, JENNIFER PONS, LORI ESTES, and ZOFIA ZUKOWSKA
Department of Physiology and Biohysics, GeorgetownUniversity Medical Center, 3900 Reservoir Rd., NW, Washington, DC, USA

1. INTRODUCTION

DPP IV was first discovered as a membrane bound, atypical serine protease, which cleaves proline in the N-terminal penultimate position[1]. The enzyme is constitutively expressed on endothelial and some epithelial cells, and is inducible in activated T and B-lymphocytes, where it is known as CD26[2]. DPP IV cleavage modifies the activity of many regulatory peptides, such as neuropeptide Y (NPY), glucagon-like peptides GLP-1 and GLP-2, as well as chemokines, and either inactivates them or changes their affinity for specific receptors[3, 4, 5, 6]. More recently, other functions of DPP IV have been discovered too. In addition to its proteolytic activity, DPP IV also acts as a binding protein mediating interactions between the cells and the extracellular matrix[7, 8] or as a co-receptor facilitating signal transmission through the plasma membrane[9]. Such multiple functions and mechanisms of actions implicate DPP IV as a potential important regulator of various physiological and pathological processes.

Dipeptidyl Aminopeptidases in Health and Disease, Edited by Hildebrandt et al.
Kluwer Academic/Plenum Publishers, New York, 2003

2. DPP IV AND THE NPY SYSTEM

One of the best DPP IV substrates is NPY. It is a 36-aa peptide primarily known as a neurotransmitter in the brain and sympathetic nerves, but also present extraneuronally, in megakariocytes, platelets, immune cells and endothelium[10, 11, 12, 13]. Interestingly, in the last two cell types, NPY is co-expressed with DPP IV. The main functions of NPY include the inhibition of neurotransmitter release, vasoconstriction and stimulation of food intake[10, 14, 15]. However, the peptide can act also as a growth factor for neurons, endothelial and vascular smooth muscle cells, as well as stimulate angiogenesis[13, 16, 17]. NPY's actions are mediated by five Gi/o-coupled receptors, designated as Y1 to Y5[18, 19, 29, 21, 22]. The receptors vary in their cellular distribution and mediate different functions of the peptide. For example, the Y1 receptor is the predominant vascular receptor mediating vasoconstriction and the major brain receptor involved in anxiety and, together with Y5, in food intake. The Y2 receptor, on the other hand, is the primary receptor responsible for the neuro-inhibitory effects of NPY in the central and peripheral nervous system[10, 15, 23].

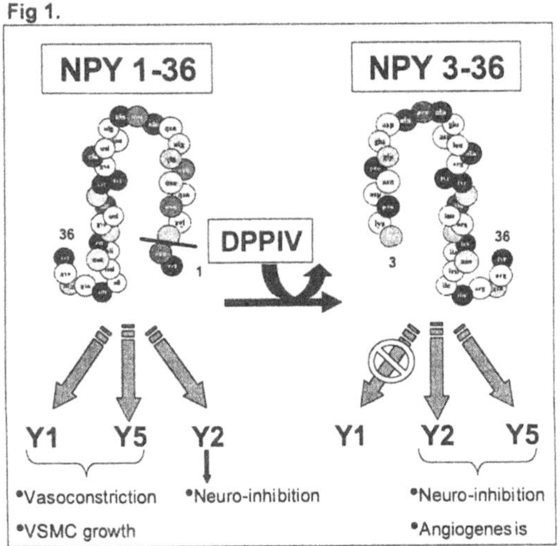

Figure 1.DPP IV-mediated processing of NPY. By cleaving Tyr-Pro off NPY's N-terminus DPP IV forms another biologically active form of the peptide, NPY_{3-36}, which is no longer able to bind to the Y1 receptor. By this way, the protease shifts the actions of the peptide from Y1- to non-Y1-receptor-mediated.

In the NPY system, DPP IV functions as the NPY-converting enzyme, which cleaves Tyr-Pro off the peptide's N-terminus and forms another biologically active form of the peptide, NPY_{3-36}[2]. The resulting shorter form of the peptide is no longer able to bind to the Y1 receptor, since it requires an intact N-terminal. However, NPY_{3-36} retains the ability to react with all other NPY receptors. Hence, DPP IV is an important regulatory molecule in the NPY system shifting actions of the peptide from Y1- to non-Y1-receptor mediated (Fig. 1).

3. ROLE OF DPP IV IN NPY-INDUCED ANGIOGENESIS

Among the many pleiotropic functions of NPY, its angiogenic activity is one of the most recently discovered. The peptide stimulates the migration and proliferation of endothelial cell, as well as aortic sprouting, vessel ingrowth into Matrigel plugs and re-vascularization of ischemic legs[17, 24]. Since endothelial cells express all the elements of the NPY system – the peptide, its receptors – Y1, Y2, Y5 and converting enzyme – DPP IV, NPY may potentially act as an autocrine growth factor in these cells (Fig 2a). Importantly, NPY_{3-36} stimulates endothelial cell proliferation with the same potency as the full-length peptide (Fig. 2a), which suggests the role of Y2 and Y5, and not Y1, receptors in this process. The question is, to what extent the conversion of NPY_{1-36} to its shorter C-terminal fragment by the abundantly present endothelial DPP IV is essential for NPY's angiogenic activities.

A critical role for DPP IV appears to be supported by several lines of evidence. First, NPY and DPP IV are co-localized in the same endothelial cell[17] and both are up-regulated during angiogenesis[25]. Secondly,. anti-DPP IV neutralizing antibody completely blocks endothelial cell migration and wound closure mediated by NPY_{1-36}, while the effect of NPY_{3-36} remains unaltered [26]. Thirdly, it appears that it is non-Y1 receptors, which are required for NPY-induced angiogenesis. For example, in the rat hind limb ischemic model local administration of NPY significantly increases vascularization of the ischemic leg. However, this effect is considerably diminished in the Y2 knockout mice (Fig. 2b). Furthermore, ischemia alone increases the release of NPY[26] and induces the expression of Y2 receptors and DPP IV in ischemic muscles (Fig. 2b). Taken together, these findings indicate the role of endogenous NPY and the Y2/DPP IV system in the process of re-vascularization.

Additional evidence comes from our aging study. Spontaneous, as well as NPY-driven angiogenesis, measured as aortic sprouting, was significantly impaired in aged mice, and this was accompanied by a loss of Y2 and DPP IV expression in spleens of old animals (Fig. 2c). Thus, NPY-induced endothelial cell proliferation and angiogenesis is Y2/Y5 and DPP IV-dependent. The system is activated during wound healing, tissue ischemia and is impaired in aging. DPP IV acts here as a protease converting NPY_{1-36} to its angiogenic, shorter form, NPY_{3-36}, which loses its ability to activate Y1 but binds to the Y2 and Y5 receptors

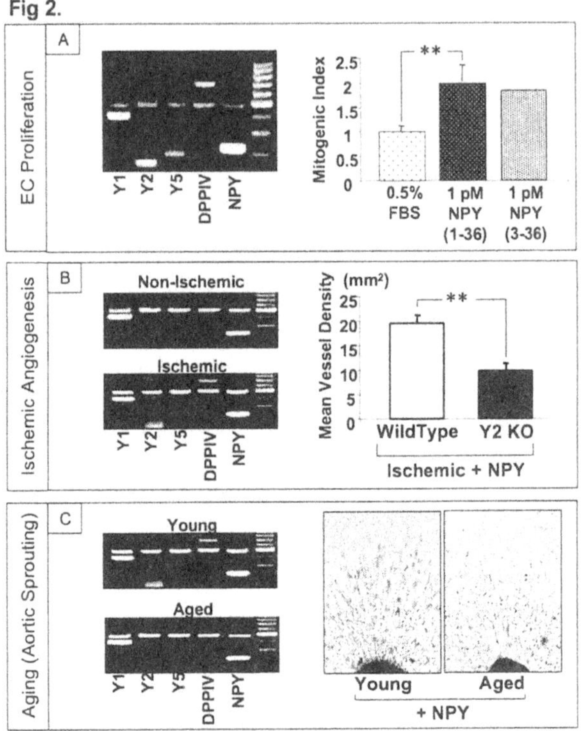

Figure 2.DPP IV and the Y2/Y5 receptor system in endothelial cell (EC) growth and angiogenesis. A. Human umbilical vein endothelial cells (HUVEC) constitutively express NPY, its Y1, Y2 and Y5 receptors and DPP IV (RT-PCR). The full length NPY_{1-36} stimulates HUVEC proliferation (measured as ^3H thymidine uptake), however, the Y2/Y5 agonist, NPY_{3-36}, which is formed by DPP IV cleavage, is as potent and efficacious. B. In the rat hind limb ischemic model, ischemia alone induces DPP IV and Y2 receptor expression (RT-PCR) in the ischemic calf muscle. Additionally, NPY-driven revascularization of the ischemic muscle (measured as vessel density) is considerably diminished in the Y2 knockout mice. C. Impairment of NPY-mediated aortic sprouting in aged mice is accompanied by a reduction of DPP IV and Y2 receptor expression in these animals, as compared to young mice (RT-PCR).

4. DPP IV IN VASCULAR REMODELING

Unlike endothelial cells, vascular smooth muscle cells, do not express high levels of NPY receptors and DPP IV constitutively. However, similarly to what happens in immune cells, DPP IV and NPY receptor synthesis is induced in vascular smooth muscle cells during their proliferation, stimulated by growth factors and NPY itself, which also promotes growth of these cells (Fig 3a). Also, in contrast to endothelial cells, which depend on Y2/Y5 receptors, the mitogenic effect of the peptide in vascular smooth muscle cells requires activation of Y1 and Y5 receptors, being blocked by combination of these antagonists (Fig. 3a). Surprisingly, this Y1/Y5-receptor-dependent mitogenic effect of NPY is blocked by the DPP IV inhibitor, Ala-Pyrr-2-CN. In addition, we have found that the enzyme itself (human recombinant, gift from Dr. Oravecz) is able to stimulate vascular smooth muscle cell proliferation, suggesting other modes of DPP IV's action leading to vascular smooth muscle cell growth, independent of the enzyme's proteolytic activity converting NPY_{1-36} to NPY_{3-36} (Fig. 3a).

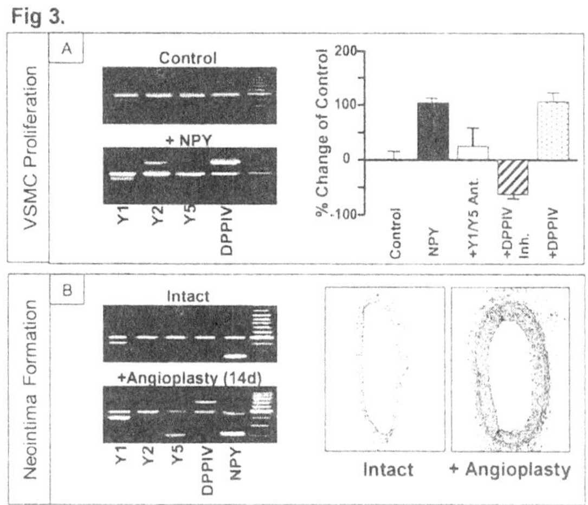

Figure 3.DPP IV and the Y1/Y5 system in vascular smooth muscle cell (VSMC) proliferation. A. The expression of NPY receptors and DPP IV is induced/up-regulated in vascular smooth muscle cells following NPY treatment (RT-PCR). The peptide stimulates proliferation of vascular smooth muscle cells (measured as ^{3}H thymidine uptake), which is prevented by combined Y1 and Y5 antagonists as well as by the DPP IV inhibitor, Ala-Pyrr-2-CN. Moreover, human recombinant DPP IV itself stimulates vascular smooth muscle cell proliferation. B. In the rat carotid artery angioplasty model, vessel injury causes vascular smooth muscle cell proliferation and neointima formation, which is accompanied by the induction/up-regulation of DPP IV and Y1 and Y5 receptor expression.

We also studied the NPY-DPP IV interactions in the rat balloon angioplasty model, where vascular injury and removal of endothelium causes vascular smooth muscle cell proliferation and formation of neointima. Administration of exogenous NPY dramatically augments this effect, leading usually to complete vessel occlusion, which resembles atherosclerotic lesions[27]. Similar to the effects *in vitro*, Y1 and Y5 antagonists prevent NPY-driven neointima formation[27]. Furthermore, angioplasty alone increases local NPY levels and up-regulates the expression of the Y1 and Y5 receptors in the vessel wall. Interestingly, DPP IV expression is also induced by angioplasty (Fig 3b). Thus, although NPY-driven vascular smooth muscle cell proliferation and neointima formation are dependent on the Y1/Y5 system, which is up-regulated during vascular injury and atherosclerosis, DPP IV paradoxically seems to be required for or involved in these processes. This raises the possibility that in the setting of vascular remodelling, DPP IV acts not as a protease, but as a co-receptor and/or binding protein, facilitating Y1/Y5 signalling. The fact that DPP IV is able to stimulate vascular smooth muscle cell proliferation in a NPY-independent manner may indicate that either DPP IV is a downstream mediator of NPY's actions or may act also via other pathways, such as chemokine activation.

5. CONCLUSION

Our studies, both *in vivo* and *in vitro*, strongly indicate a critical role of DPP IV in modifying NPY's actions in the cardiovascular system. The protein can act as an NPY converting enzyme, cleaving the full length NPY_{1-36} to its shorter form, NPY_{3-36}, and shifting the peptide's activities from Y1-mediated vasoconstriction and vascular smooth muscle cell growth to Y2/Y5-mediated angiogenesis. On the other hand, the intriguing phenomenon of DPP IV being a necessary factor in Y1/Y5-mediated vascular smooth muscle cell proliferation implicates its possible role as a co-receptor facilitating signalling of and/or ligand binding to NPY receptors. Additional studies are required to determine mechanisms of DPP IV actions in certain cellular models, e.g. why does the enzyme not inactivate NPY_{1-36} in the vascular smooth muscle cell-Y1/Y5 system by cleavage of the peptide, or whether or not other proteases, such as aminopeptidase P, can compensate for loss of DPP IV. If DPP IV proves to be indeed a critical step required for inhibiting NPY's contractile and pro-atherosclerotic effect and potentiating its angiogenic activities, abnormally low DPP IV expression and activity could be a risk factor for hypertension and ischemic cardiovascular diseases, in which NPY has already been implicated[28]. Research into this area seems particularly necessary as DPP IV inhibitors are being considered as a potential therapy for type II diabetes - a disease, which by itself increases the risk for cardiovascular consequences.

ACKNOWLEDGEMENTS

This work was supported by grants from NIH HL67357 and HL55310 and AstraZeneca to Z. Zukowska, and NIH grant AG20795 to J. Kitlinska.

REFERENCES

1. Hopsu-Havu, V.K., Glenner, G.G., 1966, A new dipeptide naphthylamidase hydrolyzing glycyl-prolyl-beta-naphthylamide. *Histochemie* **7**:197-201.
2. Mentlein, R., 1999, Dipeptidyl-peptidase IV (CD26)- role in the inactivation of regulatory peptides. *Regul. Pept.* **85**:9-24.
3. Mentlein, R., Dahms, P., Grandt, D., Kruger, R., 1993, Proteolytic processing of neuropeptide Y and peptide YY by dipeptidyl peptidase IV. *Regul. Pept.* **49**:133-44.
4. Mentlein, R., Gallwitz, B., Schmidt, W.E., 1993, Dipeptidyl-peptidase IV hydrolyses gastric inhibitory polypeptide, glucagon-likepeptide-1(7-36)amide, peptide histidine methionine and is responsible for their degradation in human serum. *Eur. J. Biochem.* **214**:829-35.
5. Drucker, D.J., Shi, Q., Crivici, A., Sumner-Smith, M., Tavares, W., Hill, M., DeForest, L., Cooper, S., Brubaker, P.L., 1997, Regulation of the biological activity of glucagon-like peptide 2 in vivo by dipeptidyl peptidase IV. *Nat. Biotechnol.* **15**:673-7.
6. Oravecz, T., Pall, M., Roderiquez, G., Gorrell, M.D., Ditto, M., Nguyen, N.Y., Boykins, R.,Unsworth, E., Norcross, M.A., 1997, Regulation of the receptor specificity and function of the chemokine RANTES(regulated on activation, normal T cell expressed and secreted) by dipeptidyl peptidase IV (CD26)-mediated cleavage. *J. Exp. Med.* **186**:1865-72.
7. Cheng, H.C., Abdel-Ghany, M., Elble, R.C., Pauli, B.U., 1998, Lung endothelial dipeptidyl peptidase IV promotes adhesion and metastasis of rat breast cancer cells via tumor cell surface-associated fibronectin. *J. Biol. Chem.* **273**:24207-15.
8. Johnson, R.C., Zhu, D., Augustin-Voss, H.G., Pauli, B.U., 1993, Lung endothelial dipeptidyl peptidase IV is an adhesion molecule for lung-metastatic rat breast and prostate carcinoma cells. *J. Cell. Biol.* **121**:1423-32.
9. Morimoto, C., Schlossman, S.F., 1998, The structure and function of CD26 in the T-cell immune response. *Immunol. Rev.* **161**:55-70.
10. Colmers, W.F., Wahlestedt, C., 1993, In The Biology of Neuropeptide Y and Related Peptides, Humana Press, Totowa, NJ p.xvi, 564.
11. Myers, A.K., Abi-Younes, S., Zukowska-Grojec, Z., 1991, Re-evaluation of the effects of neuropeptide Y on aggregation of human platelets. *Life Sci.* **49**:545-51.
12. Schwarz, H., Villiger, P.M., von Kempis, J., Lotz, M., 1994, Neuropeptide Y is an inducible gene in the human immune system. *J Neuroimmunol.* **51**:53-61.
13. Zukowska-Grojec, Z., Karwatowska-Prokopczuk, E., Fisher, T.A., Ji, H., 1998, Mechanisms of vascular growth-promoting effects of neuropeptide Y: role of its inducible receptors. Regul. Pept. **75-76**:231-8.
14. Zukowska-Grojec, Z., Wahlestedt, C., 1993, Origin and actions of neuropeptide Y in the cardiovascular system. In The Biology of Neuropeptide Y and Related Peptides (W. Colomers and C. Wahlestedt eds.), Humana Press, Totowa, NJ pp. 315-388.
15. Grundemar, L., Bloom, S.R., 1997, Neuropeptide Y and Drug Development. Academic Press, San Diego, London.

16. Hansel, D.E., Eipper, B.A., Ronnett, G.V., 2001, Neuropeptide Y functions as a neuroproliferative factor. *Nature* **410:**940-4.
17. Zukowska-Grojec, Z., Karwatowska-Prokopczuk, E., Rose, W., Rone, J., Movafagh, S., Ji, H., Yeh, Y., Chen, W.T., Kleinman, H.K., Grouzmann, E., Grant, D.S., 1998, Neuropeptide Y: a novel angiogenic factor from the sympathetic nerves and endothelium. *Circ. Res.* 83:187-95.
18. Wharton, J., Gordon, L., Byrne, J., Herzog, H., Selbie, L.A., Moore, K., Sullivan, M.H., Elder, M.G., Moscoso, G., Taylor, K.M., et al., 1993, Expression of the human neuropeptide tyrosine Y1 receptor. *Proc. Natl. Acad Sci. U S A.* **90:**687-91.
19. Larhammar, D., Blomqvist, A.G., Yee, F., Jazin, E., Yoo, H., Wahlested, C., 1992, Cloning and functional expression of a human neuropeptide Y/peptide YY receptor of the Y1 type. *J. Biol. Chem.* **267:**10935-8.
20. Bard, J.A., Walker, M.W., Branchek, T.A., Weinshank, R.L., 1995, Cloning and functional expression of a human Y4 subtype receptor for pancreaticpolypeptide, neuropeptide Y, and peptide YY. *J. Biol. Chem.* **270:**26762-5.
21. Rose, P.M., Fernandes, P., Lynch, J.S., Frazier, S.T., Fisher, S.M., Kodukula, K., Kienzle, B., Seethala, R., 1995, Cloning and functional expression of a cDNA encoding a human type 2 neuropeptideY receptor. *J. Biol. Chem.* **270:**29038.
22. Gerald, C., Walker, M.W., Criscione, L., Gustafson, E.L., Batzl-Hartmann, C., Smith, K.E., Vaysse, P., Durkin, M.M., Laz, T.M., Linemeyer, D.L., Schaffhauser, A.O., Whitebread, S., Hofbauer, K.G., Taber, R.I., Branchek, T.A., Weinshank, R.L., 1996, A receptor subtype involved in neuropeptide-Y-induced food intake. Nature 382:168-71.
23. Herzog, H., Baumgartner, M., Vivero, C., Selbie, L.A., Auer, B., Shine, J., 1993, Genomic organization, localization, and allelic differences in the gene for the human neuropeptide Y Y1 receptor. *J. Biol. Chem.* **268:**6703-7.
24. Kitlinska, J., Lee, E.W., Movafagh, S., Pons, J., Zukowska, Z., 2002, Neuropeptide Y-induced angiogenesis in aging. *Peptides* **23:**71-7.
25. Lee, E.W., 2000, Angiogenic effects of Neuropeptide Y (NPY) in rat ischemic limb model *FASEB Journal,* **14(4):** A709.
26. Ghersi, G., Chen, W., Lee, E.W., Zukowska, Z., 2001, Critical role of dipeptidyl peptidase IV in neuropeptide Y-mediated endothelial cell migration in response to wounding. *Peptides* **22:**453-8.
27. Li, L., Bojakowski, K., Kitlinska, J., Lee, E.W., Zukowska, Z., 2001, Differential effect of neuropeptide Y (NPY) on neointimal formation due to angioplasty and transplantation. *Circulation* **104:** 318.
28. Zukowska-Grojec, Z., 1997, Neuropeptide Y: Implications in vascular remodeling and novel therapeutics. *Drug News and Perspectives* **10:** 587-595.

CD26 Expression on Cutaneous Infiltrates from Patients with Cutaneous T-Cell Lymphoma (CTCL)

CD26 in cutaneous T-cell lymphoma patients

MAURO NOVELLI, ALESSANDRA COMESSATTI, PIETRO QUAGLINO, PAOLA SAVOIA, MARIA T. FIERRO, and MARIA G. BERNENGO
Department of Medical and Surgical Specialities – 1st Dermatologic Clinic, University of Turin, Via Cherasco 23, 10126, Turin, Italy.

1. INTRODUCTION

Primary cutaneous lymphomas are an heterogeneus group of rare lymphoproliferative disorders which show considerable variation in clinical presentation, histology, immunophenotype and prognosis, characterized by the development of a clonal lymphoid population of T- or B-cell origin primarily arising in the skin [1,2]. Cutaneous T-cell lymphomas (CTCL), which account for the great majority of cutaneous lymphoma cases, can be divided according to the newly proposed EORTC classification[1] into two major groups on the basis of the clinical course. Sézary syndrome (SS), the erythrodermic and leukemic variant in the CTCL spectrum, and CD30- large cell lymphoma share an aggressive clinical behaviour. On the other hand, Mycosis fungoides (MF) and CD30+ lymphoproliferative disorders (including lymphomatoid papulosis and large cell CD30+ lymphoma) are characterized by an indolent clinical course. In particular, cutaneous lesions in MF, which is the most common CTCL subtype, show a classic slow progression over years or sometimes decades, from patches to more infiltrated plaques and eventually tumours; in the later stages, there may also be lymph-node and visceral involvement. Survival of patients with early phase MF is similar to that of an age- and sex-matched population, whereas

Dipeptidyl Aminopeptidases in Health and Disease, Edited by Hildebrandt et al.
Kluwer Academic/Plenum Publishers, New York, 2003

in the presence of cutaneous tumours, or nodal/visceral involvement, the prognosis is poor and survival rates drop down to 20-30% at 5 years.

From a phenotypical point of view, atypical lymphoid cells in MF/SS bear a post thymic T helper memory phenotype (CD3+, CD4+, CD8-, CD45RO+)[3]; aberrant phenotypes with loss of T-cell lineage antigens are more commonly found in tumour-stage MF. Few data are available in literature as to the expression of CD26 on cutaneous lymphoma. CD26, a glycosilated membrane protease cleaving Xaa-Pro (or less efficiently Xaa-Ala) dipeptides from the N-terminus of poly- peptides[4], is constitutively expressed in endothelial and epithelial cells of various tissues. It is expressed on more than 50% of peripheral blood lymphocytes in healthy subjects, increasing after natural or induced activation, and has been related to a Th1 cytokine secretory pattern[5]. Among lymphomas, CD26 is detectable on neoplastic cells of most anaplastic large cell lymphomas and in a fraction of T-cell NHL[6]. In previous reports, we showed that the lack of CD26 is a constant feature of circulating Sézary cells (SC) from both SS and MF patients with peripheral blood involvement, and that the levels of the CD4+CD26- subpopulation correlate with the extent of peripheral blood involvement[7,8]. Recently, Jones et al. [9] found similar results in a cohort of 28 patients.

In this paper, immunohistochemistry and high resolution multi-parameter flow-cytometry have been applied to analyze the expression of CD26 on cutaneous lymphoid infiltrates in a series of 332 MF patients at different clinical stages and 42 SS patients, to evaluate if atypical lymphoid cells maintain the same CD26-phenotype shown in the peripheral blood. Moreover, CD26 expression has been analyzed in 46 specimens from patients with primary non MF/SS CD30+ or CD30-CTCL, to evaluate if these CTCL subtypes share with MF/SS the same CD26- phenotype.

2. PATIENTS AND METHODS

2.1 Patients

From January 1975 to December 2001, a total of 406 CTCL patients have been diagnosed, treated and followed-up at our institutions. According to the EORTC classification system[1], 332 were diagnosed as MF, 42 as SS and 46 as non MF/SS CTCL (13 with CD30+ large cell and 33 with CD30- large cell CTCL).

The diagnosis was made on the basis of clinical, immuno-pathological and molecular data. SS diagnosis was based on the following criteria, as previously reported[1,7]: erythroderma, peripheral adenopathies, peripheral blood involvement (circulating atypical SC more than 1,000/mm^3), confirmed by PCR analysis.

For patients with MF/SS, evaluation was based on detailed medical history, physical examination, complete blood cell count (including SC count on May-Grunwald stained smears) and routine laboratory tests. Staging procedures included skin biopsy and abdomen, pelvis and lung CT scan; lymph node biopsy was performed in the presence of palpable lymphadenopathies. MF patients were staged according to the TNMB classification system.

As to the patients with non MF/SS CTCL, the diagnosis of primary cutaneous lymphoma has been made on the basis of the absence of clinical evidence of extracutaneous involvement both at diagnosis and within 6 months after diagnosis.

2.2 Immunohistochemistry

Immunohistochemistry on cryostatic sections was performed in all cases by the standard streptavidin-biotin-peroxidase method (LSAB2plus Kit, Dako, Glostrup, Denmark) using the monoclonal antibodies shown in Table 1. The phenotype was evaluated predominantly on intraepithelial lymphoid atypical cells, in order to avoid any misinterpretations due to the presence of a variable amount of normal reactive lymphocytes in the dermal infiltrate; the CD26 expression on the dermal infiltrate was evaluated only when epidermotropism was absent (tumoral lesions in MF patients; non epidermotropic non MF/SS CTCL). CD26 was considered positive when expressed on more than 50% of the lymphoid cell infiltrate, negative when expressed on less than 30% of the cell infiltrate. A CD26 expression on more than 30% and less than 50% of the cell infiltrate was defined as “mixed”.

2.3 Tissue suspensions

Flow cytometry on tissue suspensions was performed in 67 patients (56 MF and 11 SS). Tissue suspensions were performed using an automated mechanical disaggregation device (Medimachine, CONSUL$^{T.S.}$, Italy, distributed by Becton-Dickinson, S.Josè, CA, USA, and Dakopatts, Denmark), as follows[10]: 1) small pieces of tissue, pretreated with 1% collagenase IA (Sigma, S.Louis, MI, USA) were placed in microblade-equipped polyethylene chambers (Medicons, CONSUL$^{T.S.}$) with 0.5 -1.5 ml of suspension buffer (RPMI supplemented with 10% FCS and antibiotics) and inserted into the Medimachine; 2) fragments were dissociated for 20 seconds, at a constant speed of about 100 rpm; 3) recovered suspension buffer was filtered on porous polyester membranes (Filcons, CONSUL$^{T.S.}$) with an effective surface area of 80 mm^2. After filtration, the cells were buffer washed and then processed for flow-cytometry.

Table 1. Monoclonal antibodies used

CD	MoAb	Isotype	Source	Reactivity
CD 2	Leu5b (S5.2)	IgG2a	BD	T cells, thymocytes, NK cells
CD 3	Leu-4 (SK7)	IgG1	BD	Thymocytes, T cells
CD 4	Leu-3 (SK3)	IgG1	BD	Thymocyte subs., T helper cells, monocytes, macrophages
CD 5	Leu-1 (L17F12)	IgG2a	BD	Thymocytes, T cells, subset of B cells
CD 7	Leu-9 (GH9)	IgG2a	BD	Pluripotential hematopoietic cells, thymocytes, T cells
CD 8	Leu-2 (SK1)	IgG1	BD	Thymocyte subsets, cytotoxic T cells
CD11a	MHM24	IgG1	DK	αL subunit of integrin LFA-1: leukocytes
CD15s	KM93	IgM	ST	Granulocytes, monocytes, NK cells, activated B and T cell
CD20	L26	IgG2a	DP	B cells
CD25	IL-2r (2A3)	IgG1	BD	IL-2 receptor: activated T cells, B cells, monocytes
CD26	L272	IgG2a	BD	DPP-IV : activated B and T cells, macrophages
CD30	Ber-H2	IgG1	DP	Activated T, B and NK cells, monocytes
CD45RA	4KB5	IgG1	DK	CD45 isoform (A exon): B cells, naive T cells, monocytes
CD45R0	UCHL1	IgG2a	BD	CD45 isoform (without A, B,C exons): T and B cells, monocytes, macrophages
CD56	Leu-19 (MY31)	IgG1	BD	NCAM: NK cells, subsets of T cells
CD62L	Leu-8 (SK11)	IgG2a	BD	L-selectin: B cells, T cells, monocytes, NK cells
CD69	L78	IgG1	BD	activated B and T cells, macrophages and NK cells
CD71	L01.1	IgG2a	BD	transferrin receptor: proliferating cells, activated leukocytes
HLA-DR	L243	IgG2a	BD	MHC-II presenting cells
TCRβ	8A3	IgG1	ST	TCR β chain: T lymphocytes
TCDδ	5A6.E9	IgG1	CA	TCR δ chain: T lymphocytes
Ki-67		IgG1	DK	Proliferating cells

AN: Ancell, Bayport, NM, USA; **BD**: Becton Dickinson, San Jose, CAS, USA; **CA**: Caltag, Burlingame, CA, USA; **CM**: Chemicon, Temecule, CA, USA; **CO**: Beckman Coulter, Fullerton, CA, USA; **DP**: Dakopatts, Copenhagen, Denmark; **IM**: Immunotech, Marseille, France; **ST**: Serotec, Oxford, United Kingdom.

Three colour immunofluorescence analysis was performed simultaneously using peridin-chlorophill-protein (PerCP) conjugated CD45, fluorescein isothiocyanate (FITC) and phycoerythrin (PE) conjugated antibodies.

Fluorescence was analyzed in a FACSCalibur cytometer (Becton-Dickinson, S.José, CA, USA). Lymphocytes were gated using a CD45 logarithmic amplified histogram, followed by a foreward and sideward gate

2.4 Determination of TCR-γ gene rearrangement

T-cell clonality was evaluated in 124 cases by the detection of a dominant TCR-γ gene rearrangement using a heteroduplex analysis (T Cell Lymphoma Kit, Experteam, Venice, Italy). Genomic DNA was extracted from 10 cryostatic sections (15 μm each) previously digested by proteinase K. The TCR-γ chain gene was amplified using the method described by McCarthy et al. [11].

3. RESULTS

The results of the CD26 expression as determined by immunohistochemistry and flow-cytometry are summarized in Table 2 and Table 3 and illustrated in Figure 1.

Marked differences were found in CD26 expression on cutaneous lymphoid infiltrates between MF and SS patients. In fact, a predominant CD26 negative phenotype was found in all the 41 SS patients. It is noteworthy that a "de novo" CD26 expression on the dermal infiltrate was found in one SS patient who developed during follow-up a transformation in high grade lymphoma. On the other hand, only 186/332 MF patients (56%) showed a lack of CD26 expression; as to the remaining patients, a variable fraction of CD26 negative cells intermingled with the CD26 positive was found in 59 cases (17.8%), whereas in 87 (26.2%) the majority of the cell infiltrate stained positively for CD26. A higher rate of CD26 negative cases was found in T_2-T_3-T_4 patients with respect to T_1 patients.

Flow-cytometry on tissue suspensions gave the same results. The percentage of CD4+CD26- cutaneous infiltrating cells was significantly higher in SS patients (60.9 ± 11.6) than in MF patients (26.1 ± 19.2; Student t test: $p<0.001$). According to the clinical stage in MF patients, the percentage of CD4+CD26- cutaneous infiltrating cells was significantly higher in T_3-T_4 than in T_1-T_2 patients (Student t test: $p=0.012$).

Table 2. CD26 expression on CTCL patients: comparison with clinico-pathologic diagnosis according to the EORTC classification, cutaneous clinical stage and PCR findings.

	N	CD26 Negative[a]		CD26 Mixed[a]		CD26 Positive[a]	
		N (%)[b]	Clone (%)[c]	N (%)[b]	Clone (%)[c]	N (%)[b]	Clone (%)[c]
MF-total	332	186 (56%)	44/83 (53%)	59 (17.8%)	6/11 (54.5%)	87 (26.2%)	21/30 (70%)
MF - T1	78	35 (44.9%)	6/15	22 (28.2%)	2/3	21 (26.9%)	5/6
MF - T2	178	102 (57.3%)	18/43	29 (16.3%)	3/6	47 (26.4%)	8/16
MF - T3	56	37 (66.1%)	15/18	7 (12.5%)	½	12 (21.4%)	5/5
MF - T4	20	12 (60%)	5/7	1 (5%)	ND	7 (35%)	3/3
SS	42	42 (100%)	13/14	0	0	0	0

[a]: CD26 was considered positive when expressed on more than 50% of the lymphoid cell infiltrate, negative when expressed on less than 20% of the cell infiltrate. A CD26 expression on more than 20% and less than 50% of the cell infiltrate was defined as "mixed".
[b]: percentage of cases within the total number of patients with that disease
[c]: percentage of patients with clonal rearrangement of TCR γ-gene within the total number of patients tested with that disease
ND: not done

Table 3. Flow cytometry on cutaneous tissue suspensions: results

	n	CD4+CD26- (%) mean ± sd
MF	56	26.11 ± 19.21
T1 MF	18	21.51 ± 16.26
T2 MF	21	23.41 ± 12.53
T3 MF	14	39.88 ± 30.71
T4 MF	3	31.44 ± 23.36
SS	11	60.86 ± 11.57

A clonal rearrangement of the γ-chain of the TCR gene was found in both CD26 negative and CD26 positive cases. No difference in the percentage of clonal cases was found according to the CD26 expression (53% clonal cases in the CD26 negative group vs 54.5% in the CD26 "mixed" and 70% in the CD26 positive group).

No difference in the clinical course was found according to the CD26 expression when stratifying patients for the T stage. The response to treatment and the relapse rate were similar between the two groups; moreover, no difference in the incidence of transformation in high grade lymphoma was found between T_3 MF patients with CD26 positive (5/12; 41.7%) or CD26 negative (18/44; 40.9%) infiltrating cells.

As to non MF/SS CTCL patients, 26/46 (56.5%) showed a predominant CD26 expression. No statistically significant difference was found in CD26 expression between CD30+ and CD30- subtypes, even if a slight prevalence of CD26 positive cases was found in the CD30+ (5/8; 61.5%) with respect to the CD30- group (18/33; 54.5%).

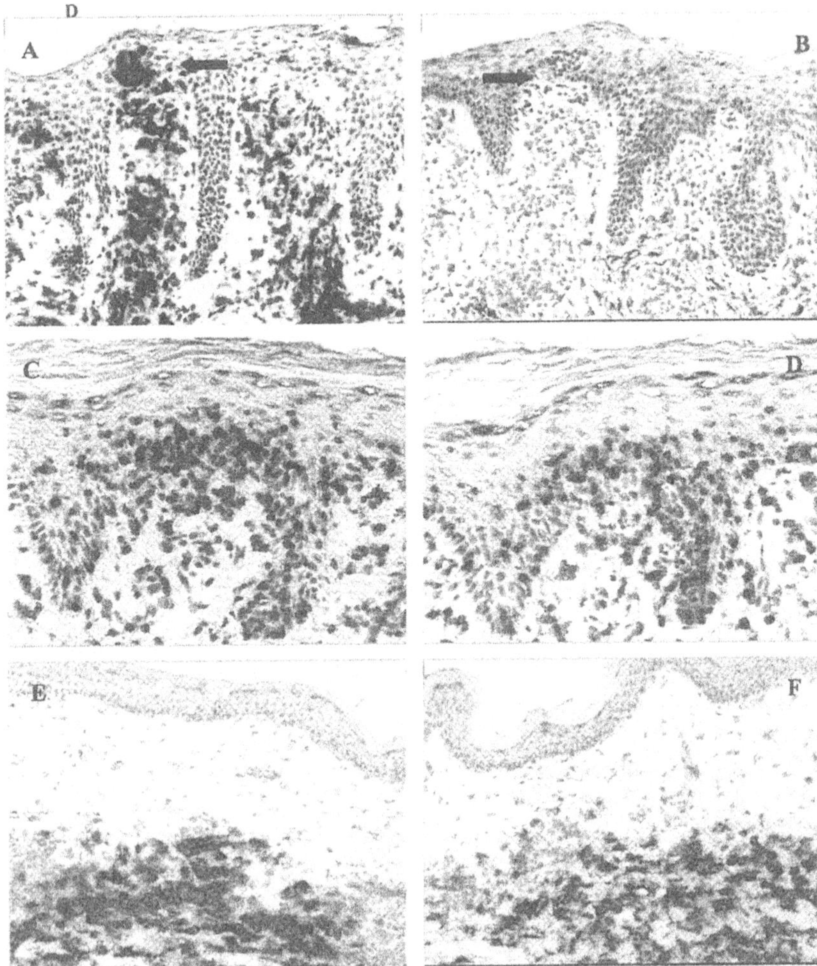

Figure 1: Immunohistochemistry on cryostatic sections (x300)
A-B: Sézary syndrome: cutaneous atypical lymphoid cells are CD3+ (A) and CD26 – (B). Arrows indicate a typical Pautrier microabscesse.
C-D: Micosis fungoides: cutaneous atypical lymphoid cells show a predominant CD3+ (C) and CD26+ (D).
E-F: Non MF/SS primary CD30+ anaplastic large-cell lymphoma: diffuse dermal infiltrate sparing the epidermis; atypical lymphoid cells are CD30+ (E) and CD26+ (F).

4. DISCUSSSION

It is known that circulating atypical lymphoid cells in MF/SS patients share the same characteristic CD26 negative pattern[7,9]. This evidence prompted us to analyze, by means of both immunohistochemistry and flow-cytometry on tissue suspensions, CD26 expression on cutaneous infiltrates in a large cohort of CTCL patients (325 MF at different clinical stages and 41 SS), to ascertain if atypical lymphoid cells maintain the same CD26-phenotype shown in the peripheral blood. Few data are available in literature as to the expression of CD26 on cutaneous infiltrates in CTCL patients. The only other study reported in literature showed that only 4/21 MF patients stained positively[12].

The results of this study clearly show that SS patients are nearly always characterized by a constant lack of CD26 expression in the skin. In fact, in all cases, CD26 was expressed on less than 30% of the cell infiltrate. It is noteworthy that the transformation in high grade lymphoma which occurred in one patient, was associated to a "de novo" CD26 expression on the dermal infiltrate. On the other hand, marked differences were found in MF patients. Even if half the cases (56%) were predominantly CD26 negative, a CD26 expression on 30% to 50% of the cell infiltrate was detected in 17.8%; moreover, in the remaining 26.2% of cases, more than 50% of the cell infiltrate stained positively for CD26 expression. The existence of CD26 positive MF was further substained by PCR studies. In fact, the percentage of clonally rearranged cases was even higher than that observed in the classic CD26 negative group. The variable degree of CD26 expression was found to be at least partially associated to the cutaneous stage. In fact, the percentage of cases with "mixed" CD26 expression was higher in early stage MF, whereas in patients with tumours or erythroderma, a predominance of CD26 negative cases was observed. We feel therefore that the differences in CD26 expression do not reflect only the difficulties in a correct identification of the "true" neoplastic population in early phase MF, but are to be considered as a characteristic feature of MF evolution.

It is noteworthy that the presence or absence of CD26 expression does not seem to correlate with a different clinical course; in fact, when stratifying MF patients for the T stage, no difference in the relapse rate, response to treatment and risk of developing transformation in high grade lymphoma were found according to the CD26 expression. All the cases with peripheral blood involvement were included in the CD26 negative or mixed group; therefore, it is unclear whether circulating atypical cells maintain always the same CD26 cutaneous phenotype or rather the presence of peripheral blood involvement could be related uniquely to a CD26 negative expression.

These data allow us to gain further insights in the pathology of CTCL. First of all, the expression of CD26 in CTCL patients does not seem to be associated to a worse prognosis, as reported in nodal lymphomas [6,12]. In fact, SS patients, who are characterized by a poor survival [7] show a constant CD26 negative phenotype, as well as half the patients with advanced MF. Moreover, even if non MF/SS CTCL show a predominant CD26 positive expression, no difference were found between the CD30+ and the CD30- cases, the former characterized by an aggressive clinical course and high risk of extracutaneous spreading, the latter associated to a relatively good prognosis. This is in contrast with the findings of Carbone et al. [6,12] who reported a predominant CD26 expression in CD30+ anaplastic lymphomas (71%) with respect to other T non Hodgkin lymphoma subtypes (47%).

Secondly, it is conceivable that the modulation in CD26 expression could reflect a different functional activation of atypical lymphoid cells. Even if the biological implications linked to the CD26 status remain to be clarified, some interesting hypotheses can be built up. CD26 represents an accessory surface molecule involved in the processes of activation and proliferation of T lymphocytes[13]. Recent studies suggested that CD26 cell surface expression correlates with the production of type 1 helper (Th1) cytokines[5], whereas CD4+CD45RO+CD26- lymphocytes are related to a Th2 secretory pattern [14]. As it is known, Th1 lymphocytes are responsible for cell-mediated immunity and macrophage-dependent immune responses, whereas Th2 cells are related to antibody production, eosinophil activation, and macrophage inhibition. In fact, all the papers confirm that the cytokine profile of Sézary cells in both skin[15] and peripheral blood[16,17] resembles that of Th2 cells. On the other hand, there is no general agreement as to the MF cytokine pattern. Saed et al. [15] found that cutaneous lesions of MF are characterized by an epidermal Th1 profile, whereas Hansen et al.[18] reported a Th1 pattern only in early phase MF lesions; more recently, Harwix et al. [19] clearly showed that T-cell clones do not show a polarized Th1 or Th2 cytokine profile. These data are in agreement with the findings of the present study, showing a predominance of CD26 negative cases only in advanced stage MF. It is conceivable, therefore, that the different expression of CD26 in MF cutaneous infiltrates could be related to a specific functional activation pattern of the atypical lymphoid cells.

REFERENCES

1. Willemze, R., Kerl, H., Sterry, W., et al., 1997, EORTC classification for primary cutaneous lymphomas: a proposal from the cutaneous lymphoma study group of the european organization for research and treatment of cancer. Blood **90**: 354.
2. Siegel, R..S., Pandolfino, T., Guitart, J., Rosen, S., Kuzel, T.M., 2000 Primary cutaneous T-cell lymphoma: review and current concepts. *J Clin Oncol* **18**(15): 2908-25.
3. Ralfkiaer, E., Wollf-Sneedorff, A., Thomsen, K., Vejlsgaard, G.L., 1993, Immunophenotypic studies in cutaneous T-cell lymphomas: clinical implications. *Br J Dermatol* **129**(6): 655-9.
4. De Meester, I., Korom, S., Van Damme, J., Scharpe, S., 1999, CD26, let it cut or cut it down. *Immunol Today* **20**: 367-375.
5. Willheim, M., Ebner, C., Baier, K., Kern, W., *et al.*, 1997, Cell surface characterization of T lymphocytes and allergen-specific T cell clones: correlation of CD26 expression with T(H1) subsets. *J Allergy Clin Immunol* **100**: 348-355.
6. Carbone, A., Cozzi, M., Gloghini, A., Pinto, A., 1994, CD26/dipeptidyl peptidase IV expression in human lymphomas is restricted to CD30-positive anaplastic large cell and a subset of T-cell non-Hodgkin's lymphomas. *Hum Pathol* **25**: 1360-1365.
7. Bernengo, M.G., Quaglino, P., Novelli, M., Cappello, N., Doveil, G.C., Lisa, F., De Matteis, A., Fierro, M.T., Appino, A., 1998, Prognostic factors in Sézary syndrome: a multivariate analysis of clinical, haematological and immunological features. *Ann Oncol* **9**: 857-863.
8. Bernengo, M.G., Novelli, M., Quaglino, P., Lisa, F., De Matteis, A., Savoia, P., Cappello, N., Fierro, M.T., 2001, The relevance of the CD4+CD26- subset in the identification of circulating Sézary cells. *Br J Dermatol* **144**: 125-135.
9. Jones, D., Dang, N.H., Duvic, M., Washington, L.T., Huh, Y.O., 2001, Absence of CD26 expression is a useful marker for diagnosis of T-cell lymphoma in peripheral blood. *Am J Clin Pathol* **115**(6):885-92.
10. Novelli, M., Savoia, P., Cambieri, I., Ponti, R., Comessatti, A., Lisa, F., Bernengo, M.G., 2000, Collagenase digestion and mechanical disaggregation as a method to extract and immunophenotype tumour lymphocytes in cutaneous T-cell lymphomas. *Clin Exp Dermatol* **25**(5): 425-431.
11. McCarthy, K.P., Sloane, J.P., Kabarowski, J.H.S., Matutes, E., Wiedemann, L.M., 1992, A simplified method of detection of clonal rearrangements of the T-cell receptor-γ chain gene. *Diagn Mol Pathol* **1**:173-9.
12. Carbone, A., Gloghini, A., Zagonel, V., Aldinucci, D., Gattei, V., Degan, M., Improta, S., Sorio, R., Monfardini, S., Pinto, A., 1995, The expression of CD26 and CD40 ligand is mutually exclusive in human T-cell non-Hodgkin's lymphomas/leukemias. *Blood* **86**(12):4617-26.
13. Kahne, T., Lendeckel, U., Wrenger; S., *et al.*, 1999, Dipeptidyl peptidase IV: a cell surface peptidase involved in regulating T cell growth. *Int J Mol Med* **4**: 3-15.
14. Scheel, D., Richter, E., Toellner, K.-M., Reiling, N., Key G., Wacker H.-H., Ulmer, A.J., Flad, H.-D, Gerdes, J., 1995, Correlation of CD26 expression with T helper (TH 1)-like reactions in granulomatous diseases. In *Leucocyte Typing V* **1**: 1111.
15. Saed, G., Fivenson, D.P., Naidu, Y., Nickoloff, B.J., 1994, Mycosis fungoides exhibits a Th1-type cell-mediated cytokine profile whereas Sezary syndrome expresses a Th2-type profile. *J Invest Dermatol* **103**(1):29-33.

16. Vowels, B.R., Cassin, M., Vonderheid, E.C., Rook, A.H., 1992, Aberrant cytokine production by Sezary syndrome patients: cytokine secretion pattern resembles murine Th2 cells. *J Invest Dermatol* **99**: 90-94.
17. Dummer, R., Heald, P.W., Nestle, F.O., *et al.*, 1996, Sézary syndrome T-cell clones display T-helper 2 cytokines and express the accessory factor-1 (interferon-gamma receptor beta-chain). *Blood* **88**: 1383-1389.
18. Hansen, E.R., 1996, Immunoregulatory events in the skin of patients with cutaneous T-cell lymphoma. *Arch Dermatol* **132**(5):554-61.
19. Harwix, S., Zachmann, K., Neumann, C., 2000, T-cell clones from early-stage cutaneous T-cell lymphoma show no polarized Th-1 or Th-2 cytokine profile. *Arch Dermatol Res* **292**(1):1-8.

Intrahepatic Expression of Collagen and Fibroblast Activation Protein (FAP) in Hepatitis C Virus Infection

MARK D. GORRELL[1], XIN M. WANG[1], MIRIAM T. LEVY[1], ELEANOR KABLE[2], GEORGE MARINOS[3], GUY COX[2], and GEOFFREY W. MCCAUGHAN[1]

[1] *A. W. Morrow Gastroenterology and Liver Centre, Royal Prince Alfred Hospital, Centenary Institute of Cancer Medicine and Cell Biology and the University of Sydney, NSW Australia.* [2] *Electron Microscope Unit, University of Sydney.* [3] *Gastroenterology Department, Prince of Wales Hospital Sydney*

DPIV is the best understood proteinase that has the rare capability of hydrolysing the prolyl bond[1]. We have suggested that DPIV, fibroblast activation protein (FAP) DP8, DP9, dipeptidyl peptidase - like protein 1 (DPL1, previously named DPX) and DPL2 form a distinct sub-class of the prolyl oligopeptidase (POP) family called the DPIV/CD26 gene family[2]. The DPIV gene family is distinguished by a pair of glutamates that is about 430 residues N terminal to the catalytic serine and are essential for DP activity[3].

FAP has 52% amino acid identity with DPIV. The FAP and DPIV genes are adjacent, suggesting recent gene duplication. FAP and DPIV exhibit different patterns of expression and substrate specificities (Reviewed in McCaughan *et al.* [4]). Both have dipeptidyl peptidase activity on Ala-Pro. FAP has a gelatinase activity that DPIV lacks[1,5]. Like DPIV, catalysis depends upon dimerisation. Considering its constitutive gelatinase activity, which is collagen type I specific[6], the tissue localisation of FAP protein is its most interesting property. In contrast to DPIV, which is widely expressed, FAP is not expressed in normal adult tissue.

FAP is strongly expressed in activated hepatic stellate cells (HSC) and myofibroblasts in cirrhotic liver[5] and other sites of tissue remodelling[7] (reviewed in Abbott and Gorrell [2]).

Dipeptidyl Aminopeptidases in Health and Disease, Edited by Hildebrandt et al.
Kluwer Academic/Plenum Publishers, New York, 2003

The HSC has an important role in the pathogenesis of cirrhosis [8]. In the normal liver HSC are quiescent, long lived cells that store vitamin A. Following liver injury, HSC undergo activation and transdifferentiation to myofibroblast-like cells. Significant functional changes accompany this phenotypic change including alterations in extracellular matrix (ECM) production and degradation and expression of various matrix metalloproteinases (MMPs) and their inhibitors. Unlike quiescent HSC, activated HSC show intense cytoplasmic alpha smooth muscle actin (SMA) immunoreactivity. Transdifferentiation of the HSC to a SMA positive phenotype is not sufficient to result in fibrosis. In chronic liver diseases such as chronic hepatitis C virus (HCV) infection, the majority of patients have considerable numbers of activated SMA-positive HSC, but a minority of patients develop cirrhosis[9]. By dual labelling we determined that subsets of HSC include many FAP single-positive and some SMA single-positive cells[5].

Here, we report FAP - positive cells in earlier stages of liver injury, where there may be inflammation but not necessarily fibrosis. We found that FAP expression by HSC correlates with the histological severity of liver disease. To further characterise the HSC subpopulations, we also studied the expression of the HSC marker Glial Fibrillary Acidic Protein (GFAP).

Certain substances have the property, when illuminated with very intense light, of generating the second harmonic (SH) - light at twice the original frequency. Recently this phenomenon has been harnessed in microscopy[10]. The ability to generate second harmonics is peculiar to molecules that are not centro-symmetric, one common biological example being collagen. The unique triple-helix structure and very high crystallinity of collagen make it exceptionally efficient in generating the second harmonic of incident light, and therefore it can provide sensitive and high-resolution information on collagen distribution, particularly the extremely crystalline type I collagen. Using a microscope optimised for SH detection[11] we found that we could detect the SH signal from collagen with much greater resolution and sensitivity than had been reported previously, typically using excitation levels lower than required for excitation of two-photon fluorescence (TPF). We present here localisation of collagen fibres along with FAP-expressing HSC at high resolution in frozen sections of human liver.

1. MATERIALS AND METHODS

Liver biopsies from patients with chronic HCV infection were used for both frozen and formalin-fixed paraffin sections. Samples from 27 patients of mean patient age 40.1 years ± SD 8.3, were analysed. Necroinflammatory activity for the portal/periportal and lobular area and the degree of fibrosis were scored by the Scheuer method. Immunoreactivity was categorised on a scale of 0 to 4, with 4 = staining of perisinusoidal cells occupying more than 30% of the sinusoidal region. Mesenchymal (fibrous septa and portal tract) cell numbers were categorised on a scale of 0 to 4 with a score of 4 = positivity of greater than 50% of mesenchymal cells. The SMA positive vascular smooth muscle cells were excluded from the scoring. Data were analysed by linear correlation analysis using GraphPad Prism® (San Diego, CA). Additional samples were obtained from three transplant donors and 16 liver transplant recipient livers.

For SH generation (SHG), ethanol-fixed cryosections of the liver explant from a patient diagnosed with primary sclerosing cholangitis Child-Pugh class C cirrhosis were immunostained for FAP using an anti-mouse Ig conjugated with Alexa 594 (Molecular Probes, Eugene, Oregon, USA). The microscope is a Leica DMIRBE inverted stand equipped with a Leica TCS2MP confocal system and Coherent Mira tunable pulsed titanium sapphire laser, tunable from 700 to 950nm, with pulses in the 100-200fs range. The microscope is equipped with dual photomultiplier transmitted light detectors, with dichroic mirrors dividing the detectable spectrum (380-680nm) at either 505nm or 560nm; further selection is accomplished by barrier filters in either or both channels. An identical dual detection unit is mounted behind the objective lens to act as a non-descanned TPF detector. A 415/10 nm narrow bandpass filter (with the laser tuned to 830nm) was used to exclude fluorescent signals in the transmission detector. (For some images a 416/30 bandpass filter was used). The SH signal was propagated almost exclusively in the forward direction and therefore was picked up only in the transmitted detector. The signal could be excited between 760 and 925nm; at shorter wavelengths the SH signal was blocked by the barrier filters at the detectors; the longer wavelength is close to the practical tuning limit of our laser. Confocal images of Alexa-stained material were collected using excitation at 543 nm and spectrometric detection in the range 590-620nm.

2. RESULTS

2.1 Correlation of FAP and lack of correlation of SMA immunoreactivities with the stage of hepatic fibrosis

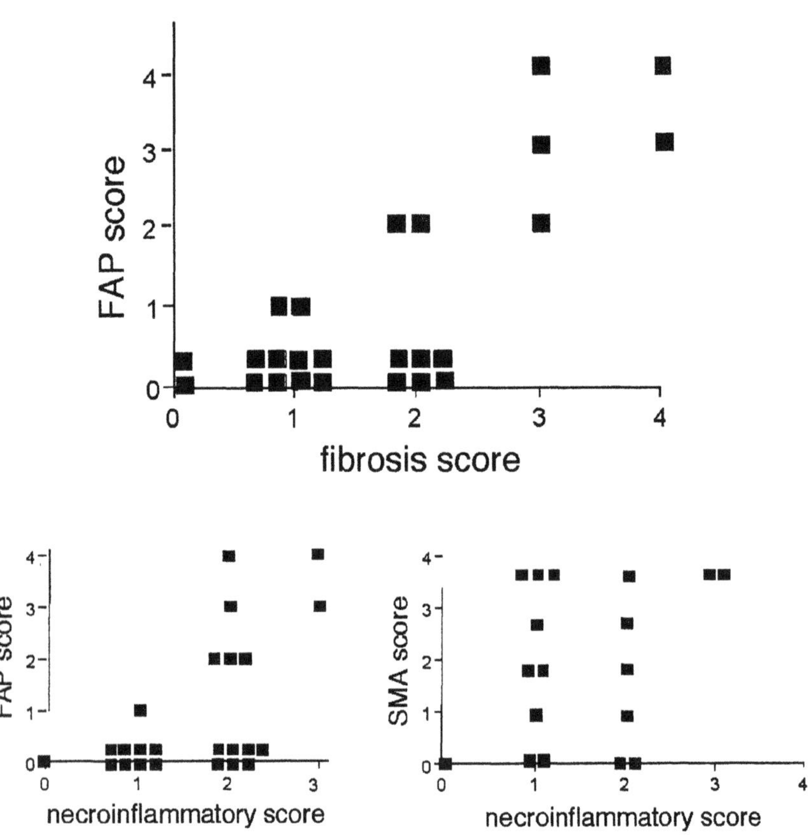

Figure 1. Correlation of FAP but not SMA expression with the stage of hepatic fibrosis and inflammation. Column scatter plots showing the periportal FAP and SMA immunoreactivity scores for each patient grouped according to the stage of liver fibrosis.

FAP protein was detected in the hepatic parenchyma in 11 of 27 patients with chronic HCV infection. The immunoreactivity was localised to the portal / periportal interface and the fibrous septa, particularly at areas of necroinflammation. Endothelial and smooth muscle cells in the walls of blood vessels were FAP negative. In 20 of 27 patients with HCV infection SMA immunoreactivity was observed in HSC diffusely throughout the liver lobule. Unlike FAP staining, there was no concentration of SMA immunoreactivity in periportal regions. FAP was detected in the mesenchymal area (the portal tracts and fibrous septa) in 19 of 27 patients. In this region SMA positive cells were detected in all 27 patients and the cells positive for FAP or SMA had spindle-shaped cell bodies with long processes consistent with the morphology of myofibroblasts.

Periportal FAP immunoreactivity was strongly correlated with the stage of liver fibrosis (r^2 = 0.77, $p < 0.0001$) (Figure 1). In contrast, SMA immunoreactivity was independent of the degree of fibrosis. Linear correlation analysis of the periportal (r^2 = 0.05), total parenchymal (r^2 = 0.01) and mesenchymal (r^2 = 0.08) scores for FAP and SMA found no significant association between these two HSC activation markers. Correlation coefficients comparing periportal SMA expression, total parenchymal scores or the mesenchymal scores with the degree of liver fibrosis were similarly significant for FAP and non-significant for SMA respectively. FAP immunoreactivity was positively correlated with the grade of necroinflammatory activity (r^2 = 0.23, p = 0.011). In contrast, there was no relationship between the level of periportal SMA expression and the grade of necroinflammatory activity (r^2 = 0.04). Similar correlation coefficients were obtained comparing all-region FAP and SMA score with grade of necroinflammatory activity.

Parenchymal GFAP immunoreactivity was observed in 10 of 25 patients. Positive cells were within the mesenchymal areas and in the periportal perisinusoidal space. GFAP positive cells were unusual within the liver lobule beyond the periportal rim. Mesenchymal GFAP immunoreactivity was more common than periportal GFAP immunoreactivity and was seen in 22 of the 25 patients. GFAP positive cell staining was usually present in up to 30% of the cells of the portal tract or fibrous septa. There was a weak but significant correlation between the immunoreactivity of GFAP and the immunoreactivity of FAP (r^2 = 0.29, p = 0.005) and the fibrosis score (r^2 = 0.18, p = 0.03).

2.2 Second Harmonics

Unlike fluorescence signals, the SH signal showed no signs of bleaching during acquisition of repeated images from a given area, showing that no damage to the collagen structure was occurring. This is to be expected since second harmonic generation is a cohererent process, unlike fluorescence, and no energy is lost. Illumination levels were typically lower than those required for two-photon excitation of fluorescent labels in the same sections. The image clarity was exceptional.

Masson's Trichrome stain was inadequate to reveal small groups of collagen fibrils. Sirius Red staining was more effective, but the SH signal was more easily distinguished and gave a much higher effective resolution; we were able to image collagen at close to the optical resolution limit. Neither stain interfered with SHG: stained sections gave a SH signal identical to that from unstained sections of the same samples.

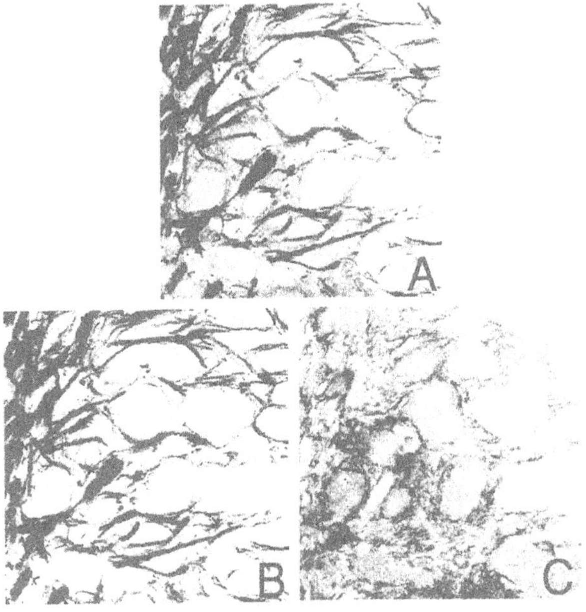

Figure 2. Ethanol-fixed, 50 micron cryosections of cirrhotic liver. SH signal (415nm) from collagen fibres in the periportal-parenchymal interface of a cirrhotic nodule and FAP immunostaining showing that fine fibrils of collagen tended to lie adjacent to activated stellate cells. A. Overlay of SH signal and single photon confocal fluorescent signal from Alexa-conjugated antibody. B. SH signal. C. Fluorescent signal. Signals are depicted in black. In this region the collagen and FAP^{+} cells tend to surround islands of hepatocytes.

In cirrhotic liver, collagen fibres through the liver were easily and effectively revealed by their SH signal. The SH signal shows both the collagen septum and proliferation of fine collagen fibres through the parenchyma of cirrhotic nodules (Figure 2). The high resolution localisation of fine fibrils of collagen by SHG shows that these fibrils generally lie alongside activated HSC. This observation is consistent with the notion that activated HSC in chronic liver disease are a net producer of fibrillar collagen. Thus, SHG has potential as a novel, rapid, high-resolution method of assessing fibrosis in patient biopsies.

2.3 CD26, CD3, CXCR4, CXCL12 and synaptophysin

In cirrhotic liver many nerve fibres and some myofibroblasts and HSC stained for synaptophysin. Using confocal microscopy, few cells were clearly double positive for synaptophysin and FAP. FAP closely co-localised with the ECM components fibronectin and collagen. FAP immunopositivity extended further periseptally towards the centre of cirrhotic nodules than did GFAP immunopositivity. Myofibroblasts were nearly all SMA+FAP+ and nearly all GFAP+FAP+. In 9 of 16 patients some myofibroblasts co-stained for both FAP and CD26. CD26 antibodies stained the bile canaliculus, bile ducts, most of the CD3+ lymphocytes and sometimes myofibroblasts. Clusters of CD3+CD26+ lymphocytes often lay near periportal areas of FAP+ HSC. Many lymphocytes were CD26+CXCR4+, as is the case in blood[12]. In addition, some CD26+CXCL12+ and some CXCR4+CXCL12+ cells were observed.

3. DISCUSSION

The HSC has a central effector role in the pathogenesis of liver fibrosis and cirrhosis. Recent discoveries of the expression by HSC of neuronal and glial cell markers are intriguing and suggest a possible neural crest origin of HSC. FAP is one such marker, being found on glial and other cell lines. FAP expression coincides with tissue remodelling. These properties suggest a functional role for FAP in the pathogenesis of liver disease. The present study strengthens this argument by showing colocalisation of FAP^{+} cells with type I collagen fibres and a strong correlation between the severity of fibrosis and the extent of FAP expression in hepatitis C. In addition, in the portal-periportal region FAP expression correlated with necroinflammatory activity. In contrast to FAP, the correlation of GFAP expression with fibrosis severity was weak and there was no correlation between fibrosis severity and SMA.

There is evidence of a distinct cell population associated with fibrosis at the tissue-remodelling interface. Collagen mRNA *in situ* studies in patients with primary biliary cirrhosis demonstrate that most collagen mRNA production occurs in cells having a similar portal/periportal location as the FAP positive cells we observed[13] and this is confirmed by the collagen localisation by SHG shown here. Other studies of HSC phenotypes in experimental and human chronic liver diseases also suggest a distinct HSC phenotype in the vicinity of the developing fibrous septa. Nestin, N-CAM, BDNF (brain-derived neurotropic factor), neurotropin 4 and nerve growth factor are expressed by subpopulations of HSC in this region[14].

FAP positive cells were topographically located near the regions of portal/periportal necroinflammatory activity and these two parameters correlated, suggesting that stimulators of necroinflammatory change might also induce FAP expression by HSC or that there is two-way cross-talk between leukocytes and HSC[15].

The role of FAP expression on HSC is currently unknown. The gelatinase activity may contribute to the damage-repair cycle that characterises ongoing fibrosis, by degrading normal basement membrane / ECM in the sinusoidal space, resulting in further HSC activation. Alternatively, FAP may be upregulated in response to the deposited collagen for the purpose of clearance of the fibrotic scar. This question requires further investigation using the FAP deficient mouse and specific enzyme inhibitors.

The second harmonic signal is only propagated forward, and hence can only be detected in a transmission detector. This provides a simple way of distinguishing it from single and two-photon excited fluorescence. Its very narrow spectral width means that it can also be separated from fluorescence by a suitable narrow-band filter. It can thus be detected quite independently of the signals from multiple fluorescent labels, and is therefore a very powerful tool in multispectral imaging. There are different ways in which SH and TPF signals can be separated. The SH signal is generated over a very wide spectral range - with our instrument we have excited it from 925nm up to 780 nm - shorter wavelengths take the signal beyond the 380nm cut-off of our current detector. This means that the wavelength can be chosen to meet the needs of TPF without compromise to the SH image. The relatively simple modification of adding a sensitive transmitted light detector and appropriate filtration will equip a two-photon microscope to image SHG.

ACKNOWLEDGEMENTS

We thank Colin Sheppard, Régis Gauderon and Phil Lukins for introducing us to SHG, Wolfgang Rettig and Thilo Kähne for antibodies, and support from the Australian National Health and Medical Research Council.

REFERENCES

1. Gorrell, M. D., Gysbers, V. and McCaughan, G. W., 2001, CD26: A multifunctional integral membrane and secreted protein of activated lymphocytes. *Scand. J. Immunol.* **54:** 249-64.
2. Abbott, C. A. and Gorrell, M. D., 2002, The family of CD26/DPIV and related ectopeptidases. In *Ectopeptidases: CD13/Aminopeptidase N and*

CD26/Dipeptidylpeptidase IV in Medicine and Biology (J. Langner and S. Ansorge ed.), Vol. ISBN 0-306-46788-7 Kluwer/Plenum, NY, p. 171-95.

3. Abbott, C. A., McCaughan, G. W. and Gorrell, M. D., 1999, Two highly conserved glutamic acid residues in the predicted beta propeller domain of dipeptidyl peptidase IV are required for its enzyme activity. *FEBS Lett.* **458:** 278-84.
4. McCaughan, G. W., Gorrell, M. D., Bishop, G. A., Abbott, C. A., Shackel, N. A., McGuinness, P. H., Levy, M. T., Sharland, A. F., Bowen, D. G., Yu, D., Slaitini, L., Church, W. B. and Napoli, J., 2000, Molecular pathogenesis of liver disease: an approach to hepatic inflammation, cirrhosis and liver transplant tolerance. *Immunol. Rev.* **174:** 172.
5. Levy, M. T., McCaughan, G. W., Abbott, C. A., Park, J. E., Cunningham, A. M., Rettig, W. J. and Gorrell, M. D., 1999, Fibroblast activation protein: A cell surface dipeptidyl peptidase and gelatinase expressed by stellate cells at the tissue remodelling interface in human cirrhosis. *Hepatology* **29:** 1768-78.
6. Park, J. E., Lenter, M. C., Zimmermann, R. N., Garin-Chesa, P., Old, L. J. and Rettig, W. J., 1999, Fibroblast activation protein: A dual-specificity serine protease expressed in reactive human tumor stromal fibroblasts. *J. Biol. Chem.* **274:** 36505-12.
7. Rettig, W. J., Su, S. L., Fortunato, S. R., Scanlan, M. J., Raj, B. K., Garin-Chesa, P., Healey, J. H. and Old, L. J., 1994, Fibroblast activation protein: purification, epitope mapping and induction by growth factors. *Int. J. Cancer* **58:** 385-92.
8. Benyon, R. C. and Arthur, M. J. P., 2001, Extracellular matrix degradation and the role of hepatic stellate cells. Sem. *Liver Dis.* **21:** 373-84.
9. Schmitt-Graff, A., Kruger, S., Bochard, F., Gabbiani, G. and Denk, H., 1991, Modulation of alpha smooth muscle actin and desmin expression in perisinusoidal cells of normal and diseased human livers. Am. J. Pathol. 138: 1233-42
10. Gauderon, R., Lukins, P. B. and Sheppard, C. J., 2001, Optimization of second-harmonic generation microscopy. *Micron* **32:** 691-700.
11. 11. Cox, G., Kable, E., Jones, A., Fraser, I., Manconi, F. and Gorrell, M. D., 2002, Three-dimensional imaging of collagen using second harmonic generation. *J. Struct. Biol.*
12. Herrera, C., Morimoto, C., Blanco, J., Mallol, J., Arenzana, F., Lluis, C. and Franco, R., 2001, Comodulation of CXCR4 and CD26 in human lymphocytes. *J. Biol. Chem.* **276:** 19532-9.
13. 13. Goddard, C. J., Smith, A., Hoyland, J. A., Baird, P., McMahon, R. F., Freemont, A. J., Shomaf, M., Haboubi, N. Y. and Warnes, T. W., 1998, Localisation and semiquantitative assessment of hepatic procollagen mRNA in primary biliary cirrhosis. *Gut* **43:** 433-40.
14. Cassiman, D., Libbrecht, L., Desmet, V., Denef, C. and Roskams, T., 2002, Hepatic stellate cell/myofibroblast subpopulations in fibrotic human and rat livers. *J. Hepatol.* **36:** 200-9.
15. Maher, J. J., 2001, Interactions between hepatic stellate cells and the immune system. *Sem. Liver Dis.* **21:** 417-26.

Expression of CD26/Dipeptidyl Peptidase IV in Endometrial Adenocarcinoma and its Negative Correlation with Tumor Grade

HIROAKI KAJIYAMA, FUMITAKA KIKKAWA, KAZUHIKO INO, KIYOSUMI SHIBATA, and SHIGEHIKO MIZUTANI

Department of Obstetrics and Gynecology, Nagoya University Graduate School of Medicine, Tsurumai-cho 65, Showa-ku, Nagoya 466-8550, Japan.

1. INTRODUCTION

Dipeptidyl peptidase IV (DPPIV) is a 110 kDa type II membrane glycoprotein has been identified as the cluster differentiation 26 (CD26), and has multiple functions, including serine protease activity. Enzymologically, DPPIV can cleave specific peptide substrates such as substance P, growth hormone releasing factor, glucagon-like peptides and some chemokines including RANTES (regulated on activation, normal T cell expressed and secreted), and reduces cellular responses to these bioactive peptides. This ectoenzyme is widely distributed in activated T-cells, epithelial cells of the small intestine, liver, prostate, renal proximal tubules, and also in melanocytes. In addition, DPPIV is expressed in the female reproductive organs such as the placenta, ovary, and endometrium. Besides its expression in normal tissues, DPPIV expression and its roles in human tumors have been reported not only in hematologic malignancies, but also in certain solid tumors including melanoma, thyroid carcinoma, prostate carcinoma, and colon carcinoma.

DPPIV expression and function in endometrial adenocarcinoma have not yet been studied. The purpose of the present study was to investigate DPPIV expression and localization by immunohistochemical analysis in normal

Dipeptidyl Aminopeptidases in Health and Disease, Edited by Hildebrandt et al.
Kluwer Academic/Plenum Publishers, New York, 2003

endometrium and different grades of endometrial adenocarcinomas, and also to determine whether its expression pattern is related to neoplastic transformation, differentiation, and disease progression. Furthermore, we also examined immunohistochemical expression of the chemokine RANTES, which is one of the DPPIV substrates, in the same tissue materials, and discuss the possible role of DPPIV in differentiation and progression of endometrial carcinoma through the regulation of this chemokine.

2. DPPIV EXPRESSION IN NORMAL ENDOMETRIUM AND ENDOMETRIAL CARCINOMA

Initially, to evaluate the immunoreactivity of TS-145 anti DPPIV monoclonal mAb kindly provided by Dr. Ryuzo Ueda, Nagoya City University on paraffin embedded tissue sections, we first immunostained the normal renal proximal tubular tissue as a positive control. Strong staining intensity was observed in the renal proximal tubular cells (data not shown). DPPIV immunoreactivity was only localized in endometrial glandular cells, but not detected in stromal cells. DPPIV expression was weakly or moderately detected in the proliferative phase of normal endometrium, while in the secretory phase of normal endometrium, its expression was strongly detected in glandular cells. These findings are well consistent with the previous report.

In endometrial adenocarcinoma, DPPIV was also localized in adenocarcinoma cells, but not in stromal cells. In grade 1, DPPIV immunoreactivity was strongly or moderately detected. However, weak or no expression of DPPIV was found in grades 2 and 3. There was an inverse correlation between DPPIV immunoreactivity and tumor grading, and comparison of the five groups showed a significant difference (Fig. 1, $p < 0.001$). However, there was no correlation with the clinical stage of endometrial adenocarcinoma (data not shown).

3. RANTES EXPRESSION IN NORMAL ENDOMETRIUM AND ENDOMETRIAL CARCINOMA

We next examined immunohistochemical staining for RANTES in normal endometrium and endometrial adenocarcinomas.

RANTES immunoreactivity was localized in the glandular cells of normal endometrium, and was strongly or moderately detected in the secretory phase of normal endometrium. Most endometrial adenocarcinoma also highly or moderately expressed RANTES. Similar moderate expression of RANTES was found in breast carcinoma tissues as a positive control (data not shown). However, there was no significant correlation between RANTES expression and tumor grade.

We also confirmed that RANTES induced cell proliferation in both HEC1A and ISHIKAWA cells in a concentration-dependent manner (data not shown).

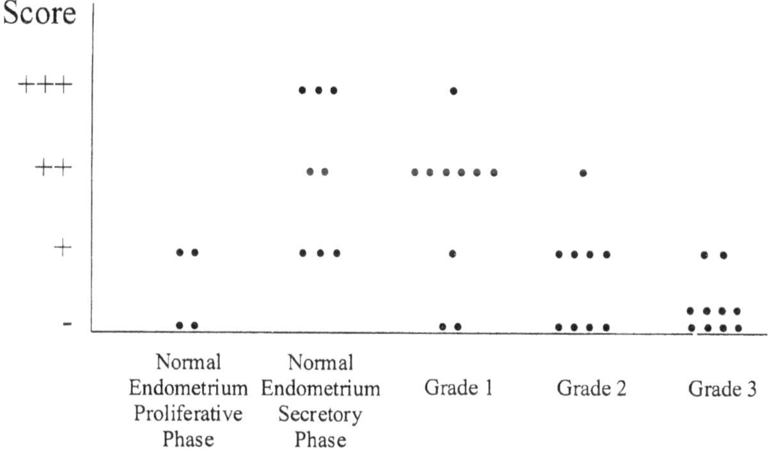

Figure 1. Staining intensity score of DPPIV immunoreactivity in the glandular cells of normal endometrium and grades 1, 2, and 3 endometrial adenocarcinoma. The score decreased with advancing tumor grade.

4. CONCLUSION

In this study, the intensity of DPPIV staining in endometrial adenocarcinoma was inversely correlated with the degree of tumor differentiation and was downregulated with advancing tumor grade. Among these various DPPIV substrates, several chemokines were shown to be involved in tumor cell proliferation and progression. Recently, Luboshits et al. have reported that the expression of RANTES is directly correlated with the advance of breast carcinoma, indicating that RANTES may be involved in breast carcinoma progression[1]. Hornung et al. previously showed RANTES expression in human endometrial and endometriosis tissues[2]. To our knowledge, our present study is the first report showing the expression of RANTES in endometrial carcinomas. Furthermore, RANTES stimulated the proliferation of HEC1A endometrial carcinoma cells in vitro. It is still unclear whether these chemokines could be involved in progression of endometrial carcinomas in vivo. However, one can speculate that loss of DPPIV in advanced endometrial carcinoma may result in the inability to degrade peptide growth factors such as RANTES. These factors may be secreted from carcinoma or stromal cells as an autocrine/paracrine growth factor, and may contribute to tumor progression.

In conclusion, this report has demonstrated the expression and localization of DPPIV in normal endometrium and endometrial adenocarcinoma. Since DPPIV expression decreased with the advancing tumor grade, reducing this enzyme may have a beneficial effect on carcinoma progression due to the loss of degrading activity of bioactive factors such as RANTES in endometrial adenocarcinoma.

REFERENCES

1. Luboshits G, Shina S, Kaplan O, Engelberg S, Nass D, Lifshitz-Mercer B, Chaitchik S., 1999, Elevated expression of the CC chemokine regulated on activation, normal T cell expressed and secreted (RANTES) in advanced breast carcinoma. *Cancer Res* **59**:4681-7.
2. Hornung D, Ryan IP, Chao VA, Vigne JL, Schriock ED, Taylor RN., 1997, Immunolocalization and regulation of the chemokine RANTES in human endometrial and endometriosis tissues and cells. *J Clin Endocrinol Metab* **82**:1621-8.

Adhesion Potency to Mesothelial Cells by Overexpression of Dipeptidyl Peptidase IV

FUMITAKA KIKKAWA, HIROAKI KAJIYAMA, KAZUHIKO INO, KIYOSUMI SHIBATA, and SHIGEHIKO MIZUTANI

Department of Obstetrics and Gynecology, Nagoya University Graduate School of Medicine, Tsurumai-cho 65, Showa-ku, Nagoya 466-8550, Japan.

1. INTRODUCTION

Dipeptidyl peptidase IV (DPPIV) is a 110 kD membrane bound peptidase on the cell surface of a wide variety of cell types and plays an important role through enzyme activity. The substrates of DPPIV are proline-containing peptides, such as certain growth factors, neuropeptides, vasoactive peptides, and chemokines. Through the digestion of these bioactive peptides, DPPIV should regulate many cellular functions. Recently, DPPIV has been reported to be expressed in tumor cells, and involved in tumor progression and invasion. However, there is no study of DPPIV in ovarian carcinoma, although this tumor is one of the most lethal malignancies of the female genital tract. Ovarian carcinoma cells have already disseminated to the peritoneum at diagnosis in most patients, and this dissemination propensity contributes in part to the difficulty in complete surgical resection. To increase survival and overcome expansion and dissemination of ovarian carcinoma cells, it is quite important to clarify the mechanism of dissemination to the mesothelium. We previously reported that human mesothelial cells expressed a high level of DPPIV compared other cell surface aminopeptidases, such as neutral endopeptidase (NEP/CD10), placental leucine aminopeptidase (P-LAP), and aminopeptidase A. In this study, we showed the presence of DPPIV in ovarian carcinoma cells for the

Dipeptidyl Aminopeptidases in Health and Disease, Edited by Hildebrandt et al.
Kluwer Academic/Plenum Publishers, New York, 2003

first time. Furthermore, we showed that the transfection of DPPIV cDNA increased the adhesion potency of carcinoma cells to mesothelial cells and this adhesion was promoted by fibronectin. Therefore, our studies partially clarified the dissemination mechanism of ovarian carcinoma to the mesothelium.

2. DPPIV EXPRESSION IN OVARIAN CARCINOMA

Initially, we performed immunohistochemical staining of DPPIV in various histologic types of ovarian carcinoma tissues and immunoreactivity of DPPIV was observed in carcinoma cells, but not in stromal cells. Next, we examined the expression levels of DPPIV by FACS in 5 ovarian carcinoma cell lines and 3 of 5 cell lines expressed high levels of endogenous DPPIV. Since SKOV3 cells expressed little endogenous DPPIV among the cell lines examined, we transfected DPPIV cDNA to these cells to define a functional role of DPPIV in ovarian carcinoma cells. Full length cDNA for DPPIV-transfected SKOV3 (SKDPIV) cells showed about 10-fold higher enzyme activity than parental SKOV3 cells, while vector-transfected SKOV3 (SKpcDNA) cells showed a level of enzyme activity similar to that of SKOV3 cells. However, the growth rate was equivalent among these 2 transfectants and SKOV3 cells.

3. ADHESION TO PLASTIC PLATES AND MESOTHELIAL CELLS

There were no clear differences in the adhesion rate on plastic plates among SKOV3, SKpcDNA, and SKDPIV cells. To investigate the effect of collagen and fibronectin on adhesion, collagen- and fibronectin-coated plates were used in adhesion experiments. The adhesion rate of SKOV3 cells on collagen- and fibronectin-coated plates increased about 1.5-fold, compared to that on uncoated plastic plates. SKpcDNA cells showed similar results as SKOV3 cells, whereas the adhesion rate of SKDPIV cells to collagen- and fibronectin-coated plates remarkably increased 2.0- and 2.5-fold compared with that to uncoated plastic plates.

Since adhesion of ovarian carcinoma cells to the peritoneum is a critical step for progression of ovarian carcinoma, we prepared plates monolayered by human mesothelial cells and performed adhesion assay (Fig. 1). The adhesion rates of SKOV3 and SKpcDNA cells were almost the same, while that of SKDPIV was significantly increased 2.0-fold compared to that of parental SKOV3 cells. Since immobilized fibronectin increased the adhesion rate of SKDPIV cells, we examined the effect of soluble fibronectin on adhesion (Fig. 1). The addition of soluble fibronectin markedly increased the adhesion rate of SKDPIV cells in a dose-dependent manner, whereas soluble fibronectin showed no effects on SKOV3 and SKpcDNA cells on adhesion.

To confirm the role of DPPIV on adhesion, we transfected antisense cDNA for DPPIV into NOS4 cells, which most expressed DPPIV by FACS and showed the highest enzyme activity. We selected 2 individual stable clones, and remarkable suppression of DPPIV expression was confirmed by FACS. The adhesion potency to mesothelial cells was also significantly decreased in these 2 transformants to 77% and 74% of that in NOS4 cells, respectively.

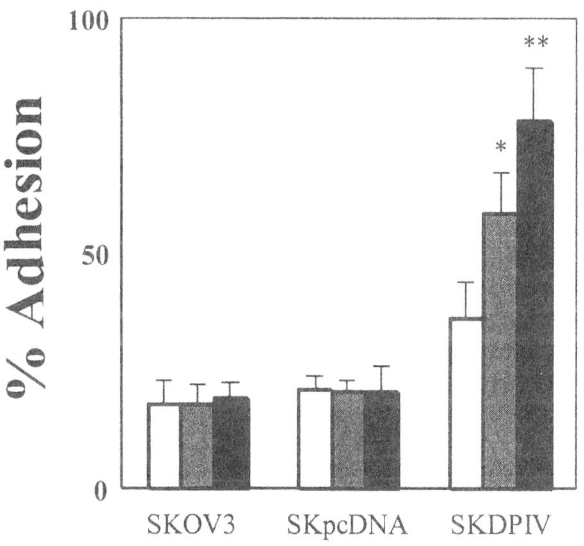

Figure 1. Adhesion of SKOV3, SKpcDNA, and SKDPIV cells to monolayered mesothelial cells with or without fibronectin. Open columns, without; shaded columns, 1 μg/ml, closed columns, 10 μg/ml fibronectin. Adhesion rate was expressed as the ratio of the count of adhesion cells to that of 2,000 labeled cells. Columns, mean; bars, SD; *, $p<0.05$ versus SKOV3 cells, **, $p<0.0005$ versus SKOV3 cells.

4. CONCLUSION

This is the first report that ovarian carcinoma cells express DPPIV and this enzyme is in part involved in adhesion of ovarian carcinoma cells to mesothelial cells.

We previously showed that human peritoneal mesothelial cells express high levels of DPPIV among several membrane bound aminopeptidases and malignant ascites stimulates DPPIV expression. Since ovarian carcinoma cells often disseminate to the mesothelium, DPPIV of mesothelial cells might be quite important in ovarian carcinoma extension and dissemination. Adhesion potency of SKDPIV on uncoated plastic plates was the almost same as those of SKOV3 and SKpcDNA cells, while the adhesion rate of SKDPIV cells significantly increased when collagen- and fibronectin-coated plates were used, suggesting that collagen and fibronectin are necessary for adhesion through DPPIV. These findings are reasonable since it was previously reported that DPPIV can bind to collagen and fibronectin.

Binding immobilized fibronectin with integrins causes many stimulatory effects on carcinoma cell invasion and proliferation. We previously reported that fibronectin secreted from the peritoneum increased MMP-9 activity and expression, and, in turn, increased the invasiveness of ovarian carcinoma cells. Furthermore, this stimulatory effect of soluble fibronectin required both FAK and c-Ras. Thus, both soluble and immobilized fibronectin can stimulate invasiveness of carcinoma cells in several steps. Although the molecular mechanisms of possible DPPIV functions have not been elucidated in detail, not only the extra-cellular serine protease activity but also the interaction between DPPIV and fibronectin may give in part an account of malignant transformation of normal cells into carcinoma cells. In conclusion, overexpression of DPPIV did not affect the growth rate on uncoated plastic plates, but did increase adhesion potency through both immobilized and soluble fibronectin in ovarian carcinoma cells. Since malignant ascites contains a concentration of soluble fibronectin, carcinoma cells floating in ascites may capture an abundance of fibronectin on cell surface DPPIV, resulting in increased adhesion to mesothelial cells.

Survival Time and Invasive Activity due to Dipeptidyl Peptidase IV Overexpression in Ovarian Carcinoma

SHIGEHIKO MIZUTANI, HIROAKI KAJIYAMA, , TAKAHIRO SUZUKI, KIYOSUMI SHIBATA, KAZUHIKO INO, and FUMITAKA KIKKAWA.

Department of Obstetrics and Gynecology, Nagoya University Graduate School of Medicine, Tsurumai-cho 65, Showa-ku, Nagoya 466-8550, Japan.

1. INTRODUCTION

Peritoneal dissemination is the main metastatic process of ovarian carcinoma besides direct extension of the carcinoma into adjacent tissues and lymphatic dissemination: Peritoneal dissemination originates from carcinoma cells released in the ascites from the ovary. Once the carcinoma cells attach to mesothelial cells, these cells may invade into mesothelial cell layer. Since extension to the peritoneum is found in over 75 percent of all cases, it is necessary to understand the mechanisms of spread by exfoliative cells that disseminate and implant throughout the peritoneal cavity. However, the biology of this dissemination is still unknown.

Dipeptidyl peptidase IV (DPPIV) is a cell surface aminopeptidase which was originally characterized as a T-cell differentiation antigen (CD26) , and has been reported to be present on epithelial cells of various tissues, including lung, liver, kidney, intestine, prostate, and placenta . It has been reported that DPPIV has a variety of functions not only serine protease activity which liberates N-terminal X-proline from peptides but also various cellular processes such as regulation of immune response, signal transduction and interaction with molecules of the extracellular matrix. A number of recent studies have provided evidence to indicate that DPPIV may play a role in tumor progression such as cell adhesion and invasion. Previous studies have also reported that DPPIV expression in melanoma cells has a

Dipeptidyl Aminopeptidases in Health and Disease, Edited by Hildebrandt et al.
Kluwer Academic/Plenum Publishers, New York, 2003

suppressing effect on malignant phenotype, and particularly, has anti-invasive function, which is related to neither protease activity locating at the extracelullar domain nor the cytoplasmic domain possibly linked to signal transduction. However, there is no evidence to indicate the association between DPPIV expression in ovarian carcinoma and its progression. In the present study, we tried the DPPIV expression in various ovarian carcinoma cell lines, and examined its effect on the progression of ovarian carcinoma in vivo. Our present data showed that DPPIV acts as a suppressor of ovarian carcinoma in peritoneal dissemination in vitro and in vivo.

2. CORRELATION AMONG DPPIV EXPRESSION, INVASIVE POTENTIAL AND CELLULAR MORPHOLOGY

The mean fluorescence intensity for DPPIV in these cell lines was shown in Table 1. NOS2, NOS4 and TAOV cells were positive for DPPIV, while SKOV3 and HRA cells were almost negative, which is consistent with data on enzyme activity analysis. Invasion assay was performed in these cell lines to compare the invasive potential and DPPIV expression. Table 1 also shows the correlation among mean fluorescence intensity of DPPIV in FACS, invasive potential and cellular morphology in these ovarian carcinoma cell lines. Furthermore, the morphology in DPPIV expressing cell lines tended to show an epithelioid pattern in contrast to that in non-DPPIV expressing cell lines, which show a long spindle/bipolar pattern like fibroblasts.

Table 1. Correlation among DPPIV expression, invasive ability and morphological characteristics in various ovarian cancer cell lines

Cell line	DPPIV Expression Mean Fluorescence Intensity[a]	Invaded cell counts[b]	Morphology
SKOV3	12	125±23	Spindle/bipolar/scattered
HRA	8	152±21	Spindle/bipolar/scattered
NOS2	684	11±4	Epitheloid/cobblestone
NOS4	421	17±6	Epitheloid/cobblestone
TAOV	293	22±8	Epitheloid/cobblestone

[a]*Mean fluorescence intensity by FACS analysis* [b]*By Matrigel invasion assay*

3. MORPHOLOGICAL AND FUNCTIONAL CHANGES BY DPPIV TRANSFECTION

To investigate the effect of the DPPIV transfection in carcinoma cells, we overexpressed DPPIV into ovarian carcinoma SKOV3 cells (SKDPIV cells). While both parental SKOV3 cells and vector-transfected SKpcDNA cells expressed little DPPIV on the cell surface, SKDPIV cells expressed a remarkably high level of DPPIV on FACS analysis. To confirm the enzyme activity of DPPIV protein expressed by the transfection, we also checked aminopeptidase activity. The DPPIV activity of SKDPIV cells was approximately 10 times higher than that of both parental SKOV3 cells and mock SKpcDNA cells in absorbance. The enzyme activity was almost completely inhibited by adding either one mM of DFP and 100 μM of diprotin A.

Furthermore, a remarkable morphological change was observed in SKDPIV cells. While both parental SKOV3 cells and SKpcDNA cells were a long-bipolar spindle-shaped morphology like fibroblasts with a scattered and unorganized growth pattern, the shape of SKDPIV cells was round with a cobblestone-like appearance. In the case of SKDPIV cells, cell-cell adhesion seemed to be tighter than that in SKOV3 or SKpcDNA cells.

We assessed the effect of DPPIV on cell proliferation, migration and invasion. No significant differences among SKOV3, SKpcDNA, and SKDPIV observed in the proliferation assay. The number of cells that had migrated on transwell cell culture assay was significantly ($p < 0.0001$) reduced in SKDPIV cells compared with that in either parental SKOV3 or SKpcDNA cells. This reduction could not be restored by the addition of either 100 μM DPA or one mM DFP. The number of invading cells on Matrigel invasion assay was also reduced in SKDPIV cells compared with that in parental SKOV3 and SKpcDNA cells ($p < 0.0001$). This reduction could not be restored by the addition of DPPIV specific inhibitors.

3.1 FUNCTIONAL ANALYSIS *IN VIVO*

We also investigated whether DPPIV suppresses the formation of peritoneal metastasis in ovarian carcinoma using nude mice. Peritonitis carcinomatosa was observed approximately 4 weeks after the inoculation of SKOV3 or SKpcDNA cells into mice. Figure 3A shows the intraabdominal appearance of the mouse 30 days after inoculation of SKpcDNA cells. A number of disseminated tumors were observed throughout the whole peritoneal cavity, especially on the omentum, mesentery and liver surface with a large amount of bloody ascites. A similar appearance was also observed in a mouse injected with parental SKOV3 cells. In contrast, a mouse injected with SKDPIV cells macroscopically had no disseminated tumor with a small amount of ascites at the same time of autopsy . In addition, the number of disseminated tumors at death in the group of mice injected with SKDPIV cells was obviously much less than the numbers in the other two groups. Only several disseminated tumors in this group of mice were observed even 60 days after the inoculation of SKDPIV cells. Figure 1 shows the survival curves among these three groups. All mice finally died of peritonitis carcinomatosa. However, the mice injected with SKDPIV cells survived significantly longer than those injected either SKOV3 cells or SKpcDNA cells {Mean survival days: 64.9±4.7 days (SKDPIV), 35.7±2.8 days (SKOV3), 36.6±1.8 days (SKpcDNA), respectively, $p < 0.0001$}

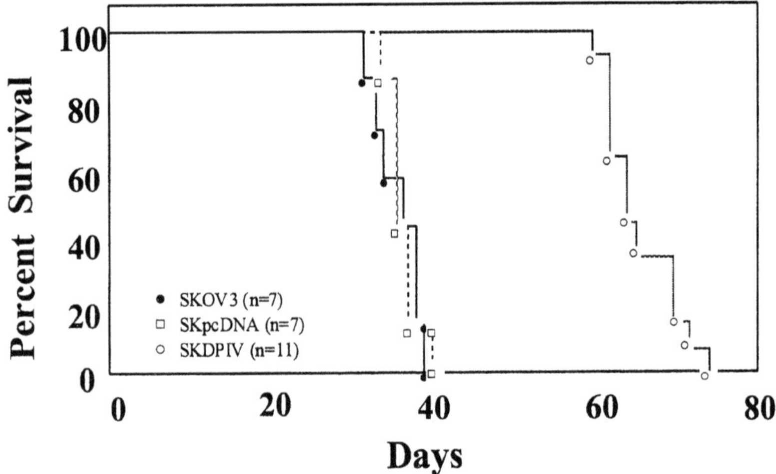

Figure 1. Effect of DPPIV on nude mice survival. The survival curves among three groups of mice treated with SKOV3, SKpcDNA and SKDPIV cells. Mice (n = 7-11) were followed for survival after inoculation with 1 x 10^7 cells.

4. CONCLUSION

the present in vitro and vivo data provide a possible link between DPPIV and decreased intraperitoneal dissemination of ovarian carcinoma due to both invasion and migration of ovarian carcinoma cells. The potential of DPPIV for treating intraperitoneal metastatic carcinoma including gene therapy seems promising, although extensive work is required before this enzyme can become available for clinical use.

Dipeptidylpeptidase IV Activities in Prostatic Secretions

MICHAEL J. WILSON[1,2,5], ROSS HALLER, JOEL W. SLATON[2,5], NEIL F. WASSERMAN[3,5], and AKHOURI A. SINHA[4,5]
VA Medical Center and Departments of Laboratory Medicine and Pathology[1], Urologic Surgery[2], Radiology[3], and Genetics, Cell Biology, & Development[4], and University of Minnesota Cancer Center[5], University of Minnesota, Minneapolis, MN, USA

1. INTRODUCTION

Dipeptidylpeptidase IV (DPP IV) is a serine type exopeptidase widely distributed in mammalian tissues that cleaves N-terminal dipeptides from polypeptides with a proline, and to a more limited extent, alanine, at the penultimate position[1,2]. This peptidase has been implicated in lymphocyte activation[3-5],binding of cells to extracellular matrix proteins[6,7] and in the metabolism of cytokines, bioactive peptides and growth factors[2].

2. DPP IV IN PROSTATE SECRETION

DPP IV activity in the seminal plasma of men[8,9] is derived nearly exclusively from the prostate gland[10] where it is concentrated in the membrane fraction or prostasomes[9,10]. It is present as a single band of 160

Dipeptidyl Aminopeptidases in Health and Disease, Edited by Hildebrandt et al.
Kluwer Academic/Plenum Publishers, New York, 2003

kDa upon histochemical detection of its activity after SDS-polyacrylamide gel electrophoresis and as 110 kDa band upon immunoblotting of heated and reduced samples, indicating the active enzyme may exist as a dimer in prostatic secretion and seminal plasma[10]. There was about 15-20% variation in DPP IV activities in repeat seminal plasma samples of vasectomized men, indicating that the level of secreted DPP IV remains relatively constant in an individual, even though there is considerable variation between individuals[10].

3. DPP IV ACTIVITIES IN PROSTATE CANCER

DPP IV activity is localized by histochemistry[10] and immunohistochemistry to the supranuclear cytoplasm of normal and benign hyperplastic prostate (BPH) cells[11,12], but is more diffuse in cancer cells[11,13]. The activity of DPP IV is elevated in prostate cancer tissues and in benign hyperplastic glands associated with cancers[13], whereas, metastases of prostate cancers and prostate cancer cell lines PC-3 and LNCaP demonstrate decreased to no DPP IV localization[11,12].

Table 1. Activities of DPP IV in extracts of prostatic cancer (CAP) and benign prostatic hyperplastic (BPH) tissues. The presence of cancer (70% of tissue piece) and BPH were verified by histology of a frozen section from each tissue piece used[13].

	N	U/mg Protein
BPH	8	167 ±23[a]
CAP	7	355 ±63

a = SEM

Table 2. The relationship of secretory Dipeptidylpeptidase IV activities and serum prostate specific antigen (PSA) and volume of the peripheral (peri) and transition (trans) zones of the prostate in men with prostate cancer (CAP) and no evidence of malignancy (NEM).

	N	Volume			Secretion
		Peri Zone	Trans Zone	PSA (ng/dl)	DPP IV Act (U/ml)
NEM	24	20.3 ±2.3a	43.5 ±6.3	7.8 ±4.4	22.2 ±2.5
CAP	26	20.9 ±1.4	21.4 ±4.4b	15.4 ±8.5	28.3 ±2.7

a = SEM b <.05

The volume of the transition zone in men with no evidence of malignancy was significantly greater than that in men with cancer, and the volume of the transition ($p<.00009$) and peripheral zones ($p<.03$) correlated to the serum PSA levels. As reported by Lepor et al.[14], the volume of the transition zone, site of origin of BPH, is associated with the rise in serum PSA seen in men with BPH. There was a negative correlation of the volume of the transition zone and the activities of DPP IV in expressed prostatic secretions. This indicates that the peripheral zone, zone of origin of cancer, may be the greater source of DPP IV in prostatic secretion. Although there was a trend for higher DPP IV activities in secretions from men with prostate cancer, this was not a statistically significant association.

4. CONCLUSIONS

Evaluation of DPP IV activities in expressed prostatic secretions indicates that the peripheral zone of the human prostate is a predominant source of DPP IV in prostatic secretions. Although, DPP IV activities are higher in prostate cancer tissues, there was not a statistically significant association of DPP IV activities in prostatic secretions and cancer as determined by biopsy.

ACKNOWLEDGEMENTS

The work from our laboratory was supported by funds of the Medical School Grant Program of Merck & Co., NIH program grant 5P30-CA77598 to the University of Minnesota Cancer Center, and the Department of Veterans Affairs.

REFERENCES

1. Yaron, A. and Naider, F., 1993, Proline-dependent structural and biological properties of peptides and proteins. *Crit. Rev. Biochem. Molec. Biol.* **28**:31-81.

2. De Meester, I., Durinx, C., Bal, G., Proost, P., Struyf, S., Goossen, F., Augustyns, K., and Scharpe, S., 2000, Natural substrates of dipeptidylpeptidase IV. *Adv. Exp. Biol. Med.* **477**:67-87.
3. Morimoto, C., Torimoto, Y., Levinson, G., Rudd, C.E., Schreiber, M., Dang, N.H., Letvin, N., and Schlossman, S.F., 1989, IF7, a novel cell surface molecule involved in helper function of CD4 cells. *J. Immunol.* **143**:3420-3439.
4. Hegen, M., Niedobitek, G., Klein, C.E., Stein, H., and Fleischer, B., 1990, The T cell triggering molecule Tp103 is associated with dipeptidyl aminopetidase IV activity. *J. Immunol.* **144**:2908-2914.
5. Schön, E., Ansorge, S., 1990, Dipeptidyl peptidase IV in the immune system. Cytofluorometric evidence for induction of the enzyme on activated T lymphocytes. Biol Chem Hoppe Seyler **371(8)**:699-705.
6. Hanski, C., Huhle, T., and Reutter, W., 1985, Involvement of plasma membrane dipeptidyl peptidase IV in fibronectin mediated adhesion of cells in collagen. *Biol. Chem. Hoppe-Seyler* **366:**1169-1176.
7. Löster, K., Zeilinger, K., Schuppan, D., and Reutter, W., 1995, The cysteine-rich region of dipeptidyl peptidase IV (CD26) is the collagen-binding site. *Biochem. Biophys. Res. Commun.* **217**:341-348.
8. Kullertz, K., Nagy, M., Fishcer, G., and Barth, A., 1986, Isolierung und Characterisierung der Dipeptidylpeptidase IV aus humanem Seminalplasma. *Biomed. Biochim. Med.* **45**:291-303.
9. Vanhoof, G., De Meester, I., van Sande, M., Scharpe, S., Yaron, A., 1992, Distribution of proline-specific aminopeptidases in human tissues and body fluids. *Eur J Clin Chem Clin Biochem.* **30(6)**:333-8.
10. Wilson, M.J., Ruhland, A.R., Pryor,J.L., Ercole, C.E., Sinha, A.A., Hensleigh, H., Kaye, K.W., Dawkins, H.J.S., Wasserman, N.F., Reddy, P., and Ahmed, K., 1998, Prostate specific origin of dipeptidylpeptidase IV (CD26) in human seminal plasma. *J. Urol.* **160:**1905-1909.
11. Dinjens, W.N.M., Kate, J.T., Kirch, J.A.J.M., Tanke, H.J., Van Der Linden, E.P.M., Van Den Inhg, H.F.G.M., Van Steenbrugge, G.J., Khan, P.M., and Bosman, F.T., 1990, Adenosine deaminase complexing protein (ADCP) expression and metastatic potential in prostatic adenocarcinomas. *J. Pathol.* **160:**195-201.
12. Bodenrieder, T., Finstad, C.L., Freeman, R.H., Papandreou, C.N., Scher, H.I., Albino, A.P., Reuter, V.E., and Nanus, D.M., 1997, Expression and localization of aminopeptidase A, aminopeptidase N, and dipeptidyl peptidase IV in benign and malilgnant human prostate tissue. *Prostate* **33:**225-232.
13. Wilson, M.J., Ruhland, A.R., Quast, B.J., Reddy, P.K., Ewing, S.L., and Sinha, A.A., 2000, Dipeptidylpeptidase IV activities are elevated in prostate cancers and adjacent benign hyperplastic glands. *J. Androl.* **21**:220-226
14. Lepor, H., Wang, B., and Shapiro, E., 1994, Relationships between prostatic epithelial volume and serum prostatic-specific antigen levels. *Urology* **44:** 199-205.

V

DIABETES AND METABOLISM

Implementation of GLP-1 Based Therapy of Type 2 Diabetes Mellitus Using DPP-IV Inhibitors

JENS JUUL HOLST
Department of Medical Physiology, University of Copenhagen, The Panum Institute, DK-2200 Copenhagen N. Tel. (45) 3552 7518. Fax: (45) 3532 7537. E-mail: holst@mfi.ku.dk

SUMMARY

GLP-1 is a peptide hormone from the intestinal mucosa. It is secreted in response to meal ingestion and normally functions in the so-called ileal brake i. e. inhibition of upper gastrointestinal motility and secretion when nutrients are present in the distal small intestine. It also induces satiety and promotes tissue deposition of ingested glucose by stimulating insulin secretion. Thus, it is an essential incretin hormone. In addition, the hormone has been demonstrated to promote insulin biosynthesis and insulin gene expression and to have trophic effects on the beta cells. The trophic effects include proliferation of existing beta cells, maturation of new cells from duct progenitor cells and inhibition of apoptosis. Furthermore glucagon secretion is inhibited. Because of these effects, the hormone effectively improves metabolism in patients with type 2 diabetes mellitus. However, continuous administration of the peptide is necessary because of an exceptionally rapid rate of degradation catalyzed the enzyme dipeptidyl peptidase IV. With inhibitors of this enzyme, it is possible to protect the endogenous hormone and thereby elevate both fasting and postprandial levels of the active hormone. This leads to enhanced insulin secretion and glucose turnover. But will DPP-IV inhibition enhance all effects of the endogenous peptide? The mode of action of GLP-1 is complex involving also interactions with sensory

Dipeptidyl Aminopeptidases in Health and Disease, Edited by Hildebrandt et al.
Kluwer Academic/Plenum Publishers, New York, 2003

neurons and the central nervous system, where a DPP-IV mediated degradation does not seem to occur. Therefore, it is as yet uncertain wether DDP-IV inhibitors will affect gastrointestinal motility, appetite and food intake. Even the effects of GLP-1 effects on the pancreatic islets may be partly neurally mediated and therefore uninfluenced by DPP-IV inhibition.

1. WHAT IS GLP-1(GLUCAGON-LIKE PEPTIDE-1)?

GLP-1 is a product of the glucagon gene[1]. This gene is expressed not only in the pancreatic alpha-cells, but also in the L-cells of the intestinal mucosa, one of the most abundant endocrine cells of the gut[2]. Here, the primary translation product, proglucagon, is cleaved, not to produce glucagon as in the islets, but to release from its C-terminal part the two glucagon-like peptides GLP-1 and GLP-2[3], which show about 50 % sequence homology with glucagon. In spite of the structural homology, GLP-2 shares few of the biological actions of GLP-1, but rather acts as a regulator of adaptive growth in the gut[4;5]. GLP-1 secretion is stimulated by the presence of nutrients in the lumen of the gut (but additional neural or endocrine mechanisms may also operate)[6], and the secretion of GLP-1 throughout the day is highly correlated to the release of insulin[7]. This is particularly evident in patients with reactive hyperglycemia after gastric operations and accelerated gastric emptying of meals. These patients have grossly exaggerated GLP-1 responses to test meals, which correlate with the gastric emptying rate as well as with postprandial insulin levels[8]. Indeed, it has been shown that a pronounced reactive hyperglycaemia can be provoked in healthy subjects, if the meal-induced glucose excursions and exaggerated GLP-1 responses of the gastrectomized patients are mimicked by intravenous infusions of glucose and GLP-1[9]. In healthy subjects, GLP-1 levels are elevated 2-4 fold in response to ingestion of each meal[7], but also in the interdigestive state there appears to be a certain basal secretion. Thus, fasting levels of GLP-1 may be significantly suppressed during infusions of high doses of somatostatin[10], a peptide shown to exert a pronounced constraining effect on GLP-1 secretion[11].

The biological effects of GLP-1 comprise not only an effect on insulin secreting cells, but also on other cells of the islets as well as effects on several extrapancreatic sites.

Insulinotropic effects

Firstly, GLP-1 is one of the most potent insulinotropic substances known, with half maximal effective concentrations for its effects on the beta cells around 10 pmol/l[12]. Its insulinotropic effect is strictly glucose dependent and there is no effect on insulin secretion at glucose concentrations below approximately 4.5 mmol/l[13]. GLP-1's insulinotropic activity is exerted via interaction with a specific receptor located on the cell membrane of the ß-cells[14;15]. Binding of GLP-1 to the receptor causes activation - via a stimulatory G-protein - of adenylate cyclase resulting in the formation of cAMP[16;17]. There is agreement that all of the actions of GLP-1 are secondary to the formation of cAMP. Subsequent activation of protein kinase A leads to a plethora of events including altered ion channel activity, intracellular calcium^{2+} handling and enhanced exocytosis of insulin containing granules[17]. As mentioned a certain level of glucose must be present for GLP-1 to have any effect on insulin secretion. In addition, GLP-1 potentiates strongly the insulinotropic actions of glucose itself. Conversely, it seems that GLP-1 (or perhaps any hormone that can causes sufficient cAMP accumulation in the ß-cells?) is required for glucose to exert its activity. Thus, in experiments in single beta cells, neither glucose nor GLP-1 alone affect intracellular calcium levels or membrane potential, whereas together they bring about a strong activation[18;19]. In other words, GLP-1 conveys "glucose competence" to the ß-cells[18]. The clinical implication of the dependence on blood glucose concentrations at or above normal fasting glucose levels is, of course, that GLP-1 is incapable of causing profound hypoglycaemia[20]. The reactive hypoglycaemia alluded to above[9], is due to an extremely exaggerated release of insulin, which, due to the rather slow inactivation time of insulin, may lead to a short-lasting overshoot of insulin's hypoglycaemic effects extending beyond the stage where insulin secretion (and the GLP-1 effect) has ceased because of falling glucose concentrations.

Secondly, GLP-1 stimulates all steps of insulin biosynthesis as well as insulin gene transcription[21], thereby providing continued and augmented supplies of insulin for secretion. In addition, it upregulates the genes for the cellular machinery involved in insulin secretion, such as the glucokinase and GLUT-2 genes[22].

And finally and most importantly, GLP-1 has been shown to have trophic effects on ß-cells. Not only does it stimulate ß-cell proliferation[23;24], it also enhances the differentiation of new ß-cells from progenitor cells in the pancreatic duct epithelium[25]. A proliferation was also induced in aging glucose-intolerant rats with a resulting improvement of glucose tolerance[26]. This indicates that GLP-1 may be capable of providing new ß cells in individuals with an insufficient number of functioning cells[26] such as 2DM patients (although it is not yet established to what extent this process occurs in humans).

Most recently, but of no less interest, GLP-1 has been shown to be capable of inhibiting both cytokine and fatty acid induced apoptosis in beta cells[27;28]. Since the normal number of beta cells is maintained in a balance between apoptosis and proliferation[29], this observation is of considerable interest, and also raises the possibility that GLP-1 could be useful in conditions with beta cell apopotosis. Thus a GLP-1 analogue, NN 2211 (NovoNordisk, Denmark), has been demonstrated to markedly reduce and delay the development of diabetes in ZDF rats[30], the diabetes of which is known to occur as apoptosis reduces the number of functioning beta cells[31].

Inhibition of glucagon secretion

In addition to its effects on the ß-cells, GLP-1 also strongly inhibits glucagon secretion[32]. The mechanism by which this occurs is unclear. Both inhibition via stimulated insulin release and via stimulated somatostatin secretion have been suggested[12], but a direct effect on the glucagon cells which seem to express GLP-1 receptors[33;34] is also possible. It is of considerable clnical importance that the inhibition exhibits glucose dependency. Thus, in a recent study in healthy volunteers, glucagon secretion was studied under the conditions of a stepwise, hyperinsulinemic, hypoglycaemic clamp, with and without additional infusion of GLP-1. At basal glycaemia and at the earliest steps of the clamp, glucagon secretion was clearly suppressed during GLP-1 infusion compared to controls, but at lower glucose levels glucagon secretion was stimulated by hypoglycaemica and to identical levels in the two groups[35]. Thus it is unlikely that a GLP-1-based treatment will impede the glucagon-mediated defence against hypoglycemia in the clinical setting.

The importance of the inhibition of glucagon secretion at hyperglycaemic levels, i.e. for diabetes treatment is perhaps best illustrated in studies of GLP-1 infusion in patients with type 1 (insulin-dependent) diabetes and no residual ß-cell secretory capacity[36]. In these patients, GLP-1 retains substantial glucose lowering activity, in spite of undetectable C-peptide responses (i. e. there is no insulin effect), while glucagon secretion is strongly inhibited. We assume that the glucose-lowering effect is a consequence of the inhibition of glucagon secretion, since we have shown that the glucose lowering effect of GLP-1 is probably mainly due to inhibition of hepatic glucose production as a result of its combined action on the secretion of the two pancreatic glucoregulatory hormones[37].

Effects on the gastrointestinal tract

Further important effects of GLP-1 include inhibition of gastrointestinal secretion and motility, notably gastric emptying[38;39].

A considerable amount of recent evidence seems to indicate that, physiologically, one of the main functions of GLP-1 is to act as one of the hormones of the so-called "ileal brake", an endocrine mechanism by which the presence of nutrients in the distal small intestine causes inhibition of upper gastrointestinal motor and secretory activity[40]. Thus, if the ileum is perfused luminally with solutions containing lipids or carbohydrates in amounts corresponding to what has been called physiological malabsorption, GLP-1 secretion is markedly stimulated. At the same time gastric and pancreatic secretion is inhibited[41]. If the plasma concentrations observed in these experiments are reproduced by intravenous infusion of exogenous GLP-1, gastric motility and secretion as well as pancreatic secretion are markedly inhibited[42]. Physiologically, it may be that these effects of GLP-1 are more important than the insulinotropic activity. Thus, when GLP-1 is infused intravenously during ingestion of a meal, the insulin responses are diminished dose-dependently, rather than being enhanced[39]. At the same time gastric emptying is being progressively retarded, so that the explanation of the reduced insulin secretion is the reduced gastric emptying of and reduced subsequent absorption of insulinotropic nutrients; recall, that GLP-1 acts by potentiating nutrient-stimulated insulin secretion. Thus the physiological role of GLP-1 may be to adjust the delivery of chyme to the digestive and absorptive capacity of the gut by retarding propulsion and digestion of the gastric contents[39;42;43].

The inhibitory effect of GLP-1 on gastric emptying is desirable in patients with diabetes because the slower gastric emptying rate reduces postprandial glucose excursions. This is evident from the use of another potent gastric inhibitor, amylin, with similar effects on gastric emptying for diabetes treatment[44]. Amylin reduces postprandial glucose excursions sufficiently to lower haemoglobin A1c levels significantly in patients with diabetes mellitus.

Effects on appetite and food intake

Nutrients in the ileum are thought to have a satiating effect, curtailing food intake[45]. Does GLP-1 play a physiological role as a satiating agent? Indeed, recent studies have shown that infusions of slightly supraphysiological amounts of GLP-1 significantly enhance satiety and reduce food intake in normal subjects[46;47]. The effect on food intake and satiety is preserved in obese subjects also[48] as well as in obese subjects with type 2 diabetes[49;50].

In addition to the effects of peripherally administered GLP-1, direct injections of GLP-1 into the cerebral ventricles also inhibit food intake[51;52]. Here the peptide presumably interacts with GLP-1 receptors that are expressed at numerous locations in the brain, but in particular in the hypothalamus around the paraventricular nucleus and the arcuatus[53;54]. The cerebral GLP-1 receptors are likely to be targets for GLP-1 released from nerve fibers ascending from cell bodies in the nucleus of the solitary tract in the brain stem, in which the proglucagon gene is expressed and in which proglucagon is processed in the same way as in the gut[55]. The question arises whether these neurones are linked to meal-induced satiety. Rinaman et al[56] analysed the neurons of the nucleus of the brain stem that were activated by various procedures designed to model enteroceptive stress (lithium chloride administration, CCK injection, lipopolysaccharide) and observed C-fos expression in cell bodies that also stained for GLP-1, whereas neurons showing c-fos expression after meal ingestion were distinct from the GLP-1 neurons. In further experiments, they administered the GLP-1 receptor antagonist, exendin 9-39, intracebroventricularly to rats given lithium chloride as above, and found that the antagonist could completely reverse the anorexigenic effect of systemic lithium chloride[57]. Thus, is seems clear that GLP-1 from brain stem functions as a mediator of the anorexic effects of enteroceptive stress, whereas its role in meal-induced satiety is less clear.

The effect of GLP-1 on appetite and food intake would support attempts at weight reduction in 2DM patients and, if effective, would be considered most desirable. It should be noted though that GLP-1 receptor knock out mice do not become obese[58], but this may reflect the redundancy of the appetite regulating mechanisms rather than ineffectiveness of the signal.

1.1 Effect of GLP-1 in type 2 diabetes

Taken together all of these effects render GLP-1 unusually attractive as a therapeutic agent. The full hypoglycaemic effect of GLP-1 in type 2 diabetic patients was first demonstrated by Nauck et al[59] who were able to completely normalize plasma glucose levels by a 4-hour iv GLP-1 infusion in patients admitted to hospital for initiation of insulin treatment because of secondary failure of oral antidiabetic agents. These patients had high haemoglobin A1c levels and fasting plasma glucose levels of approximately 13 mmol/l indicating that GLP-1 would be effective even in patients with limited insulin reserve. Recently, Zander et al[60] used continuous subcutaneous infusion of GLP-1 for 6 weeks in order to explore in a subchronic design the potential of GLP-1 in diabetes treatment. After a 3 weeks wash-out period 20 patients were allocated to continuous infusion of either saline or GLP-1 at a rate of 4.8 pmol/kg/min using Minimed[R] pumps. The patients were evaluated before, after 1 week and after 6 weeks of treatment. No changes were observed in the saline treated group, whereas in the GLP-1 group fasting and average plasma glucose concentrations were lowered by approximately 5 mmol/l, haemoglobin A1c decreased by 1.2 %, free fatty acids were significantly lowered, and the patients had a significant weight loss of approximately 2 kg. In addition, insulin sensitivity as determined by a hyperinsulinaemic euglycaemic clamp almost doubled, and insulin secretion capacity (measured using a 30 mmol/l glucose clamp + arginine) greatly improved. There was no significant difference between results obtained after one and 6 weeks treatment, but there was a tendency towards further improvement of plasma glucose as well as insulin secretion. There were very few side effects and no differences between saline and GLP-1 treated patients in this respect. Thus, GLP-1 seems to live up to all the expectations as a therapeutic agent for type 2 diabetes. The burning question therefore is how its therapeutic potential can be utilised in clinical praxis.

1.2 Limitation of the effects of GLP-1 because of enzymatic degradation

It spite of the impressive effects of intravenous or subcutaneous infusions of GLP-1, it turned out that simple subcutaneous injections of GLP-1 only lead to small lowering of blood glucose and a short-lived stimulation of insulin secretion[61]. The reason is that GLP-1 is degraded extremely rapidly after subcutaneous injection. In fact, less than 10 % of the peptide survives in intact bioactive form[62]. The rapid initial degradation is due to the ubiquitously expressed enzyme, dipeptidyl-peptidase IV (DPP-IV)[63], which cleaves off the two N-terminal amino acid residues from GLP-1 whereby the molecule not only is inactivated, but actually may be turned into an antagonist at the GLP-1 receptor[64]. Physiologically this is an interesting process with important implications for the normal functions of GLP-1. Using isolated perfused preparations of porcine small intestine Hansen et al[65] were able to show that only about 1/3 of the GLP-1 that is secreted from the gut leaves the gut in the intact biologically active form; the remainder is constituted by GLP-1 9-36amide. Immunohistochemical studies showed that

DPP-IV is localised to the endothelium of the capillaries of the villi. Newly released GLP-1 therefore gets degraded upon entry into the capillaries. The implications of this is that GLP-1 must have exerted most of its actions before it enters the villous capillary, i. e. in the lamina propria. The hypothesis is that here GLP-1 interacts with sensory afferent neurons and that, therefore, most of its effects are exerted in a reflex like manner. A number of studies support this notion[66], including studies involving deafferentation[67], and studies in which impulse activity were detected in the vagal trunks as well as vagal pancreatic efferents after GLP-1 administration in rats[68] and in studies where reflex activity was inhibited by ganglionic blockers, which also abolished GLP-1 induced insulin secretion[69]. In a series of studies of gastric acid secretion in man, Wettergren et al similarly demonstrated that GLP-1 inhibited vagally stimulated secretion but was ineffective after vagotomy[70;71]. In conclusion, it seems that, physiologically, an important part of the activity of endogenous GLP-1 is exerted via interaction with sensory nerve fibres relaying in the brain and modulating efferent probably mainly vagal fibres that regulate gastrointestinal secretion and motility as well as pancreatic endocrine secretion. During large meals with large GLP-1 responses larger amounts of intact GLP-1 may reach the islets and other tissues via the circulation and contribute to stimulation of insulin secretion. In fact, it has been shown that the amount of intact GLP-1 that reaches the circulation postprandially may be particularly low in type 2 diabetic subjects[72].

2. CIRCUMVENTING THE METABOLISM OF GLP-1 USING DPP-IV INHIBITORS

The extensive and rapid degradation of GLP-1 in the body implies that the peptide cannot be immediately employed for clinical treatment of type 2 diabetes. A number of strategies have, therefore, been explored including development of small molecule agonists for the GLP-1 receptor; development of DPP-IV resistant analogues; and finally inhibition of DPP-IV, in analogy with the use of angiotensin converting enzyme inhibitors for the treatment of hypertension[63;73]. The degradation occurs in the circulation at such a high rates that steady state cannot be established. In this situation the usual parameters of elimination T½ and metabolic clearance rate have no meaning, but calculated as apparent values the half-life is around 1-1½ min and the clearance rate exceeds cardiac output by a factor of 2[74,75].

It was also demonstrated that with available inhibitors it was possible to completely protect exogenous and endogenous GLP-1 from DPP-IV mediated degradation and thereby to enhance greatly its insulinotropic activity[76]. Numerous subsequent studies have indicated that administration of orally active DPP-IV inhibitors markedly improve metabolism and glucose regulation in animal models of glucose intolerance. For example, in mice rendered glucose intolerant by high fat diets, the inhibitor valine pyrrolidide, almost doubled the plasma levels of undegraded bioactive GLP-1, augmented insulin secretion and virtually normalised the otherwise considerably impaired glucose tolerance[77]. Using another inhibitor, Pospisilik et al[78] reported sustained improvements in fasting glucose, glucose tolerance, insulin sensitivity, and ß-cell responsiveness to glucose in Zucker fatty rats (a specific Vancouver strain), and finally, in a human study over 4 weeks, very significant improvements of metabolic control resulted from two or three times daily administration of an inhibitor developed by Novartis[79]. There are many other substrates for DPP-IV than GLP-1[80], but it seems that the extreme degradation of GLP-1 makes it a preferential target, so that significant protection of GLP-1 can be accomplished without significantly compromising the other functions of DPP-IV. It should be noted that Fisher rats with mutations in the catalytic part of DPP-IV have no malfunctions[81]. Recently, mice with a targeted deletion of the gene encoding DPP-IV (CD 26) have been generated, and except for showing greater than normal glucose tolerance, these mice also seem to be normal[84].

3. GLP-1 OR DPP-IV INHIBITORS?

Although the potential of DPP-IV inhibitors in diabetes treatment is amply supported by experimental evidence, it is worthwhile to consider to what extent treatment with GLP-1 (analogues) and DPP-IV inhibitors will be similar and also whether the two treatments are applicable to identical or separate segments of the diabetic population.

The foremost question is efficacy. Will DPP-IV inhibition be as efficacious as GLP-1 analogues? The question involves consideration of the circulating levels of intact GLP-1 obtained after inhibitor administration. During continuous intravenous infusion of GLP-1 only 20 % of the hormone survives in intact form[62]. One would therefore expect an up to five-fold elevation of circulating levels of intact GLP-1 during effective inhibitor treatment. Alternatively, since it has been demonstrated that DPP-IV inhibition may completely protect the hormone from N-terminal truncation, the levels of intact GLP-1 might approach the levels of "total GLP-1" (intact + metabolite) observed without inhibition[76]. If true, it should be possible to elevate the concentrations of intact GLP-1 to 20-30 pmol/L, which is similar to concentrations obtained in diabetic patients during infusion of GLP-1 at a rate that results in complete normalisation of (fasting) hyperglycaemia[62]. However, recent investigations have shown, that the protection of GLP-1 during DPP-IV inhibition causes a down regulation of the secretion of GLP-1 (and of the other incretin hormone, GIP (glucose dependent insulinotropic polypeptide)), so that meal stimulated levels of intact GLP-1 do increase during inhibitor treatment, but not to the levels obtained without inhibitor treatment[83]. Possibly, a negative feed-back cycle involving increased paracrine somatostatin secretion from D-cells in the vicinity of the GLP-1 secreting L-cells is responsible for the lower meal-induced secretion[11]. This negative feed-back cycle obviously limits the increases in intact GLP-1 levels one can expect during inhibitor treatment, and, in turn, the insulinotropic and blood glucose lowering effects of DPP-IV inhibitors.

Another question is to what extent inhibitor treatment will result in inhibition of gastric motility, appetite and food intake. As discussed above, these effects of GLP-1 apparently involve activation of sensory afferent nerve fibres in the lamina propria of the gut wall[65], and this interaction is likely to occur *before* GLP-1 enters the capillaries and gets degraded. These actions of *endogenous* GLP-1, therefore, are unlikely to be affected by DPP-IV inhibition. In agreement with this concept, 8 weeks of extensive DPP-IV inhibition did not result in changes in body weight in glucose intolerant rats[84]. DPP-IV inhibition in Zucker Diabetic Fatty rats for 3 months resulted in slightly lower weight gain than in control animals but no change in food intake[78]. The reason for the difference is difficult to unravel in these leptin receptor deficient animals, which during treatment exhibited significantly improved glycaemic control. Similar findings were made by Sudre et al[85], who noted that no anorectic effects of the inhibitor employed could be demonstrated in fully developed ZDF rats.

As discussed above, even GLP-1 induced enhancement of insulin secretion may involve reflex pathways[69]. It cannot be excluded, therefore, that it is mainly the endocrine actions of GLP-1 (as opposed to the neurally mediated) that are enhanced by DPP-IV inbibition. Interestingly, this may also indicate that DPP-IV inhibition will not result in the nausea and vomiting that results from administration of large amounts of GLP-1, and which may involve interaction with the central nervous system[56;86].

Taken together these limitations possibly indicate that DPP-IV inhibition will be less efficacious than administration of GLP-1 analogues, a notion that finds some support from the clinical studies reported by Ahrén et al[79], where DPP-IV inhibition certainly improved glycaemic control in type 2 diabetic patients, but not to the extent observed with continuous administration of GLP-1[87;88].

The most important potential of the treatment with DPP-IV inhibitors[73] should be found in their oral availability and lack of side effects. This means that they can be offered to subjects at risk for developing diabetes, e.g. persons with IGT, with genetic disposition, obese subjects and subjects with mild diabetes. In these subjects it can be expected, because of the amplification of the GLP-1 effects, that glucose tolerance will be improved and that the beta cells may be protected; and the long term benefits may turn out to be prevention of progression of disease and prevention of complications as indicated in the studies of ZDF rats alluded to above. Thus, DPP-IV inhibitors may be particularly useful for the *treatment of prediabetes*, presently considered a necessity if an explosive development of the type 2 diabetes epidemic is to be prevented.

REFERENCES

1. Bell, G.I., Sanchez-Pescador, R., Laybourn, P.J., Najarian, R.C., 1983, Exon duplication and divergence in the human preproglucagon gene. *Nature,* **304(5924):**368-371.
2. Mojsov, S., Heinrich, G., Wilson, I.B., Ravazzola, M., Orci, L., Habener, J.F., 1986, Preproglucagon gene expression in pancreas and intestine diversifies at the level of post-translational processing. *J Biol Chem*, **261(25):**11880-11889.
3. Orskov, C., Holst, J.J., Knuhtsen, S., Baldissera, F.G., Poulsen, S.S., Nielsen, O.V., 1986, Glucagon-like peptides GLP-1 and GLP-2, predicted products of the glucagon gene, are secreted separately from pig small intestine but not pancreas. *Endocrinology*, **119(4):**1467-1475.
4. Drucker, D.J., Erlich, P., Asa, S.L., Brubaker, P.L., 1996, Induction of intestinal epithelial proliferation by glucagon-like peptide 2. *Proc Natl Acad Sci U S A*, **93(15):**7911-7916.
5. Hartmann, B., Thulesen, J., Hare, K.J., Kissow, H., Orskov, C., Poulsen, S.S., et al., 2002, Immunoneutralization of endogenous glucagon-like peptide-2 reduces adaptive intestinal growth in diabetic rats. *Regul Pept*,**105(3):**173-179.
6. Holst, J.J., 1999, Glucagon-like Peptide 1(GLP-1): an intestinal hormone signalling nutritional abundance, with an unusual therapeutic potential. *Trends Endocrinol Metab*, **10(6):**229-234.
7. Orskov, C., Wettergren, A., Holst, J.J., 1996, Secretion of the incretin hormones glucagon-like peptide-1 and gastric inhibitory polypeptide correlates with insulin secretion in normal man throughout the day. *Scand J Gastroenterol*, **31(7):**665-670.

8. Miholic, J., Orskov, C., Holst, J.J., Kotzerke, J., Meyer, H.J., 1991, Emptying of the gastric substitute, glucagon-like peptide-1 (GLP- 1), and reactive hypoglycemia after total gastrectomy. *Dig Dis Sci*, **36(10):**1361-1370.
9. Toft-Nielsen, M., Madsbad, S., Holst, J.J., 1998, Exaggerated secretion of glucagon-like peptide-1 (GLP-1) could cause reactive hypoglycaemia. *Diabetologia*, **41(10):**1180-1186.
10. Toft-Nielsen, M., Madsbad, S., Holst, J.J., 1996, The effect of glucagon-like peptide I (GLP-I) on glucose elimination in healthy subjects depends on the pancreatic glucoregulatory hormones. *Diabetes*, **45(5):**552-556.
11. Hansen, L., Hartmann, B., Bisgaard, T., Mineo, H., Jørgensen, P.N., Holst, J.J., 2000, Somatostatin restrains the secretion of glucagon-like peptide-1 and 2 from isolated perfused porcine ileum. *Am J Physiol*, **278(6):** E1010-E1018.
12. Fehmann, H.C., Goke, R., Goke, B., 1995, Cell and molecular biology of the incretin hormones glucagon- like peptide-I and glucose-dependent insulin releasing polypeptide. *Endocr Rev*, **16(3):**390-410.
13. Weir, G.C., Mojsov, S., Hendrick, G.K., Habener, J.F., 1989, Glucagonlike peptide I (7-37) actions on endocrine pancreas. Diabetes, **38(3):**338-342.
14. Thorens, B., 1992, Expression cloning of the pancreatic beta cell receptor for the gluco-incretin hormone glucagon-like peptide 1. *Proc Natl Acad Sci U S A*, **89(18):**8641-8645.
15. Thorens, B., Porret, A., Buhler, L., Deng, S.P., Morel, P., Widmann, C., 1993, Cloning and functional expression of the human islet GLP-1 receptor. Demonstration that exendin-4 is an agonist and exendin- (9-39) an antagonist of the receptor. *Diabetes*, **42(11):**1678-1682.
16. Fehmann, H.C., Goke, R., Goke, B., 1995, Cell and molecular biology of the incretin hormones glucagon- like peptide-I and glucose-dependent insulin releasing polypeptide. *Endocr Rev*, **16(3):**390-410.
17. Gromada, J., Holst, J.J., Rorsman, P., 1998, Cellular regulation of islet hormone secretion by the incretin hormone glucagon-like peptide 1. *Pflugers Arch*, **435(5):**583-594.
18. Holz, G.G., Kuhtreiber, W.M., Habener, J.F., 1993, Pancreatic beta-cells are rendered glucose-competent by the insulinotropic hormone glucagon-like peptide-1(7-37). *Nature*, **361(6410):**362-365.
19. Gromada, J., Dissing, S., Bokvist, K., Renstrom, E., Frokjaer Jensen, J., Wulff, B.S., et al., 1995, Glucagon-like peptide I increases cytoplasmic calcium in insulin- secreting beta TC3-cells by enhancement of intracellular calcium mobilization. *Diabetes*, **44(7):**767-774.
20. Qualmann, C., Nauck, M.A., Holst, J.J., Orskov, C., Creutzfeldt, W., 1995, Insulinotropic actions of intravenous glucagon-like peptide-1 (GLP-1) [7-36 amide] in the fasting state in healthy subjects. *Acta Diabetol*, **32(1):**13-16.
21. Fehmann, H.C., Habener, J.F., 1992, Insulinotropic hormone glucagon-like peptide-I(7-37) stimulation of proinsulin gene expression and proinsulin biosynthesis in insulinoma beta TC-1 cells. *Endocrinology*, **130(1):**159-166.
22. Buteau, J., Roduit, R., Susini, S., Prentki, M., 1999, Glucagon-like peptide-1 promotes DNA synthesis, activates phosphatidylinositol 3-kinase and increases transcription factor pancreatic and duodenal homeobox gene 1 (PDX-1) DNA binding activity in beta (INS-1)-cells. *Diabetologia*, **42(7):**856-864.
23. Xu, G., Stoffers, D.A., Habener, J.F., Bonner-Weir, S., 1999, Exendin-4 stimulates both beta-cell replication and neogenesis, resulting in increased beta-cell mass and improved glucose tolerance in diabetic rats. *Diabetes*, **48(12):**2270-2276.

24. Stoffers, D.A., Kieffer, T.J., Hussain, M.A., Drucker, D.J., Bonner-Weir, S., Habener, J.F., et al., 2000, Insulinotropic glucagon-like peptide 1 agonists stimulate expression of homeodomain protein IDX-1 and increase islet size in mouse pancreas. *Diabetes*, **49(5):**741-748.
25. Zhou, J., Wang, X., Pineyro, M.A., Egan, J.M., 1999, Glucagon-like peptide 1 and exendin-4 convert pancreatic AR42J cells into glucagon- and insulin-producing cells. *Diabetes*, **48(12):** 2358-2366.
26. Perfetti, R., Zhou, J., Doyle, M.E., Egan, J.M., 2000, Glucagon-like peptide-1 induces cell proliferation and pancreatic- duodenum homeobox-1 expression and increases endocrine cell mass in the pancreas of old, glucose-intolerant rats. *Endocrinology*, **141(12):**4600-4605.
27. Bregenholt, S., Moldrup, A., Blume, N., Knudsen, L.B., Petersen, J.S., 2001, The GLP-1 analogue, NN2211, inhibits free fatty acid induced apoptosis in primary rat beta cells. *Diabetologia*, **44[suppl. 1]:** A19.
28. Bregenholt, S., Moldrup, A., Knudsen, L.B., Petersen, J.S., 2001, The GLP-1 derivative NN2211 inhibits cytokine-induced apoptosis in primary rat beta cells. *Diabetes*, **50[suppl.2]:** A31.
29. Bonner-Weir, S., 2000, Life and death of the pancreatic beta cells. *Trends Endocrinol Metab,* **11(9):**375-378.
30. Sturis, J., Jappe, M.B., Knudsen, L.B., Wilken, M., Gjedsted, A., Primdahl, S., et al., 2000, Long-acting GLP-1 derivative NN2211 markedly attenuates diabetes development in the male Zucker fatty rat. *Diabetologia*, **43[suppl.1]:** A145.
31. Pick, A., Clark, J., Kubstrup, C., Levisetti, M., Pugh, W., Bonner-Weir, S., et al., 1998, Role of apoptosis in failure of beta-cell mass compensation for insulin resistance and beta-cell defects in the male Zucker diabetic fatty rat. *Diabetes*, **47(3):**358-364.
32. Orskov, C., Holst, J.J., Nielsen, O.V., 1988, Effect of truncated glucagon-like peptide-1 [proglucagon-(78- 107) amide] on endocrine secretion from pig pancreas, antrum, and nonantral stomach. *Endocrinology*, **123(4):**2009-2013.
33. Orskov, C., Poulsen, S.S., 1991, Glucagonlike peptide-I-(7-36)-amide receptors only in islets of Langerhans. Autoradiographic survey of extracerebral tissues in rats. *Diabetes,* **40(10):**1292-1296.
34. Heller, R.S., Kieffer, T.J., Habener, J.F., 1997, Insulinotropic glucagon-like peptide I receptor expression in glucagon-producing alpha-cells of the rat endocrine pancreas. *Diabetes*, **46(5):**785-791.
35. Nauck, M.A., Heimesaat, M.M., Behle, K., Holst, J.J., Nauck, M.S., Ritzel, R., et al., 2002, Effects of glucagon-like peptide 1 on counterregulatory hormone responses, cognitive functions, and insulin secretion during hyperinsulinemic, stepped hypoglycemic clamp experiments in healthy volunteers. *J Clin Endocrinol Metab*, **87(3):**1239-1246.
36. Creutzfeldt, W.O., Kleine, N., Willms, B., Orskov, C., Holst, J.J., Nauck, M.A., 1996, Glucagonostatic actions and reduction of fasting hyperglycemia by exogenous glucagon-like peptide I(7-36) amide in type I diabetic patients. *Diabetes Care*, **19(6):**580-586.
37. Hvidberg, A., Nielsen, M.T., Hilsted, J., Orskov, C., Holst, J.J., 1994, Effect of glucagon-like peptide-1 (proglucagon 78-107amide) on hepatic glucose production in healthy man. *Metabolism*, **43(1):**104-108.
38. Wettergren, A., Schjoldager, B., Mortensen, P.E., Myhre, J., Christiansen, J., Holst, J.J., 1993, Truncated GLP-1 (proglucagon 78-107-amide) inhibits gastric and pancreatic functions in man. *Dig Dis Sci*, **38(4):**665-673.

39. Nauck, M.A., Niedereichholz, U., Ettler, R., Holst, J.J., Orskov, C., Ritzel, R., et al., 1997, Glucagon-like peptide 1 inhibition of gastric emptying outweighs its insulinotropic effects in healthy humans. *Am J Physiol*, **273(5 Pt 1):**E981-E988.
40. Holst, J.J., 1997, Enteroglucagon. *Annu Rev Physiol*, **59:**257-71:257-271.
41. Layer, P., Holst, J.J., Grandt, D., Goebell, H., 1995, Ileal release of glucagon-like peptide-1 (GLP-1). Association with inhibition of gastric acid secretion in humans. *Dig Dis Sci*, **40(5):**1074-1082.
42. Wettergren, A., Schjoldager, B., Mortensen, P.E., Myhre, J., Christiansen, J., Holst, J.J., 1993, Truncated GLP-1 (proglucagon 78-107-amide) inhibits gastric and pancreatic functions in man. *Dig Dis Sci*, **38(4):**665-673.
43. Layer, P., Holst, J.J., Grandt, D., Goebell, H., 1995, Ileal release of glucagon-like peptide-1 (GLP-1). Association with inhibition of gastric acid secretion in humans. *Dig Dis Sci*, **40(5):**1074-1082.
44. Young, A., Denaro, M., 1998, Roles of amylin in diabetes and in regulation of nutrient load. *Nutrition*, **14(6):**524-527.
45. Read, N., French, S., Cunningham, K., 1994, The Role of the Gut in Regulation Food Intake in Man. *Nutrition Reviews*, **52(1):**1-10.
46. Flint, A., Raben, A., Astrup, A., Holst, J.J., 1998, Glucagon-like Peptide 1 Promotes Satiety and Suppresses Energy Intake in Humans. *J Clin Invest*, **101(3):**515-520.
47. Gutzwiller, J.P., Goke, B., Drewe, J., Hildebrand, P., Ketterer, S., Handschin, D., et al., 1999, Glucagon-like peptide-1: a potent regulator of food intake in humans. *Gut, **44(1)**:81-86.*
48. Naslund, E., Barkeling, B., King, N., Gutniak, M., Blundell, J.E., Holst, J.J., et al., Energy intake and appetite is supressed by glucagon-like peptide-1 (GLP-1) in obese men. *Int J Obes*. In press.
49. Gutzwiller, J.P., Drewe, J., Goke, B., Schmidt, H., Rohrer, B., Lareida, J., et al., 1999, Glucagon-like peptide-1 promotes satiety and reduces food intake in patients with diabetes mellitus type 2. *Am J Physiol*, **276(5 Pt 2):**R1541-R1544.
50. Verdich, C., Flint, A., Gutzwiller, J.P., Naslund, E., Beglinger, C., Hellstrom, P.M., et al., 2001, A meta-analysis of the effect of glucagon-like peptide-1 (7-36) amide on ad libitum energy intake in humans. *J Clin Endocrinol Metab*, **86(9):**4382-4389.
51. Turton, M.D., O'Shea, D., Gunn, I., Beak, S.A., Edwards, C.M., Meeran, K., et al., 1996, A role for glucagon-like peptide-1 in the central regulation of feeding [see comments]. *Nature*, **379(6560):**69-72.
52. Tang-Christensen, M., Larsen, P.J., Goke, R., Fink-Jensen, A., Jessop, D.S., Moller, M., et al., 1996, Central administration of GLP-1-(7-36) amide inhibits food and water intake in rats. *Am J Physiol*, **271(4 Pt 2):**R848-R856.
53. Goke, R., Larsen, P.J., Mikkelsen, J.D., Sheikh, S.P., 1995, Distribution of GLP-1 binding sites in the rat brain: evidence that exendin-4 is a ligand of brain GLP-1 binding sites. *Eur J Neurosci*, **7(11):**2294-2300.
54. Tang-Christensen, M., Vrang, N., Larsen, P.J., 1998, Glucagon-like peptide 1(7-36) amide's central inhibition of feeding and peripheral inhibition of drinking are abolished by neonatal monosodium glutamate treatment. *Diabetes*, **47(4):**530-537.
55. Larsen, P.J., Tang-Christensen, M., Holst, J.J., Orskov, C., 1997, Distribution of glucagon-like peptide-1 and other preproglucagon- derived peptides in the rat hypothalamus and brainstem. *Neuroscience*, **77(1):**257-270.
56. Rinaman, L., 1999, Interoceptive stress activates glucagon-like peptide-1 neurons that project to the hypothalamus. *Am J Physiol*, **277(2 Pt 2):**R582-R590.

57. Rinaman, L., 1999, A functional role for central glucagon-like peptide-1 receptors in lithium chloride-induced anorexia. *Am J Physiol*, **277(5 Pt 2)**:R1537-R1540.
58. Scrocchi, L.A., Brown, T.J., MaClusky, N., Brubaker, P.L., Auerbach, A.B., Joyner, A.L., et al., 1996, Glucose intolerance but normal satiety in mice with a null mutation in the glucagon-like peptide 1 receptor gene. *Nat Med*, **2(11)**:1254-1258.
59. Nauck, M.A., Kleine, N., Orskov, C., Holst, J.J., Willms, B., Creutzfeldt, W., 1993, Normalization of fasting hyperglycaemia by exogenous glucagon- like peptide 1 (7-36 amide) in type 2 (non-insulin-dependent) diabetic patients. *Diabetologia*, **36(8)**:741-744.
60. Zander, M., Madsbad, S., Madsen, J.L., Holst, J.J., 2002, Effect of 6-week course of glucagon-like peptide 1 on glycaemic control, insulin sensitivity, and beta-cell function in type 2 diabetes: a parallel-group study. *Lancet*, **359(9309)**:824-830.
61. Nauck, M.A., Wollschlager, D., Werner, J., Holst, J.J., Orskov, C., Creutzfeldt, W., et al., 1996, Effects of subcutaneous glucagon-like peptide 1 (GLP-1 [7-36 amide]) in patients with NIDDM. *Diabetologia*, **39(12)**:1546-1553.
62. Deacon, C.F., Nauck, M.A., Toft-Nielsen, M., Pridal, L., Willms, B., Holst, J.J., 1995, Both subcutaneously and intravenously administered glucagon-like peptide I are rapidly degraded from the NH2-terminus in type II diabetic patients and in healthy subjects. *Diabetes*, **44(9)**:1126-1131.
63. Deacon, C.F., Johnsen, A.H., Holst, J.J., 1995, Degradation of glucagon-like peptide-1 by human plasma in vitro yields an N-terminally truncated peptide that is a major endogenous metabolite in vivo. *J Clin Endocrinol Metab*, **80(3)**:952-957.
64. Knudsen, L.B., Pridal, L., 1996, Glucagon-like peptide-1-(9-36) amide is a major metabolite of glucagon-like peptide-1-(7-36) amide after in vivo administration to dogs, and it acts as an antagonist on the pancreatic receptor. *Eur J Pharmacol*, **318(2-3)**:429-435.
65. Hansen, L., Deacon, C.F., Orskov, C., Holst, J.J., 1999, Glucagon-like peptide-1-(7-36)amide is transformed to glucagon-like peptide-1-(9-36)amide by dipeptidyl peptidase IV in the capillaries supplying the L cells of the porcine intestine. *Endocrinology*, **140(11)**:5356-5363.
66. Wettergren, A., Wojdemann, M., Holst, J.J., 1998, Glucagon-like peptide-1 inhibits gastropancreatic function by inhibiting central parasympathetic outflow. *Am J Physiol*, **275(5 Pt 1)**:G984-G992.
67. Imeryuz, N., Yegen, B.C., Bozkurt, A., Coskun, T., Villanueva-Penacarrillo, M.L., Ulusoy, N.B., 1997, Glucagon-like peptide-1 inhibits gastric emptying via vagal afferent- mediated central mechanisms. *Am J Physiol*, **273(4 Pt 1)**:G920-G927.
68. Nakabayashi, H., Nishizawa, M., Nakagawa, A., Takeda, R., Niijima, A., 1996, Vagal hepatopancreatic reflex effect evoked by intraportal appearance of tGLP-1. *Am J Physiol*, **271(5 Pt 1)**:E808-E813.
69. Balkan, B., Li, X., 2000, Portal GLP-1 administration in rats augments the insulin response to glucose via neuronal mechanisms. *Am J Physiol Regul Integr Comp Physiol*, **279(4)**:R1449-R1454.
70. Wettergren, A., Wojdemann, M., Meisner, S., Stadil, F., Holst, J.J., 1997, The inhibitory effect of glucagon-like peptide-1 (GLP-1) 7-36 amide on gastric acid secretion in humans depends on an intact vagal innervation. *Gut*, **40(5)**:597-601.
71. Wettergren, A., Petersen, H., Orskov, C., Christiansen, J., Sheikh, S.P., Holst, J.J., 1994, Glucagon-like peptide-1 7-36 amide and peptide YY from the L- cell of the ileal mucosa are potent inhibitors of vagally induced gastric acid secretion in man. *Scand J Gastroenterol*, **29(6)**:501-505.

72. Vilsboll, T., Krarup, T., Deacon, C.F., Madsbad, S., Holst, J.J., 2001, Reduced postprandial concentrations of intact biologically active glucagon-like peptide 1 in type 2 diabetic patients. *Diabetes*, **50(3)**:609-613.
73. Holst, J.J., Deacon, C.F., 1998, Inhibition of the activity of dipeptidyl-peptidase IV as a treatment for type 2 diabetes. *Diabetes*, **47(11)**:1663-1670.
74. Deacon, C.F., Pridal, L., Klarskov, L., Olesen, M., Holst, J.J., 1996, Glucagon-like peptide 1 undergoes differential tissue-specific metabolism in the anesthetized pig. *Am J Physiol*, **271(3 Pt 1)**:E458-E464.
75. Vilsboll, T., Agersoe, H., Krarup, T., Holst, J.J., Similar elimination rates of GLP-1 in obese type diabetic patients and healthy subjects. *J Clin Endocrinol Metab.,* in press.
76. Deacon, C.F., Hughes, T.E., Holst, J.J., 1998, Dipeptidyl peptidase IV inhibition potentiates the insulinotropic effect of glucagon-like peptide 1 in the anesthetized pig. *Diabetes*, **47(5)**:764-769.
77. Ahren, B., Holst, J.J., Martensson, H., Balkan, B., 2000, Improved glucose tolerance and insulin secretion by inhibition of dipeptidyl peptidase IV in mice. *Eur J Pharmacol*, **404(1-2)**:239-245.
78. Pospisilik, J.A., Stafford, S.G., Demuth, H.U., Brownsey, R., Parkhouse, W., Finegood, D.T., et al., 2002, Long-term treatment with the dipeptidyl peptidase IV inhibitor P32/98 causes sustained improvements in glucose tolerance, insulin sensitivity, hyperinsulinemia, and beta-cell glucose responsiveness in VDF (fa/fa) Zucker rats. *Diabetes*, **51(4)**:943-950.
79. Ahren, B., Simonsson, E., Larsson, H., Landin-Olsson, M., Torgeirsson, H., Jansson, P.A., et al., 2002, Inhibition of dipeptidyl peptidase IV improves metabolic control over a 4-week study period in type 2 diabetes. *Diabetes Care*, **25(5)**:869-875.
80. Mentlein, R., 1999, Dipeptidyl-peptidase IV (CD26)--role in the inactivation of regulatory peptides. *Regul Pept*, **85(1)**:9-24.
81. Pederson, R.A., Kieffer, T.J., Pauly, R., Kofod, H., Kwong, J., McIntosh, C.H., 1996, The enteroinsular axis in dipeptidyl peptidase IV-negative rats. *Metabolism*, **45(11)**:1335-1341.
82. Marguet, D., Baggio, L., Kobayashi, T., Bernard, A.M., Pierres, M., Nielsen, P.F., et al., 2000, Enhanced insulin secretion and improved glucose tolerance in mice lacking CD26. *Proc Natl Acad Sci U S A*, **97(12)**:6874-6879.
83. Deacon, C.F., Wamberg, S., Bie, P., Hughes, T.E., Holst, J.J., 2002, Preservation of active incretin hormones by inhibition of dipeptidyl peptidase IV suppresses meal-induced incretin secretion in dogs. *J Endocrinol*, **172(2)**:355-362.
84. Li, X., Kwasnik, L., Miserendino, R., Mone, M., Hughes, T.E., Vilhauer, E.B., et al., 1997, Improved insulin secretion and oral glucose tolerance after in vivo inhibition of DPP-IV in insulin resistant rats. *Diabetes*, **46[suppl.1]**: 237A.
85. Sudre, B., Broqua, P., White, R.B., Ashworth, D., Evans, D.M., Haigh, R., et al., 2002, Chronic inhibition of circulating dipeptidyl peptidase IV by FE 999011 delays the occurrence of diabetes in male zucker diabetic fatty rats. *Diabetes,* **51(5)**:1461-1469.
86. Ritzel, R., Orskov, C., Holst, J.J., Nauck, M.A., 1995, Pharmacokinetic, insulinotropic, and glucagonostatic properties of GLP-1 [7-36 amide] after subcutaneous injection in healthy volunteers. Dose-response-relationships. *Diabetologia*, **38(6)**:720-725.
87. Rachman, J., Barrow, B.A., Levy, J.C., Turner, R.C., 1997, Near-normalisation of diurnal glucose concentrations by continuous administration of glucagon-like peptide-1 (GLP-1) in subjects with NIDDM. *Diabetologia*, **40(2)**:205-211.

88. Larsen, J., Hylleberg, B., Ng, K., Damsbo, P., 2001, Glucagon-like peptide-1 infusion must be maintained for 24 h/day to obtain acceptable glycemia in type 2 diabetic patients who are poorly controlled on sulphonylurea treatment. *Diabetes Care*, **24(8)**:1416-1421.

Dipeptidyl Peptidase IV Inhibition in Animal Models of Diabetes

J. ANDREW POSPISILIK[*], JAN A. EHSES[*], TIMOTHY DOTY[*], CHRISTOPHER H.S. M[C]INTOSH[*], HANS-ULRICH DEMUTH[#], and RAYMOND A. PEDERSON[*]
[*] *Department of Physiology, University of British Columbia, Vancouver, Canada*
[#] *Probiodrug Research, Biocenter, Halle (Saale), Germany*

1. INTRODUCTION

Dipeptidyl Peptidase IV (DP IV; EC 3.4.14.5; CD26) is a membrane bound ectopeptidase found on numerous cell types including vascular endothelium and certain functional leukocyte subsets[1]. A soluble form of the enzyme, lacking a short transmembrane sequence, can also be found circulating in the plasma. With a substrate specificity favouring oligopeptides with a penultimate prolyl-, analyl-, or seryl- residue at the N-terminus, DP IV rapidly cleaves the N-terminal dipeptide from a number of metabolic hormones and neuroendocrine factors including (in order of catalytic efficiency) NPY, PYY, and the gluco-regulatory peptides glucagon-like peptide-1 (GLP-1), glucose-dependent insulinotropic polypeptide (GIP) and glucagon[1-3].

Released post-prandially from the intestinal mucosa, GIP and GLP-1 are responsible for >50% of meal-stimulated insulin release. Their potent insulinotropic actions are the result of direct interaction of these "incretin" hormones with their respective G-protein coupled receptors on the surface of the pancreatic β-cell[4]. Like all members of the glucagon/secretin superfamily

Dipeptidyl Aminopeptidases in Health and Disease, Edited by Hildebrandt et al.
Kluwer Academic/Plenum Publishers, New York, 2003

of peptide hormones, the integrity of the N-terminus of the GIP and GLP-1 molecules is requisite for activation of their respective receptors and stimulation of their downstream biological effects[4]. The extensive presence of DP IV in the circulation therefore limits incretin function and comprises a major regulatory mechanism for their individual actions.

1.1 Why inhibit DP IV?

Type-2 diabetes mellitus (T2DM) is a gluco-regulatory disorder for which susceptibility is believed to be both genetically and environmentally determined. Onset of T2DM results from the establishment of two general metabolic abnormalities: insulin resistance and pancreatic β-cell dysfunction.

Studies spanning more than three decades have revealed a pleiotropy of incretin effects including stimulation of insulin secretion and biosynthesis, β-cell growth and differentiation, β-cell glucose competence, glucose-uptake in the periphery, inhibition of glucagon secretion, and most recently the stimulation of β-cell cytoprotection[4-6]. It is this breadth of anti-diabetic actions that has fueled intense research into the application of GIP and GLP-1 as potential therapeutic tools. Unfortunately, two major obstacles have hampered clinical development of these compounds: rapid inactivation in the circulation by DP IV ($T_{1/2} \sim$ 1-2 min) and poor oral bioavailability.

An alternative incretin-based approach currently under investigation uses low molecular weight, specific DP IV-inhibitors to bypass both afore-mentioned caveats, stimulating the GIP/GLP-1 axis simply by protecting endogenously secreted hormone from DP IV-mediated inactivation. By prolonging the circulating half-lives of GIP and GLP-1, DP IV inhibitors increase the fraction of these peptides reaching the pancreatic β-cells in the intact form and in doing so enhance all of their beneficial actions (Fig. 1)[7-13].

2. DP IV-INHIBITION IN TYPE-2 MODELS

On the heels of the first proof of concept of DP IV inhibition *in vivo* by Pauly et al. (1996)[8], our group began a systematic examination of the effects of both acute and chronic DP IV inhibition in a number of animal models of diabetes. The following is a short review of our investigations in this area and is in no way intended to comprise a complete review of all contributors to the field.

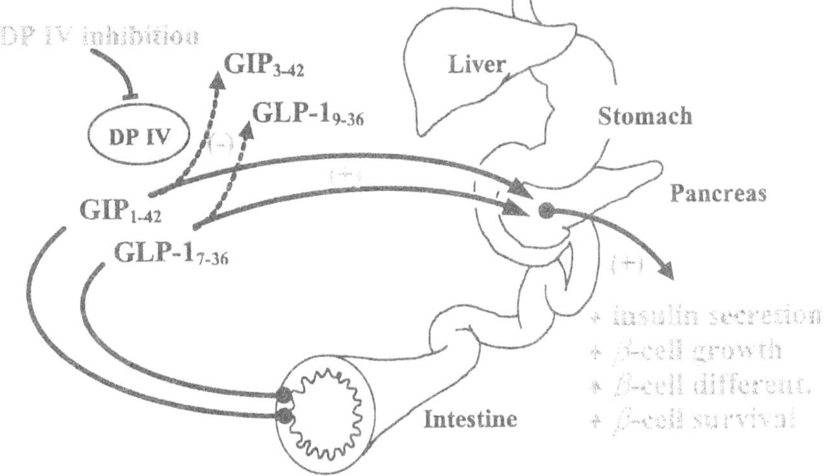

Figure 1. Schematic representation of the use of DP IV inhibitors to enhance GIP and GLP-1 action. Copyright © 2002 American Diabetes Association. From *Diabetes*, Vol 51, 2002, 2677-2683. Reprinted with permission from *The American Diabetes Association.*

2.1 Effects on glucose tolerance and β-cell function

Many groups have since confirmed Pauly's initial findings of acute DP IV inhibitor-mediated improvements in oral glucose tolerance *in vivo*, including a study by Pederson et al. conducted in the Vancouver Diabetic Fatty rat (*fa/fa*; fatty Zucker)[10]. This sub-strain of the obese *fa/fa* Zucker rat has been inbred at the Vancouver facility for over twenty years and provides an effective model of type-2 diabetes displaying both fasting and fed hyperglycemia, hyperinsulinemia, and hyperlipidemia concomitant with severe insulin resistance. Interestingly, acute administration of the DP IV inhibitor P32/98 (isoleucine thiazolidide) produced a much more profound reduction in post-prandial glucose levels (OGTT) in the obese diabetic animals than in their lean, normoglycemic littermates[10].

The pleiotropic glucose-lowering effects of incretins that would not be exploited by acute inhibitor dosing coupled with the results of the initial studies provided a strong rationale for a long-term investigation into the effects of chronic DP IV inhibitor therapy. A treatment regimen of twice-daily P32/98 (10 mg/kg *p.o.*) was initiated in a group of 12 week old obese VDF rats (N=6), with the control group receiving only injection vehicle[14]. Long-term DP IV inhibition showed little effect on gross metabolic indicators such as body weight, morning blood glucose, and nutrient intake, with the notable exception of a 12.5 % reduction in body weight gain in the treated animals (vs. controls). In addition to these bidaily measures, a 24-hour profile of plasma glucose, insulin and DP IV-activity performed 6 weeks into the investigation revealed: 1. effective DP IV-inhibition for >9 hours after dosing; 2. a reduction in meal-stimulated blood glucose levels (>2.5 mM during the feeding phase; and 3. a reduction in basal and meal-stimulated insulinemia concomitant with an increased insulin response to a meal in the treated animals[14].

As an adjunct to bidaily glucose measurements and the 24-hour profile, oral glucose tolerance tests (OGTT) performed after drug washout were used to define the long-term changes in glucose tolerance initiated by chronic DP IV-inhibitor treatment. Absent of acute incretin enhancement, fasting plasma glucose levels were shown to fall progressively from diabetic levels (~8 mM) at the outset to near normal levels (~6 mM) by twelve weeks of treatment. Similarly, glucose tolerance was seen to improve in the face of an ever-worsening condition in the untreated animals. Further, by twelve weeks of treatment, insulin secretory profiles (unaltered at 4 and 8 weeks) revealed an increased early phase insulin secretory response to oral glucose (Fig. 2). Active GLP-1 levels, measured using an antibody specific for the intact N-terminus of the molecule, showed no difference between the two groups, indirectly confirming drug washout and reaffirming the results as lasting changes in insulin secretory function[14].

Were these changes the manifestation of an increased β-cell mass or improved glucose competence on a cellular level? Pancreas perfusion studies and β-cell mass determinations revealed a marked elevation in glucose-responsiveness in the absence of any changes in islet size, morphology, or in β-cell mass in the treated animals, data suggestive of a DP IV-inhibitor induced enhancement of glucose responsiveness at the level of the β-cell[14]. These data do not, however, preclude the possibility that additional mechanisms (e.g. increased incretin sensitivity) played a causative role in the *in vivo* results mentioned above.

Recently a number of reports have provided support for DP IV-inhibitor induced improvements in β-cell function. Ahren et al. showed qualitative increases in GLUT 2 expression, as well as enhanced *in vitro* glucose responsiveness in islets isolated from both normal and high fat fed C57BL/6J mice[15]. Further the authors provided evidence for a reduced tendency towards increased islet size in the high-fat fed model of insulin resistance. Unfortunately the authors of that study did not provide evidence of any lasting effects of DP IV-inhibitor therapy on glucose tolerance, only an acute improvement in the presence of inhibitor when assessed in the face of a pharmacological dose of oral glucose (7.5 g/kg). The positive effects observed in the study were, however, in agreement with our own observations of lasting improvements in glucose tolerance and β-cell glucose responsiveness after three weeks of P32/98 treatment in high-fat fed Wistar rats (unpublished observations). Also worthy of mention is a recent report of delayed diabetes onset, reduced hypertriglyceridemia and stabilization of nutrient intake in the Zucker Diabetic Fatty rat, a strain related to the VDF that displays more rapid and robust diabetic progression[16].

2.2 Effects on insulin sensitivity

Data from the initial long-term study provided a strong suggestion of heightened insulin sensitivity in DP IV-inhibitor treated animals and thus the rationale for an in-depth examination of the effects of long-term DP IV-inhibitor treatment on insulin-sensitivity. Once again using the VDF rat as a model, two groups of eight animals were treated for 12 weeks as described in section 2.1, after which indwelling carotid and jugular cannulae were inserted under anaesthesia[17]. After two consecutive days of bodyweight gain following surgery, the animals were fasted and subjected to a 3-stage conscious euglycemic-hyperinsulinemic clamp (in the absence of DP IV-inhibitor). The protocol consisted of three sequential 90-min periods with insulin infusion rates of 0, 5, and 15 mU/kg/min (representative of fasting, basal fed, and peak fed plasma insulin levels). Inclusion of [^{3}H]-glucose into the infusion protocol allowed calculation of both hepatic and peripheral contributions to the circulating glucose pool and thus an estimation of total body insulin sensitivity. The exogenous glucose infusion rate (GIR) required to maintain fasting "euglycemia" in the treatment group was significantly elevated during both stages of insulin infusion (Fig.2). Analysis of the hepatic and peripheral contributions towards these changes revealed a reduction in hepatic glucose output (HGO) at all tested insulin concentrations as well as a left-shift and return of responsiveness of peripheral glucose uptake[17]. An *in vitro* examination of insulin mediated inhibition of isoproterenol-stimulated lipolysis in abdominal adipocytes from the same animals showed a left shift in the treated group suggesting a role for fat, in addition to skeletal muscle (soleus strip experiments), in the altered peripheral sensitivity revealed by the clamp[17].

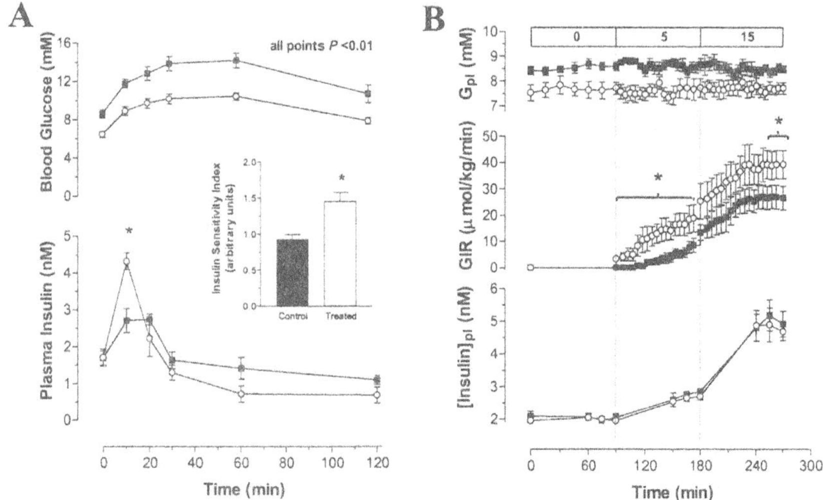

Figure 2. Long-term DP IV-inhibitor treatment improves oral glucose tolerance (A), and insulin sensitivity (B) in the VDF rat. Animals were treated either with (open circles) or without (closed squares) the DP IV-inhibitor P32/98 (10 mg/kg twice daily; N=7) for a period of twelve weeks. Animals were fasted for 16 hours prior to either protocol, sufficient time for complete drug washout. Both the OGTT (1 g/kg) and the clamp study were performed according to conventional techniques. Values in the header of panel B represent insulin infusion rates in mU/kg/min. G_{pl} = plasma glucose concentration; GIR = glucose infusion rate; $[Insulin]_{pl}$ = plasma insulin concentration * *P* <0.05. Copyright © 2002 American Diabetes Association. From *Diabetes*, Vol 51, 2002, 2677-2683. Reprinted with permission from *The American Diabetes Association.*

3. DP IV-INHIBITION IN TYPE-1 MODELS

Interest in the potential of GLP-1 in the treatment of type-1 diabetes goes back as far as 1992 when Gutniak attempted to harness the glucagonostatic actions of the peptide in type-1 patients[18, 19]. Recently, the stimulation of β-cell growth, differentiation, proliferation and perhaps most importantly the stimulation of β-cell survival, have been added to the growing list of GIP and GLP-1 actions. These effects alone are sufficient to warrant investigation into the applicability of DP IV inhibitors to models of type-1 diabetes, a autoimmune disorder in which destruction of pancreatic β-cells reduces the body's insulin secretory capacity.

In addition, extensive research (including some of the first studies on DP IV-inhibition) have shown that inhibition of surface DP IV activity suppresses T-lymphocyte (T-cell) proliferation, T-cell stimulation of B-lymphocyte IgG release, and Th_1 cytokine production, processes intimately involved in the progression of autoimmune diabetes[20]. *In vivo* proof of the immunomodulatory potential of DP IV-inhibitors was recently provided by Steinbrecher et al. in the form of prevention (over the short-term) and amelioration of experimental autoimmune encephalomyelitis (EAE) in mice[21]. Coupled with stimulation of β-cell growth, cell survival and function (through incretin enhancement), the immunosuppressive potential of DP IV-inhibitors casts them as a potentially unique, multi-faceted approach to the treatment and perhaps prevention of type-1 diabetes.

3.1 DP IV inhibition in the streptozotocin (STZ) diabetic rat

Though it lacks an appreciable autoimmune component, the STZ rat is widely accepted as an effective model of type-1 diabetes. A high single dose of the β-cell specific toxin streptozotocin (STZ) initiates a rapid phase of cell death involving a number of apoptotic as well as necrotic mechanisms. STZ induced diabetes models not only the rapid apoptotic phase of cell death postulated to be responsible for disease manifestation in humans but also the insulin deficiency that is established thereafter. In fact, the lack of ongoing autoimmune attack after the toxic insult actually provides us with a useful model for the β-cell regeneration, differentiation and neogenesis that occur following massive β-cell destruction.

To satisfy the rationale mentioned above, we initiated a study into the long-term effects of DP IV-inhibition in the STZ model[22]. Three groups of rats were injected with a single high dose of STZ (50 mg/kg) and started on DP IV inhibitor therapy (20 mg/kg daily *p.o.*) either one week before or after STZ administration. Over the seven week course of the study P32/98 treated animals displayed increased weight gain (230%) and nutrient intake, a reduction in fed blood glucose levels (~21 mM vs. 26 mM for control animals) and a return of plasma insulin levels towards normalcy (0.12 nM vs. 0.07 nM respectively). Marked improvements in glucose tolerance were observed concomitant with elevated fasting and glucose stimulated insulin secretory values (33 % reduction in glucose AUC; 177 % increase in insulin

AUC). This indirect evidence of enhanced insulin secretory capacity was confirmed by a 2-fold rise in glucose-stimulated insulin secretion during pancreas perfusion and an 8-fold increase in total pancreatic insulin content when compared to untreated STZ-control animals. To provide further insight into the mechanism underlying these data, an immunohistochemical survey of pancreatic sections was performed. The number of very small islets with a normal β-cell complement was shown to be increased (+35 %) in the treated animals, as was total β-cell number (+120 %) and islet β-cell fraction (doubling). Together the results provided strong evidence of DP IV-inhibitor stimulation of islet neogenesis, β-cell survival, and insulin biosynthesis or β-cell growth[22].

3.2 DP IV inhibition in the BB rat

The chief disadvantage of using the STZ-rat as a model of type-1 diabetes is the lack of a significant autoimmune component to the disease progression. The biobreeding (BB) rat, on the other hand, spontaneously develops extensive insulitis (mononuclear cell infilitration of the islet), and in approximately 60 % of animals, virtually complete β-cell destruction. The majority of these animals turn diabetic between 60 and 120 days of age. Despite several key differences from the human autoimmune disorder (eg. peripheral lymphopenia), the BB rat is considered one of the best models of type-1 diabetes.

The positive results of the STZ study prompted a similar investigation in the BB rat examining the potential immunosuppressive and β-cell enhancing effects of DP IV inhibition. Preliminary results from the study show a delay of onset of approximately 10 days associated with DP IV inhibitor treatment, as well as minor improvements in glucose tolerance in both pre-diabetic and diabetic animals. DP IV-inhibitor treated diabetic animals displayed a mild amelioration in severity of diabetes, however insulin therapy was still required (unpublished observations).

4. CONCLUSIONS

The success of DP IV-inhibitors as a therapeutic strategy in the treatment of diabetes is owed in great part to the pleiotropic nature of its primary effectors, the incretins GIP and GLP-1. The findings of the studies outlined above further exemplify the importance of the non-insulinotropic effects of GIP and GLP-1 in the regulation of glucose homeostasis, and in the maintenance of β-cell survival. The recent addition of improved hepatic and peripheral insulin sensitivity, β-cell protection, and delay of onset of autoimmune diabetes to the list of beneficial metabolic effects of long-term DP IV-inhibitor therapy, provides strong support for the use of these compounds in the treatment of type-2 diabetes while opening the door for further study into their applicability to type-1 diabetes.

ACKNOWLEDGEMENTS

The work outlined in this review was funded through grants from the Canadian Institutes for Health Research (CIHR; #MOP-13192) and the Department of Science and Technology of Sachsen Anhalt (9704/00116). Further JAP and JAE are indebted to the CIHR and the Michael Smith Foundation for Health Research (MSFHR) for scholarship support.

REFERENCES

1. Mentlein, R.: 1999. Dipeptidyl-peptidase IV (CD26)--role in the inactivation of regulatory peptides. *Regul Pept* **85:** 9-24
2. Lambeir, A.-M., Durinx, C., Proost, P., Van Damme, J., Scharpe, S., De Meester, I.: 2001. Kinetic Study of the Processing by Dipeptidyl-Peptidase IV/CD26 of Neuropeptides Involved in Insulin Secretion. *FEBS Lett* **507:** 327-30
3. Pospisilik, J.A., Hinke, S.A., Pederson, R.A., Hoffmann, T., Rosche, F., Schlenzig, D., Glund, K., Heiser, U., McIntosh, C.H., Demuth, H.-U.: 2001. Metabolism of glucagon by dipeptidyl peptidase IV (CD26). *Regul Pept* **96:** 133-41
4. Kieffer, T.J., Habener, J.F.: 1999. The glucagon-like peptides. *Endocr Rev* **20:** 876-913
5. Ehses, J., Casilla, V., Pospisilik, J., Doty, T., Demuth, H.-U., Pederson, R., McIntosh, C.: 2002. Glucose-dependent Insulinotropic Polypeptide (GIP) Stimulates Cell Growth and Promotes Survival of INS-1 Cells (Abstract). *Diabetes* **51:** A339
6. Perry, T., Haughey, N.J., Mattson, M.P., Egan, J.M., Greig, N.H.: 2002. Protection and reversal of excitotoxic neuronal damage by glucagon-like peptide-1 and exendin-4. *J Pharmacol Exp Ther* **302:** 881-8
7. Kieffer, T.J., McIntosh, C.H., Pederson, R.A.: 1995. Degradation of glucose-dependent insulinotropic polypeptide and truncated glucagon-like peptide 1 in vitro and in vivo by dipeptidyl peptidase IV. *Endocrinology* **136:** 3585-96

8. Pauly, R., Demuth, H.-U., Rosche, F., Schmidt, J., White, H., McIntosh, C., Pederson, R.: 1996. Inhibition of dipeptidyl peptidase IV (DP IV) in rat results in improved glucose tolerance. *Regul Pept* (abstract) **64:** 148
9. Pauly, R.P., Demuth, H.U., Rosche, F., Schmidt, J., White, H.A., Lynn, F., McIntosh, C.H., Pederson, R.A.: 1999. Improved glucose tolerance in rats treated with the dipeptidyl peptidase IV (CD26) inhibitor Ile-thiazolidide. *Metabolism* **48:** 385-9
10. Pederson, R.A., White, H.A., Schlenzig, D., Pauly, R.P., McIntosh, C.H., Demuth, H.U.: 1998. Improved glucose tolerance in Zucker fatty rats by oral administration of the dipeptidyl peptidase IV inhibitor isoleucine thiazolidide. *Diabetes* **47:** 1253-8
11. Holst, J.J., Deacon, C.F.: 1998. Inhibition of the activity of dipeptidyl-peptidase IV as a treatment for Type 2 Diabetes. *Diabetes* **47:** 1663-70
12. Balkan, B., Kwasnik, L., Miserendino, R., Holst, J.J., Li, X.: 1999. Inhibition of dipeptidyl peptidase IV with NVP-DPP728 increases plasma GLP-1 (7-36 amide) concentrations and improves oral glucose tolerance in obese Zucker rats. *Diabetologia* **42:** 1324-31
13. Deacon, C.F., Nauck, M.A., Meier, J., Hucking, K., Holst, J.J.: 2000. Degradation of endogenous and exogenous gastric inhibitory polypeptide in healthy and in type 2 diabetic subjects as revealed using a new assay for the intact peptide. *J Clin Endocrinol Metab* **85:** 3575-81
14. Pospisilik, J., Stafford, S., Demuth, H.-U., Brownsey, R., Parkhouse, W., Finegood, D., McIntosh, C., Pederson, R.: 2002. Long-term treatment with the DP IV inhibitor P32/98 causes sustained improvements in glucose tolerance, insulin sensitivity, hyperinsulinemia and b-cell glucose responsiveness in VDF (fa/fa) Zucker Rats. *Diabetes* **51(4):**943-50.
15. Ahren, B., Simonsson, E., Larsson, H., Landin-Olsson, M., Torgeirsson, H., Jansson, P.-A., Sandqvist, M., Bavenholm, P., Efendic, S., Eriksson, J., Dickinson, S., Holmes, D.: 2002. Inhibition of Dipeptidyl Peptidase IV Improves Metabolic Control Over a 4-Week Study Period in Type 2 Diabetes. *Diabetes Care* **25:** 869-75
16. Sudre, B., Broqua, P., White, R., Ashworth, D., Evans, D., Haigh, R., Junien, J.-L., Aubert, M.: 2002. Chronic Inhibition of Circulating Dipeptidyl Peptidase IV by FE 999011 Delays the Occurrence of Diabetes in Male Zucker Diabetic Fatty Rats. *Diabetes* **51:** 1461-69
17. Pospisilik, J.A., Stafford, S.G., Demuth, H.U., McIntosh, C.H., Pederson, R.A.: 2002. Long-term treatment with dipeptidyl peptidase IV inhibitor improves hepatic and peripheral insulin sensitivity in the VDF Zucker rat: a euglycemic-hyperinsulinemic clamp study. *Diabetes* **51:** 2677-83
18. Gutniak, M., Orskov, C., Holst, J.J., Ahren, B., Efendic, S.: 1992. Antidiabetogenic effect of glucagon-like peptide-1 (7-36)amide in normal subjects and patients with diabetes mellitus. *N Engl J Med* **326:** 1316-22
19. Dupre, J., Behme, M.T., Hramiak, I.M., McFarlane, P., Williamson, M.P., Zabel, P., McDonald, T.J.: 1995. Glucagon-like peptide I reduces postprandial glycemic excursions in IDDM. *Diabetes* **44:** 626-30
20. De Meester, I., Korom, S., Van Damme, J., Scharpe, S.: 1999. CD26, let it cut or cut it down. *Immunol Today* **20:** 367-75
21. Steinbrecher, A., Reinhold, D., Quigley, L., Gado, A., Tresser, N., Izikson, L., Born, I., Faust, J., Neubert, K., Martin, R., Ansorge, S., Brocke, S.: 2001. Targeting dipeptidyl peptidase IV (CD26) suppresses autoimmune encephalomyelitis and up-regulates TGF-beta 1 secretion in vivo. *J Immunol* **166:** 2041-8

22. Pospisilik, J., Martin, J., Doty, T., Ehses, J., Pamir, N., Lynn, F., Piteau, S., Demuth, H.-U., McIntosh, C., Pederson, R.: 2002. Long-Term DP IV-Inhibitor Treatment Reduces the Severity of Streptozotocin-Induced Diabetic Rats Through Improvements in Glucose Tolerance, Islet Size and Function and Promotion of b-cell Survival (Abst.). *Diabetes* **51:** A343

Glucose-dependent Insulinotropic Polypeptide (GIP): Development of DP IV-Resistant Analogues with Therapeutic Potential

SIMON A. HINKE[A], FRANCIS LYNN[A], JAN EHSES[A], NATHALIE PAMIR[A], SUSANNE MANHART[B], KERSTIN KÜHN-WACHE[B], FRED ROSCHE[B], HANS-ULRICH DEMUTH[B], RAYMOND A. PEDERSON[A] and CHRISTOPHER H.S. MCINTOSH[A]
[a]Department of Physiology, University of British Columbia, Vancouver, Canada, [b]Probiodrug AG, Biocenter, Halle (Saale), Germany

1. INTRODUCTION

Glucose-dependent Insulinotropic Polypeptide (GIP_{1-42}; Figure 1) is a gastrointestinal hormone that is released in response to nutrient intake and stimulates insulin secretion in a glucose-dependent manner[1,2.] Although GIP was the first established hormonal component (incretin) of the enteroinsular axis identified, intestinal products of the proglucagon gene, GLP-1_{7-36} and GIP-1 $_{7-37}$ (GLP-1), were subsequently shown to share its ability of stimulating insulin secretion, and the dual actions of GIP and GLP-1 are widely considered to be the major hormonal contributors to intestinal regulation of pancreatic endocrine function.

There are two major defects in type 2 diabetes that contribute to the hyperglycemia that is characteristic of the disorder: defective responsiveness to glucose and insulin resistance. Current therapies for type 2 diabetics generally involve dietary control plus stimulants of insulin secretion (e.g. sulfonylureas) and/or insulin sensitizers (e.g. metformin, thiazolidinediones). However considerable numbers of patients become resistant to sulfonylurea action and many eventually become insulin-dependent. As a consequence,

Dipeptidyl Aminopeptidases in Health and Disease, Edited by Hildebrandt et al.
Kluwer Academic/Plenum Publishers, New York, 2003

there is increasing interest in developing alternative methods for stimulating endogenous insulin secretion. Among these, there has been a major emphasis on long-acting analogues of GLP-1[3] and, more recently, DP IV-inhibitors, with little interest shown in GIP[4]. The current review focuses on the underlying reasons for this disinterest and attempts to provide a balanced view on the potential for DP IV-resistant analogues of GIP in the treatment of diabetes.

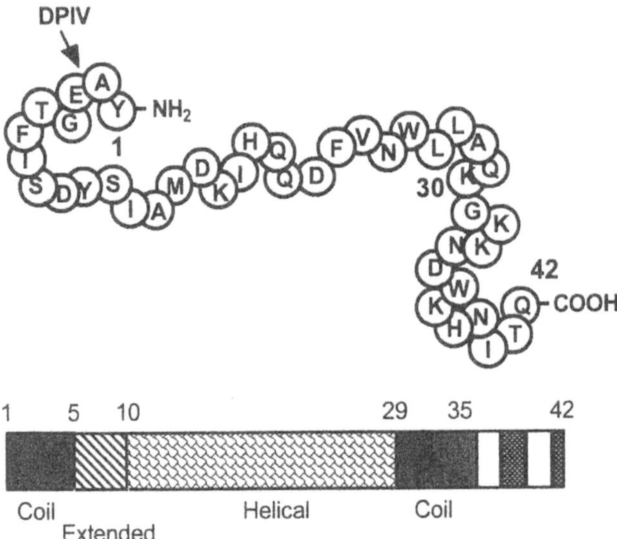

Figure 1. Predicted Secondary Structure of GIP. Secondary structure was predicted by the Gascuel and Golmard Basic Statistical Method using PCGENE. The site of DP IV-cleavage is shown.

2. WHY DEVELOP GIP ANALOGUES?

GIP potently stimulates insulin secretion in humans in a glucose-dependent manner[2]. Additionally, recent studies have demonstrated that it stimulates beta cell mitogenesis and inhibits apoptosis[5,6]. GIP_{1-42} and GLP-1 are equally insulinotropic[7] and conclusions of a much greater potency of GLP-1 can be attributed to comparative studies with batches of synthetic

GIP with low biological activity[7,8]. In addition to the common physiological actions exhibited by GIP and GLP-1, GIP also exhibits discrete actions. For example, GIP demonstrates both lipolytic and lipogenic activity on fat cells[4,9]. Although the significance is not clear, one possibility is that GIP stimulates lipolysis during fasting conditions, thus providing tissues, including the β-cell, with fatty acids that are essential for their normal function[9]. During a meal, incretins stimulate insulin secretion and GIP may then act in a lipogenic manner, as insulin inhibits the lipolytic pathway.

3. PROBLEMS WITH GIP-ANALOGUE DEVELOPMENT

The two major arguments brought against developing analogues of GIP for therapeutic use are that it is rapidly degraded in the bloodstream and that type 2 diabetic patients exhibit resistance to GIP action. The first indication that GIP was a target for DP IV degradation was obtained in HPLC studies on partially purified fractions from porcine intestinal extracts[10]. A peptide corresponding to GIP_{3-42} was identified and it was suggested that enzymatic cleavage resulted in formation of the N-terminally truncated molecule[11]. GIP_{3-42} was shown to be non-insulinotropic in the perfused rat pancreas[10] and isolated islets. Mentlein et al[12] first reported that GIP_{1-42} was a substrate for DP IV, with release of Tyr^1-Ala^2. Serum-degradation of GIP by DP IV was established using the inhibitor, Lys-Pyrrolidide[12]. The *in vivo* importance of DP IV-mediated degradation of GIP was demonstrated independently by following ^{125}I-GIP_{1-42} conversion to $^{125}IGIP_{3-42}$ following i.v. injection into anaesthetized rats[13].

Identification of the second problem, GIP resistance, has its origins in studies by Perley and Kipnis[14] in which they observed a greatly reduced incretin effect in type 2 diabetes. Studies were later performed to determine whether pathophysiological changes in GIP secretion contributed to this defect[1,2] and whether responsiveness to incretins was compromised. Several groups have now shown blunted responses to GIP infusion in type 2 diabetes, although the degree of resistance observed differs between groups[15-17]. It is important to note that all of the studies to date have employed infusion conditions designed to produce circulating levels in the physiological range.

4. DEVELOPMENT OF DP IV-RESISTANT GIP-ANALOGUES

The protocol that we have used for screening for DP IV-resistant analogues over the past few years can be summarized[18,19]:

1. Solid phase peptide synthesis and screening for DP IV-resistance using Maldi-TOF mass spectrometry,
2. Competitive binding assays using CHO-K1 cells transfected with the GIP receptor (GIP-R1 cells) to establish binding constants (IC_{50} values).
3. Measurement of analogue-induced cyclic AMP production in CHO-K1 cells and determination of maximal response and EC_{50} values.

   Analogues with promising characteristics in 1-3:

4. Determination of analogue-induced insulin secretion from insulinoma cells (βTC-3, INS-1) and/or isolated perfused rat pancreas.
5. In vivo bioassay during an oral (OGTT) or intraperitoneal (IPGTT) glucose tolerance test in normal and diabetic rats.

Design of the GIP-analogues was based on previous studies with other members of the secretin-glucagon family of peptides and on the basis of knowledge regarding the substrate specificity requirements of DP IV. The N-terminus of GIP is extremely sensitive to change and removal of the first two amino-terminal residues (GIP_{3-42}) results in a peptide that exhibits reduced receptor affinity in competitive binding studies and is completely devoid of the ability to stimulate cAMP production or affect glucose excursions and insulin profiles *in vivo*. Therefore, to generate DPIV-resistant GIP analogues, peptides were generated with modifications or substitutions of amino acids in positions 2 and 3. Additionally, for the majority of syntheses, the GIP_{1-30} backbone was used, rather than the intact peptide. Using the above screening procedure the following GIP analogues were identified as being completely or moderately DPIV-resistant and worthy of further study[18]:

COMPLETELY RESISTANT
MODERATELY RESISTANT*

[D-Ala^2]GIP_{1-42} and [D-Ala^2] GIP_{1-30}
[Pro^2] GIP_{1-30}
[Tyr^1-Ala^2 ψ(CH_2NH)Glu^3] GIP_{1-30}[#]
[(P)Ser^2]$GIP_{1-30NH2}$
[Pro^3]$GIP_{1-30NH2}$
[N-MeGlu^3] GIP_{1-30}
Val^2] GIP_{1-30}

[Gly^2] GIP_{1-30}
[Ser^2] GIP_{1-30}
[D-Tyr^1] GIP_{1-30}
[PPa^1] GIP_{1-30} ##
[D-Glu^3] GIP_{1-30}

*$t_{1/2}$ = 137-298 min with purified enzyme # [Tyr^1-$Ala^2\psi(CH_2NH)Glu^3$]$GIP_{1-30NH2}$: reduced peptide bond between Ala^2 and Glu^3 ##PPa = ([Desamino Tyr^1(phenylproprionic acid)

Competitive binding and cyclic AMP stimulation studies were performed on GIP-R1 cells with all peptides. In general, when examining binding affinity, the amino-terminus of GIP was fairly tolerant of amino acid substitution or modification. With [PPa^1]-, [D-Tyr^1]-, [D-Ala^2]- and [D-Glu^3]-GIP little or no change in binding affinity was observed. Substitutions of [$Y^1A^2\psi(CH_2NH)$], [Gly^2], and [Ser^2] were well tolerated with respect to binding affinity, and binding of [Val^2]-, [Pro^3]- and [N-$MeGlu^3$]-GIP was only modestly reduced. However all peptides exhibiting complete DP IV resistance, apart from [D-Ala^2]GIP, displayed dramatically reduced cyclic AMP stimulating ability. [$Y^1A^2\psi(CH^2NH)$]-, [Pro^3]- and [N-$MeGlu^3$]-GIP were not even capable of achieving maximal levels. [(P)Ser^2]GIP showed both greatly reduced binding affinity and cAMP production.

From these binding and cyclic AMP data, [D-Ala^2]GIP was concluded to have the greatest potential for further peptide development[20]. Both [D-Ala^2]GIP_{1-30} and [D-Ala^2]GIP_{1-42} were completely resistant to DP IV degradation for over 24 hours, and had minimal changes in receptor binding. When tested *in vivo*, [D-Ala^2]GIP reduced glycaemic excursions in Wistar rats and both lean and obese Vancouver diabetic Zucker (VDZ) rats to a greater extent than native GIP. This was associated with enhanced early phase insulin release in lean animals, and in diabetic rats where first phase insulin release is compromised, augmentation over the entire insulin time-course was observed. This is of particular interest since, in contrast to the lack of insulin response in fatty VDZ rats when GIP was administered to approximate physiological levels[21], insulin responses could be induced with pharmacological doses.

Although [Ser^2]$GIP_{1-30NH2}$ was only moderately resistant to DP IV degradation (~137-686 min), it had favourable binding and signalling characteristics at the cloned GIP receptor. In studies on the degradation of glucagon[22], which normally has a serine in position 2, serine phosphorylation resulted in complete resistance to purified DP IV, and this molecule was rapidly dephosphorylated in serum. We therefore examined the effect of phosphorylation of Ser^2 on the characteristics of $GIP_{1-30NH2}$. [(P)Ser^2]$GIP_{1-30NH2}$ was completely resistant to DP IV. Although the affinity of

[(P)Ser[2]]GIP was reduced 21-fold and the concentration-response curve for cAMP production was right shifted (433-fold), when [Ser[2]]$GIP_{1-30NH2}$ was administered sc during a OGTT, it resulted in a slightly more pronounced reduction in the glycaemic profile than $GIP_{1-30NH2}$, and an enhanced insulin time-course. Higher doses of [(P)Ser[2]] GIP resulted in greater reduction in the glycemic profile than $GIP_{1-30NH2}$, and significantly enhanced insulin responses. Both [D-Ala[2]]GIP and [(P)Ser[2]]GIP are therefore promising analogues and, along with [Tyr[1]-Glucitol]GIP[23], may well prove useful for improving glucose tolerance in humans.

5. DEVELOPMENT OF TRUNCATED GIP ANALOGUES

GIP is the largest polypeptide of the glucagon-secretin family, with 42 amino acids, as compared to 30 in GLP-1_{7-36}. It would clearly be advantageous to develop truncated forms of the molecule that retain biological activity, and we performed structure-activity studies with the view to designing such analogues.

Evidence has been presented for the existence of multiple domains in GIP_{1-42}[19]. Early studies showed that it is possible to truncate the C-terminus with retention of biological activity; synthetic GIP_{1-38} and GIP_{1-42} were shown to be equally insulinotropic[1,2]. Equal potency to GIP_{1-42} has been demonstrated for GIP_{1-31} and GIP_{1-30} in stimulating cyclic AMP production in insulinoma cells[8] and insulin release from the perfused pancreas. However $GIP_{1-30NH2}$ lacks gastric effects, although it is unclear whether this is due to the existence of a second GIP receptor, an alternatively spliced receptor, or differential ligand recognition and coupling of the receptor in gastric cells.

The insulinotropic domain of GIP has been localized to residues 19 to 30, consistent with partial retention of insulinotropic activity of GIP_{19-30}, GIP_{15-42} and GIP_{17-42} [1,19]. Residues 27-30 may be important for biological activity, as GIP_{1-27} and GIP_{1-28} are devoid of insulinotropic potency. Computer assisted secondary structure analysis of GIP predicts an alpha helical region between residues 10 and 29 (Fig. 1). Hence, it is possible that this helical structure is important for biological activity. Antagonism of the GIP receptor has been demonstrated with N-terminally truncated peptides, $GIP_{6-30NH2}$, $GIP_{7-30NH2}$, and $GIP_{10-30NH2}$, and the complete high affinity binding domain of GIP resides between residues 6 and 30[24]. Recently, we obtained evidence for a bioactive domain of GIP residing in residues 1-14[20]. In binding studies, bioactivity of N-terminal GIP fragments was restricted to GIP_{1-14}, with weak responses to amidated forms of GIP_{1-13} and GIP_{1-15}. $GIP_{1-14NH2}$ exhibits low, but significant, insulinotropic activity. Using a bioassay, both $GIP_{1-14NH2}$ and, to a lesser extent, $GIP_{19-30NH2}$ reduced excursions in glycaemia in an equivalent manner to GIP_{1-42OH}, although 100-fold greater doses were needed[20].

In recent studies, an alanine scan of GIP_{1-14OH} was performed to identify key residues contributing to biological activity. Substitution of any residue of the 1-14 primary sequence resulted in significantly reduced binding affinity, with the exception of a glucagon-substitution analogue, $[Tyr^{13}]GIP_{1-14OH}$. Apart from the Ala^1 and Ala^4 analogues, most of the substituted alanine analogues were devoid of biological activity indicating that these residues are particularly important for conferring structure optimal for binding. $[D-Ala^2]$ substitution of GIP_{1-14OH} was also not well tolerated.

Given that GIP_{1-14} and GIP_{19-30} both demonstrate receptor binding ability, and that the high affinity binding domain of GIP resides within residues 6 to 30[24], it is likely that multiple contact residues contribute to high affinity receptor binding. Additionally, the two N-terminal residues may either interact with or be in close proximity to the core region (possibly indicating the presence of a functional hinge in the alpha helices), resulting in receptor activation. Using this information it may be possible to develop more potent analogues.

6. CONCLUSION

Although type 2 diabetic patients exhibit resistance to GIP when the peptide is administered in doses that result in circulating levels approximating those found physiologically, it is likely that DP IV-resistant forms of the peptide administered in pharmacological doses will prove to be effective in improving glucose tolerance. Additionally, in view of recent studies showing that GIP receptor knockout mice are resistant to diet induced obesity[25], it is possible that GIP-antagonists will prove useful in obesity treatment.

ACKNOWLEDGEMENTS

Work by the authors described in this review was funded by the Canadian Institutes of Health Research (RAP and CHSM Grant #MOP-13192), the Department of Science and Technology of Saxony Anhalt (HUD # 9704/ 00116) and the Federal Department of Science and Technology (HUD. #0312302).

REFERENCES

1. Brown J.C., Buchan, A.M.J., McIntosh, C.H.S., and Pederson R.A. 1989. Gastric inhibitory polypeptide. In *Handbook of Physiology, Section 6 the Gastrointestinal System* (S.G. Schultz, G.M. Makhlouf, and B.B. Rauner), Am. Physiol. Soc., Bethesda pp. 403-430.
2. Pederson, R.A. 1994, GIP. In *Gut Peptides: Biochemistry and Physiology* (J.H. Walsh and G.J. Dockray, eds.), Raven Press, New York, pp. 217-259.
3. Holst, J.J. 1999. Glucagon-like peptide 1 (GLP-1): an intestinal hormone signaling nutritional abundance, with an unusual therapeutic potential. *Trends Endocrinol. Metab.* **10**: 229-234.
4. Meier, J.J., Nauck, M.A., Schmidt, W.E. and Gallwitz, B. 2002. Gastric inhibitory polypeptide: the neglected incretin. *Reg. Peptides* **107**: 1-13.
5. Trümper, A., Trümper, K., Trusheim, H., Arnold, R., et al. 2001.Glucose-dependent insulinotropic polypeptide is a growth factor for β(INS-1) cells by pleiotropic signaling. *Mol. Endocrinol.* **15**: 1559-1570.
6. Ehses, J., Casilla, V., Doty, T., Pospisilik, J.A., Demuth, H.-U., Pederson, R.A. and McIntosh, C.H.S. 2002. Glucose-dependent Insulinotropic Polypeptide (GIP) stimulates cell growth and promotes cell survival in INS-1 cells. *Diabetes* **51**:A339.
7. Jia, X., Brown, J.C., Ma P., Pederson, R.A., and McIntosh, C.H.S. 1995. The effects of glucose dependent insulinotropic polypeptide and glucagon-like peptide-1(7-36) on insulin secretion. *Am. J. Physiol* **268**: E645-E651.
8. Wheeler, M.B., Gelling, R.W., McIntosh, C.H.S., Georgiou, J., Brown, J.C. and Pederson, R.A. 1995 Functional expression of the rat pancreatic islet glucose-dependent insulinotropic polypeptide (GIP) receptor: Ligand binding and intracellular signaling properties. *Endocrinology* **136**: 4629-4639.
9. McIntosh, C.H.S., Bremsak, I., Lynn, F.C., Gill R., Hinke, S.A., Gelling, R., McKnight, G., Jaspers, S., Pederson, R.A. 1999. Glucose dependent insulinotropic polypeptide stimulation of lipolysis in differentiated 3T3-L1 cells: wortmannin-sensitive inhibition by insulin. *Endocrinology* **140**: 398-404.
10. Brown, J.C., Dahl, M., Kwauk, S., McIntosh, C.H.S., Ottte, S.C. and Pederson, R.A., 1981, Actions of GIP. *Peptides* **2 (Suppl. 2)**: 241-245.
11. Jörnvall, H., Carlquist, M., Kwauk, S., Otte, S.C., McIntosh, C.H.S., Brown, J.C. and Mutt, V. 1981. Amino acid sequence and heterogeneity of gastric inhibitory polypeptide (GIP). *FEBS Lett.* **123**: 205-210.
12. Mentlein, R., Gallwitz, B. and Schmidt, W.E. 1993. Dipeptidyl peptidase IV hydrolyses gastric inhibitory polypeptide, glucagon-like peptide-1 (7-36)amide, peptide histidine methionine and is responsible for their degradation in human serum. *Eur. J. Biochem.* **214**: 829-835.
13. Kieffer, T.J. McIntosh, and Pederson, R.A. 1995. Degradation of glucose-dependent insulinotropic polypeptide and truncated glucagon-like peptide 1 *in vitro* and *in vivo* by dipeptidyl peptidase IV. *Endocrinology* **136**: 3585-3596.
14. Perley, M.J., and Kipnis, D.M. 1967. Plasma insulin responses to oral and intravenous glucose: studies in normal and diabetic subjects. *J. Clin. Invest.* **46**: 1954-1962.
15. Jones, I.R., Owens, D.R., Moody, A.J., Luzio, S.D., et al. 1987. The effects of glucose dependent insulinotropic polypeptide infused at physiological concentrations in normal subjects and type 2 (non-insulin-dependent) diabetic patients on glucose tolerance and B-cell secretion. *Diabetologia* **30**: 707-712.

16. Meneilly, G.S., Bryer-Ash, M. and Elahi, D. 1993. The effect of glyburide on ß-cell sensitivity to glucose-dependent insulinotropic polypeptide. *Diabetes Care* **16**: 110-114.
17. Nauck, M., Heimesaat, M.M., Ørskov, C., Holst, J.J., Ebert, R. and Creutzfeldt, W. Preserved incretin activity of glucagon-like peptide 1 [7-36 amide] but not of synthetic human gastric inhibitory polypeptide in patients with type-2 diabetes mellitus. *J. Clin. Invest.* **91**: 301-307.
18. Kühn-Wache, K., Manhart, S., Hoffmann, T., Hinke, S.A., Gelling, R., Pederson, R.A., McIntosh, C.H.S. and Demuth, H-U. 2000. Analogs of glucose-dependent insulinotropic polypeptide with increased dipeptidyl peptidase resistance. *Adv. Exp. Med. Biol.* **477**: 187-195.
19. Hinke, S.A., Manhart, S., Pamir, N., Demuth, H.-U., Gelling, R.W., Pederson, R.A. and McIntosh. C.H.S. 2001. Identification of a bioactive domain in the amino-terminus of glucose-dependent insulinotropic polypeptide (GIP) *Biochim. Biophys. Acta* **1547**: 143-155.
20. Hinke, S.A., Gelling, R.W., Pederson, R.A., Manhart, S, Nian, C., Demuth, H.-U. and McIntosh, C.H.S. Dipeptidyl peptidase IV-resistant [D-Ala^2]glucose-dependent insulinotropic polypeptide (GIP) improves glucose tolerance in normal and obese diabetic rats. *Diabetes* **51**: 652-661, 2002.
21. Lynn, F.C., Pamir, N., Ng, E.H.C., McIntosh, C.H.S., and Pederson, R.A. 2001. Defective glucose-dependent insulinotropic polypeptide receptor expression in diabetic fatty Zucker rats. *Diabetes* **50**:1004-1011.
22. Hinke, S. A., Pospisilik,J.A., Demuth, H.-U., Mannhart, S., Kühn-Wache, K., Hoffmann, T., Nishimura, E., Pederson, R. A. and McIntosh, C.H.S. 2000. Dipeptidyl peptidase IV (DP IV/CD26) degradation of glucagon. *J. Biol. Chem.* **275**: 3827-3834.
23. O'Harte, F., Mooney, M. and Flatt, P. 1999. NH_2-terminally modified gastric inhibitory polypeptide exhibits amino-peptidase resistance and enhanced antihyperglycemic activity. *Diabetes* **48**: 758-765.
24. Gelling R.W., Coy D., Pederson R.A., Wheeler M.B., O'Dorisio T., Hinke S. and McIntosh C.H.S. 1997. GIP_{6-30} amide contains the high affinity binding region of GIP and is a potent inhibitor of GIP_{1-42} action *in vitro*. *Reg. Peptides* **69**: 151-154.
25. Miyawaki, K., Yamada, Y., Ban, N., Ihara, Y., Tsukiyama, K., Zhou, H., Fujimoto, S., Oku, A., Tsuda, K. et al. Inhibition of gastric inhibitory polypeptide signaling prevents obesity. *Nature Med.* **8**: 738-742.

Neutral Endopeptidase 24.11 and Dipeptidyl Peptidase IV are Both Involved in Regulating the Metabolic Stability of Glucagon-like Peptide-1 *in vivo*

ASTRID PLAMBOECK*, JENS J.HOLST*, RICHARD D. CARR#, and CAROLYN F. DEACON*
**Department of Medical Physiology, Panum Institute, Blegdamsvej 3, DK-2200 Copenhagen N, Denmark; #Pharmacology Research 1, Novo Nordisk A/S, Novo Allé, DK-2880 Bagsværd, Denmark*

1. INTRODUCTION

The incretin hormone, glucagon-like peptide-1 (GLP-1) has a spectrum of effects which makes it an attractive target in the search for new therapies for type 2 diabetes. However, the main factor limiting the clinical application of GLP-1 is its metabolic instability, since the peptide is rapidly degraded and inactivated *in vivo*. A number of strategies have been proposed to take advantage of beneficial effects of GLP-1, including the development of degradation-resistant analogues and the use of selective enzyme inhibitors to potentiate the effects of the endogenously released peptide. Much recent research has, therefore, focussed upon the metabolic stability of GLP-1 *in vivo*, in an attempt to uncover which enzymes are important physiologically in regulating the peptide's metabolic stability.

Dipeptidyl Aminopeptidases in Health and Disease, Edited by Hildebrandt et al.
Kluwer Academic/Plenum Publishers, New York, 2003

2. GLUCAGON-LIKE PEPTIDE-1

2.1 Physiology

Glucagon-like peptide-1 arises from specific post-translational processing of the glucagon precursor, proglucagon, within the intestinal mucosal L-cells[1,2]. It is released in response to orally ingested nutrients[3] and has effects on the endocrine pancreas and the gastrointestinal tract as well as in the brain. These effects are mediated by activation of a specific GLP-1 receptor linked to adenylyl cyclase[4]. Structure-activity studies have demonstrated that while residues in the C-terminal region of the molecule are important for receptor binding, the penultimate two N-terminal residues are crucial for receptor activation[5]. GLP-1 is a highly potent insulin secretagogue[6], an effect which is dependent upon the prevailing blood glucose concentration, so that the insulinotropic action is greater at elevated glucose concentrations and declines as normal fasting glucose levels are approached[7]. As well as stimulating insulin secretion, the peptide also increases insulin gene transcription[8,9]. GLP-1 confers glucose competence to the β-cell *in vitro*[10] and improves its ability to sense and respond to glucose *in vivo*[11]. More recently, it has been shown to promote β-cell growth and survival[12,13]. In addition to effects on the β-cell, GLP-1 also inhibits glucagon secretion[6]. GLP-1 reduces gastrointestinal motility and gastric acid secretion[14], helping to reduce post-prandial blood glucose excursions by delaying the passage of further nutrients to the small intestine. Moreover, it even affects feeding behaviour, reducing appetite and promoting the sense of satiety[15].

2.2 Pharmacology

GLP-1 potentiates meal-induced insulin in both healthy subjects and patients with type 2 diabetes, and many of the other actions of GLP-1 are also preserved in diabetes and will additionally contribute to lowering blood glucose and minimising glucose excursions. However, it seems that exogenous GLP-1 must be continuously administered in order to be effective[16]. Thus, when given as a continuous intravenous infusion, GLP-1 actually normalises blood glucose in diabetic subjects[17,19,16], but there is no sustained improvement once the infusion is discontinued[16,19]. The same pattern is found when GLP-1 is given as single subcutaneous injections. It is effective at curtailing post-prandial glucose excursions when administered before each meal, but the blood glucose-lowering effect is lost between successive injections[20]. However, the effects of repeated subcutaneous injections are as effective as intravenous infusion[18], and continuous subcutaneous infusions of GLP-1 improve glucose profiles in type 2 diabetic patients[21].

3. DEGRADATION OF GLP-1

The need for exogenous GLP-1 to be administered continuously in order to be effective is explained by the peptide's metabolic stability. After intravenous administration, when measured with a non-discriminating assay, the half-life corresponds to the renal clearance, being around 3 minutes in pigs[22], and slightly longer (4-5 minutes) in humans[23]. However, when determined using methodology specific for the intact, biologically active peptide, it is eliminated with an apparent plasma half-life of only 1-2 minutes[24,23]. Many recent studies have focussed upon the mechanisms responsible for the N-terminal truncation of GLP-1, since this appears to be the primary inactivating step. Rather fewer studies have, so far, examined the involvement of other enzymes.

3.1 GLP-1 and Dipeptidyl Peptidase IV

GLP-1 was indicated to be susceptible to N-terminal degradation *in vitro* by Buckley and Lundquist[25], and later, dipeptidyl peptidase IV (DPP IV) was shown to be capable of mediating such cleavage[26]. This enzyme is a serine protease which is prevalent in sites including the kidney and intestinal brush-border membranes, hepatocytes, the capillary endothelium, and in plasma[27]. It cleaves a dipeptide from the N-terminal end of the GLP-1 molecule (Fig. 1), generating a metabolite which is non-insulinotropic[28]. Studies *in vivo* have demonstrated that DPP IV is likely to have a physiological role in regulating the metabolic stability of GLP-1[29,30,24].

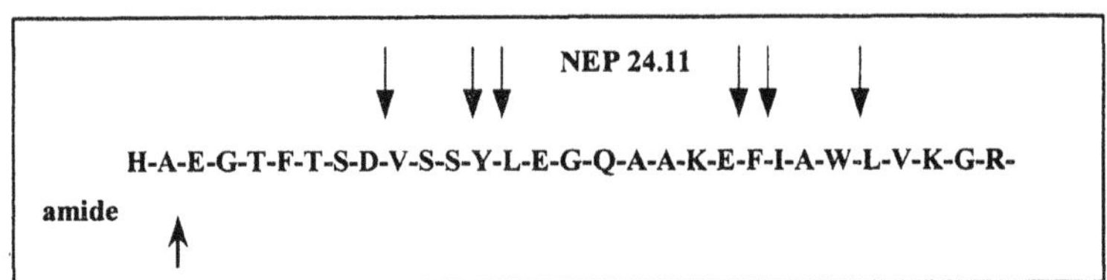

Figure 1. The amino acid sequence of GLP-1 (7-36)amide, showing the cleavage sites of DPP IV and NEP 24.11 (from Mentlein *et al.*[26] and Hupe-Sodmann *et al.*[31])

3.2 GLP-1 and Neutral Endopeptidase 24.11

Neutral endopeptidase 24.11 (NEP 24.11) is a membrane-bound zinc metallopeptidase which cleaves peptides at the N-terminal side of aromatic or hydrophobic amino acids (Fig. 1). It is capable of degrading members of the glucagon/secretin/glucose-dependent insulinotropic polypeptide (GIP) family of peptides in *in vitro* studies, with GLP-1 being a reasonably good substrate[31,32]. However, the physiological significance of NEP 24.11 has not been examined *in vivo*. NEP 24.11 has a widespread tissue distribution and is found in particularly high concentration in the kidneys[33], where it could be speculated to be involved in the renal clearance of peptide hormones.

4. ENZYME INHIBITION AND GLP-1 STABILITY IN *VIVO*

Although *in vitro* studies provide compelling evidence that GLP-1 is a substrate for DPP IV and NEP 24.11, it is only by selective elimination of the enzymes' effects *in vivo* that the physiological relevance of these enzymes can be demonstrated. Early reports using a strain of rat with a mutated, enzymatically inactive DPP IV showed that N-terminal degradation of GLP-1 (and GIP) was reduced[24]. Later, studies in mice with a targeted disruption of CD26, in which DPP IV is not expressed, also demonstrated that the metabolic stability of GLP-1 was increased and, furthermore, that these animals had improved glucose tolerance[34]. A number of studies have now examined the effect of inhibiting DPP IV activity *in vivo* using selective inhibitors. The conclusions of these studies may, perhaps, be viewed as more meaningful, in terms of potential therapeutic use, since the animals inherently possess active DPP IV, the action of which is then selectively eliminated. This, therefore, avoids the possibility of compensatory adaptive mechanisms which may occur in the animals genetically devoid of DPP IV activity.

4.1 DPP IV Inhibition

Early studies examining the effect of valine-pyrrolidide (a selective DPP IV inhibitor[35]) on the pharmacokinetics and pharmacodynamics of exogenously infused GLP-1 showed that DPP IV inhibition totally prevents N-terminal degradation of GLP-1 *in vivo*. This resulted in a 3-fold increase in the half-life of the intact, biologically active peptide, and an improvement of its insulinotropic activity[36]. DPP IV inhibitors also reduce N-terminal degradation of endogenous incretin hormones, increasing nutrient-induced concentrations of intact GLP-1[37,38]. Subsequently, a number of studies have examined the effect of several different DPP IV inhibitors on glucose tolerance. Thus, in rodents, DPP IV inhibition improves glucose tolerance and enhances insulin secretion[39,37,40,41]. A recent clinical trial of 4-weeks treatment with a DPP IV inhibitor shows that it is also effective in reducing glucose profiles and HbA1c levels in patients with type 2 diabetes[42].

4.2 Combined DPP IV and NEP 24.11 Inhibition

In order to see whether NEP 24.11 plays a physiological role *in vivo*, the effect of valine-pyrrolidide, given either alone or together with candoxatril (a selective NEP 24.11 inhibitor[43]) was examined in anaesthetised pigs. The administration of both inhibitors increased the plasma half-life of exogenously infused GLP-1 by more than 3-fold (from 2.5±0.2 to 8.3±0.9 min, $P<0.002$) and halved the metabolic clearance rate (from 25.1±3.9 to 11.7±1.2 ml/kg/min, $P<0.008$) compared to DPP IV inhibition alone (Fig.2). At least part of this is accounted for by a reduction in the renal clearance of GLP-1, presumably due to the inhibition of NEP 24.11 in the kidney.

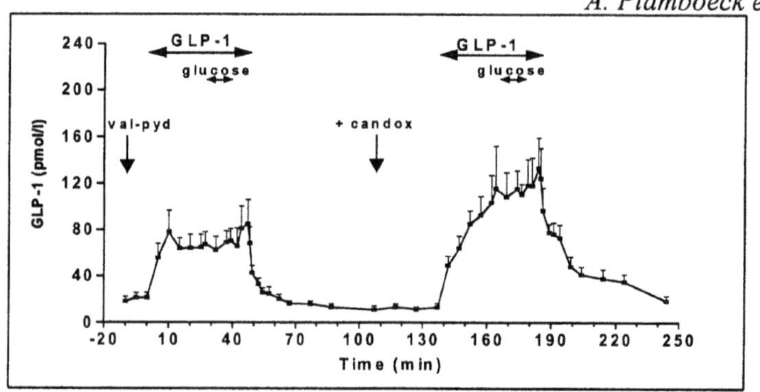

Figure 2. Plasma GLP-1 profile after administration of a DPP IV inhibitor (val-pyd) alone or together with a NEP 24.11 inhibitor (candox)

The glucose excursion following an intravenous glucose load in the presence of valine-pyrrolidide alone ($AUC_{27-67\ min}$, 59±4 mmol/l min) was further reduced ($P<0.016$) by co-administration of candoxatril (to 28±10mmol/l min) and the glucose elimination rate was increased (from 6.6±0.5 to 11.6±1.3 %/min). Determination of insulin concentrations revealed that the combined enzyme inhibition potentiated ($P<0.008$) insulin secretion ($AUC_{27-67\ min}$, 6486±1064 pmol/l min) even more than DPP IV inhibition alone (3606±668 pmol/l min).

5. CONCLUSION

DPP IV is the enzyme responsible for the initial inactivation of GLP-1, but NEP 24.11 plays a role in the subsequent elimination of both the intact peptide and the truncated metabolite. DPP IV inhibition protects intact GLP-1 from N-terminal truncation, leading to improved insulinotropic and anti-hyperglycaemic activity. However, NEP 24.11 inhibition also contributes to improving the metabolic stability of GLP-1 *in vivo.* Combined NEP 24.11 and DPP IV inhibition is superior to DPP IV inhibition alone in reducing clearance and improving the anti-hyperglycaemic and insulinotropic activity of GLP-1, providing the first evidence that inhibition of NEP 24.11 may also have therapeutic potential in diabetes treatment.

ACKNOWLEDGEMENTS

These studies were supported by Danish Medical Research Council, the Novo Nordisk Foundation and the Danish Biotechnology Programme.

REFERENCES

1. Ørskov, C., Holst, J.J., Knuhtsen, S., Baldissera, F.G., Poulsen, S.S., and Nielsen, O.V., 1986, Glucagon-like peptides GLP-1 and GLP-2, predicted products of the glucagon gene, are secreted separately from pig small intestine but not pancreas. *Endocrinology* **119**: 1467-1475
2. Mojsov, S., Heinrich, G., Wilson, I.B., Ravazzola, M., Orci, L., and Habener, J.F., 1986, Preproglucagon gene expression in pancreas and intestine diversifies at the level of post-translational processing. *J Biol Chem* **261**: 11880-11889
3. Elliott, R.M., Morgan, L.M., Tredger, J.A., Deacon, S., Wright, J., and Marks, V., 1993, Glucagon-like peptide-1 (7-36)amide and glucose-dependent insulinotropic polypeptide secretion in response to nutrient ingestion in man: acute post-prandial and 24-h secretion patterns. *Endocrinology* **138**: 159-166
4. Thorens, B., 1992, Expression cloning of the pancreatic beta cell receptor for the gluco-incretin hormone glucagon-like peptide 1. *Proc Natl Acad Sci U S A* **89**: 8641-8645
5. Adelhorst, K., Hedegaard, B.B., Knudsen, L.B., and Kirk, O., 1994, Structure-activity studies og glucagon-like peptide-1. *J Biol Chem* 269: 6275-6278
6. Ørskov, C., Holst, J.J., and Nielsen, O.V., 1988, Effect of truncated glucagon-like peptide-1 [proglucagon-(78-107)amide] on endocrine secretion from pig pancreas, antrum and non-antral stomach. *Endocrinology* **123**: 2009-2013
7. Weir, G.C., Mojsov, S., Hendrick, G.K., and Habener, J.F., 1989, Glucagon-like peptide I (7-37) actions on the endocrine pancreas. *Diabetes* **38**: 338-342
8. Drucker, D.J., Philippe, J., Mojsov, S., Chick, W.L., and Habener, J.F., 1987, Glucagon-like peptide I stimulates insulin gene expression and increases cyclic AMP levels in rat islet cell line. *Proc Natl Acad Sci USA* **84**: 3434-3438
9. Fehmann, H.C., and Habener, J.F., 1992, Insulinotropic hormone glucagon-like peptide-I (7-37) stimulation of proinsulin gene expression and proinsulin biosynthesis in insulinoma βTC-1 cells. *Endocrinology* **130**: 159-166
10. Holz, G.G., Kühtreiber, W.M., and Habener, J.F., 1993, Pancreatic beta cells are rendered glucose-competent by the insulinotropic hormone glucagon-like peptide-1 (7-36). *Nature* **361**: 362-365
11. Byrne, M.M., Gliem, K., Wank, U., Arnold, R., Katschinski, M., Polonsky, K.S., and Göke, B., 1998, Glucagon-like peptide 1 improves the ability of the beta-cell to sense and respond to glucose in subjects with impaired glucose tolerance. *Diabetes* **47**: 1259-1265
12. Edvell, A., and Lindström, P., 1999 Initiation of increased pancreatic islet growth in young normoglycemic mice (Umeå +/?). Endocrinology **140**: 778-783

13. Hansotia, T., Yusta, B., and Drucker, D.J., 2001, Activation of GLP-1 receptor signalling is coupled to inhibition of apoptosis in heterologous cell types. *Diabetes* **50** (Suppl 2): A350 (abstract)
14. Wettergren, A., Schjoldager, B., Mortensen, P.E., Myhre, J., Christiansen, J., and Holst, J.J., 1993, Truncated GLP-1 (proglucagon 78-107-amide) inhibits gastric and pancreatic functions in man. *Dig Dis Sci* **38**: 665-673
15. Flint, A., Raben, A., Rehfeld, J.F., Holst, J.J., and Astrup, A., 2000, The effect of glucagon-like peptide-1 on energy expenditure and substrate metabolism in humans. *Int J Obes Relat Metab Disord* **24**: 288-298
16. Larsen, J., Hylleberg, B., Ng, K., and Damsbo, P., 2001, Glucagon-like peptide-1 infusion must be maintained for 24 h/day to obtain acceptable glycemia in type 2 diabetic patients who are poorly controlled on sulphonylurea treatment. *Diabetes Care* **24**: 1416-1421
17. Nauck, M.A., Kleine, N., Ørskov, C., Holst, J.J., Willms, B., and Creutzfeldt, W., 1993, Normalization of fasting hyperglycemia by exogenous GLP-1 (7-36amide) in type 2 diabetic patients. *Diabetologia* **36**: 741-744
18. Nauck, M.A., Wollschläger, D., Werner, J., Holst, J.J., Ørskov, C., Creutzfeldt, W., and Willms, B., 1996, Effects of subcutaneous glucagon-like peptide 1 (GLP-1 [7-36 amide] in patients with NIDDM. *Diabetologia* **39**: 1546-1553
19. Rachman, J., Barrow, B.A., Levy, J.C., and Turner, R.C., 1997, Near-normalisation of diurnal glucose concentrations by continuous administration of glucagon-like peptide-1 (GLP-1) in subjects with NIDDM. *Diabetologia* **40**: 205-211
20. Juntti-Berggren, L., Pigon, J., Karpe, F., Hamsten, A., Gutniak, M., Vignati, L., and Efendic, S., 1996, The antidiabetogenic effect of GLP-1 is maintained during a 7-day treatment period and improves diabetic dyslipoproteinemia in NIDDM patients. *Diabetes Care* **19**: 1200-1206
21. Zander M, Madsbad S, Madsen JL, Holst JJ., 2002, Effect of 6-week course of glucagon-like peptide 1 on glycaemic control, insulin sensitivity, and beta-cell function in type 2 diabetes: a parallel-group study. *Lancet* **359**: 824-830
22. Deacon, C.F., Pridal, L., Klarskov, L., Olesen, M, and Holst, J.J., 1996, Glucagon-like peptide-1 undergoes differential tissue-specific metabolism in the anesthetized pig. *Am J Physiol* **271**: E458-E464
23. Ørskov, C., wettergren, A., and Holst, J.J., 1993, Biological effects and metabolic rates of glucagonlike peptide-1 7-36 amide and glucagonlike peptide-1 7-37 in healthy subjects are indistinguishable. *Diabetes* **42**: 658-661
24. Kieffer, T.J., McIntosh, C.H., and Pederson, R.A., 1995, Degradation of glucose-dependent insulinotropic polypeptide and truncated glucagon-like peptide 1 in vitro and in vivo by dipeptidyl peptidase IV. *Endocrinology* **136**: 3585-3596
25. Buckley, D.I., and Lundquist, P., 1992, Analysis of the degradation of insulinotropin [GLP-1 (7-37)] in human plasma and production of degradation resistant analogs. *Regul Pept* **40**: 117 (abstract)
26. Mentlein, R., Gallwitz, B. and Schmidt, W.E., 1993, Dipeptidyl peptidase IV hydrolyses gastric inhibitory polypeptide, glucagon-like peptide-1 (7-36)amide, peptide histidine methionine and is responsible for their degradation in human serum. *Eur. J. Biochem.* **214**: 829-835
27. Mentlein, R., 1999, Dipeptidyl-peptidase IV (CD26)--role in the inactivation of regulatory peptides. *Regul Pept.* **85**: 9-24

28. Deacon, C.F., Plamboeck, A., Møller, S., and Holst, J.J., 2002a, GLP-1 (9-36)amide reduces blood glucose in anesthetized pigs by a mechanism that does not involve insulin secretion. *Am J Physiol* **282**: E873-E879
29. Deacon, C.F., Johnsen, A.H., and Holst, J.J., 1995, Degradation of glucagon-like peptide-1 by human plasma in vitro yields an N-terminally truncated peptide which is a major endogenous metabolite in vivo. *J Clin Endocrinol Metab* **80**: 952-957
30. Deacon, C.F., Nauck, M.A., Toft-Nielsen, M., Pridal, L., Willms, B., and Holst, J.J., 1995, Both subcutaneously and intravenously administered glucagon-like peptide-1 are rapidly degraded from the NH_2-terminus in type II diabetic patients and in healthy subjects. *Diabetes* **44**: 1126-1131
31. Hupe-Sodmann, K., McGregor, G.P., Bridenbaugh, R., Göke, R., Göke, B., Thole, H., Zimmermann, B., and Voigt, K., 1995, Characterisation of the processing by human neutral endopeptidase 24.11 of GLP-1(7-36) amide and comparison of the substrate specificity of the enzyme for other glucagon-like peptides. *Regul Pept.* **58**: 149-156.
32. Hupe-Sodmann, K., Göke, R., Göke, B., Thole, H.H., Zimmermann, B., Voigt, K., and McGregor, G.P., 1997, Endoproteolysis of glucagon-like peptide (GLP)-1 (7-36) amide by ectopeptidases in RINm5F cells. *Peptides* **18**: 625-632.
33. Gee, N.S., Bowes, M.A., Buck, P., and Kenny, A.J., 1985, .An immunoradiometric assay for endopeptidase-24.11 shows it to be a widely distributed enzyme in pig tissues. *Biochem J* **228**: 119-126
34. Marguet, D., Baggio, L., Kobayashi, T., Bernard, A.M., Pierres, M., Nielsen, P.F., Ribel, U., Watanabe, T., Drucker, D.J., and Wagtmann, N., 2000, Enhanced insulin secretion and improved glucose tolerance in mice lacking CD26. *Proc Natl Acad Sci U S A* **97**:6874-6879
35. Neubert, K., Born, I., Faust, J., Heins, J., Barth, A., Demuth, H.U., Rahfeld, J.U., and Steinmetzer, T., 1983, Verfahren zur Herstellung neuer Inhibitoren der Dipeptidyl Peptidase IV. *German Patent Application* Number DD 296 075 A5
36. Deacon, C.F., Hughes, T.E., and Holst, J.J., 1998, Dipeptidyl peptidase IV inhibition potentiates the insulinotropic effect of glucagon-like peptide-1 in anesthetized pigs. *Diabetes* **47**: 764-769
37. Balkan, B., Kwasnik, L., Miserendino, R., Holst, J.J., and Li, X., 1999, Inhibition of dipeptidyl peptidase IV with NVP-DPP728 increases plasma GLP-1 (7-36 amide) concentrations and improves oral glucose tolerance in obese Zucker rats. *Diabetologia* **42**: 1324-1331
38. Deacon, C.F., Wamberg, S., Bie, P., Hughes, T.E., and Holst, J.J., 2002b, Preservation of active incretin hormones by inhibition of dipeptidyl peptidase IV suppresses meal-induced incretin secretion in dogs. *J Endocrinol* **172**: 355-362
39. Pederson, R.A., White, H.A., Schlenzig D., Pauly, R.P., McIntosh, C.H., and Demuth, H.U., 1998, Improved glucose tolerance in Zucker fatty rats by oral administration of the dipeptidyl peptidase IV inhibitor isuleucine thiazolidide. *Diabetes* **47**: 1253-1258
40. Pospisilik, J.A., Stafford, S.G., Demuth, H.U., Brownsey, R., Parkhouse, W., Finegood, D.T., McIntosh, C.H., and Pederson, R.A., 2002, Long-term treatment with the dipeptidyl peptidase IV inhibitor P32/98 causes sustained improvements in glucose tolerance, insulin sensitivity, hyperinsulinemia, and beta-cell glucose responsiveness in VDF (fa/fa) Zucker rats. *Diabetes* **51**: 943-950
41. Sudre, B,, Broqua, P., White, R.B., Ashworth, D., Evans, D.M., Haigh, R., Junien, J.L., and Aubert, M.L., 2002, Chronic inhibition of circulating dipeptidyl peptidase IV by FE 999011 delays the occurrence of diabetes in male zucker diabetic fatty rats. *Diabetes* **51**: 1461-1469

42. Ahrén, B., Simonsson, E., Larsson, H., Landin-Olsson, M., Torgeirsson, H., Jansson, P.A., Sandqvist, M., Bavenholm, P., Efendic, S., Eriksson, J.W., Dickinson, S., and Holmes, D., 2002, Inhibition of dipeptidyl peptidase IV improves metabolic control over a 4-week study period in type 2 diabetes. *Diabetes Care* **25**: 869-875
43. Kaye, B., Brearley, C.J., Cussans, N.J., Herron, M., Humphrey, M.J., and Mollatt, A.R., 1997, Formation and kinetics of the active drug candoxatrilat in mouse, rabbit, dog and man following administration of the prodrug candoxatril. *Xenobiotica* **27**: 1091-1102

DPP IV, Immune Parameters, and Psychometrical Variables in Patients with Eating Disorders

HERBERT FLIEGE, MATTHIAS ROSE, BURGHARD F. KLAPP, and MARTIN HILDEBRANDT
[*] *Department of Internal Medicine/Psychosomatics, University Hospital Charité, Humboldt University Berlin, Luisenstrasse 13 a, D-10117 Berlin, Germany*

1. INTRODUCTION

Dipeptidyl peptidase IV (DPP IV) is involved in the metabolism of peptide hormones, T-cell-activation and -proliferation. It has been under study in various medical contexts, especially in diseases in which the immune system plays a crucial role, in the nutritional area, and in the field of psychoneuroendocrine processes including research on the assumedly stress-triggered diseases[1-3]. Still, fairly little is known about a possible role of psychological factors in these processes.

In the past decade, a couple of studies have emerged that could show alterations in serum activity of DPP IV in certain mental disorders. In particular, serum DPP IV activity was found to be significantly lower in patients with major depression or chronic depression than in normal controls[4,5]. A trend to a higher DPP IV activity was found in patients with schizophrenia[5].

Furthermore, DPP IV activity was found to be increased compared to healthy controls in the sera of patients with anorexia nervosa and, to a lesser degree, those with bulimia nervosa[3]. Yet, for none of the groups of behavioural or mental disturbances, a well-defined factor for elevated or decreased DPP IV serum activity is known.

Dipeptidyl Aminopeptidases in Health and Disease, Edited by Hildebrandt et al.
Kluwer Academic/Plenum Publishers, New York, 2003

The reported results are few and should still be considered preliminary. But from a theoretical perspective of a possible body-mind interaction they appear tentative. For all the mental disorders that have been under investigation, mood disturbances are considered an outstanding feature. But there is no convincing evidence for a direct correlation between moods and DPP IV activity.

Maes *et al.*[6] could show that in patients with chronic hepatitis C a suppression of serum DPP IV activity was significantly correlated to interferon alpha-induced increases in depression, as measured on the *Montgomery Asberg Depression Rating Scale* and the *Hamilton Anxiety Scale*. The authors had previously investigated levels of DPP IV in healthy students who did or did not respond to a stress condition with an increase in the *Spielberger State Anxiety Inventory*[7]. There were no significant effects of stress on serum DPP IV activity and no significant differences between anxiety responders and non-responders. Rose *et al.*[8] examined whether a relationship exists between T-cell activation (including DPP IV) and depressive personality characteristics in patients with Crohn's disease. DPP IV activity was significantly lower in patients with an active disease, as was their subjective health status, assessed on the *Short Inflammatory Bowel Disease Quality of Life Questionnaire*. But a connection with measured parameters of mental depression could not be found.

In the face of these scarce and contradictory results, the focus of this study is to explore the possible connections between subjective well-being indicators on one side and DPP IV activity and immune parameters on the other. We investigated this in patients with eating disorders, because it has been postulated that DPP IV modulates nutrition control by modifying or inactivating peptide hormones in the enteroinsular axis.

2. METHODS

2.1 Patients

Thirty-four female hospital in-patients with eating disorders participated in the study. The mean age was 28 years (SD 8.8, range 17 to 55 ys.). 11 were diagnosed with Anorexia nervosa (AN) and 23 with Bulimia nervosa (BN) according to the clinical guidelines of Part F the WHO's International

Classification of Diseases ICD-10. At study entry AN patients had an average body mass index (BMI) of 14.6, BN patients an average BMI of 19.1. The mean duration of treatment was 39 days. There was a small but significant increase of 0.6 points in BMI in AN patients under psychotherapy ($t=2.7$, $p=.02$). Fasting blood samples were analysed at hospital admittance and at discharge. Healthy controls were 28 female volunteers.

2.2 Assessment instruments

Determination of DPP IV activity was performed as described[1]. 20 μL of serum was incubated with 10 μL of 2mM of Gly Pro p-nitroanilide (Sigma Chemie, Germany) in 170 μL of 0.1 M Tris-HC1 (pH 8.0) for 30 min. The reaction was stopped by the addition of 800 μL of sodium acetate buffer (1 M, pH 4.5). The DPP IV activity was deduced from the increase of extinction at 405 mm due to the amount of chromogenic substrate metabolised by DPP IV. Mononuclear cells were isolated from whole-blood samples by centrifugation (Seromed, Germany). Immunophenotyping was performed with a FACscan flow cytometer (Becton-Dickinson, USA). The results were recorded as the percentage of cells positive for the respective antibody (or two different antibodies) CD2, CD3, and CD26. Determination of immunoglobulin (IgA, IgG, IgM) in serum was performed with standard techniques.

Mood dimensions were assessed by the *"Berlin Mood Questionnaire BSF"*[9], a validated self-rating questionnaire with the scales *elevated mood*, *commitment*, *anger*, *anxious depression*, *fatigue*, and *listlessness*. Complaints were assessed by the *"Giessen Complaints Questionnaire" GBB*[10], also a validated self-rating instrument. The overall score sums up basic dimensions of subjective complaints, like abdominal complaints, exhaustion, muscular-skeletal pain, or symptoms of the cardiovascular system.

Statistical tests comprise oneway analyses for multiple group comparisons, t-tests for independent samples to compare median-split groups, product-moment correlational and stepwise multiple regression analyses for associations at one point in time, and stepwise multiple regression analyses for predicting immune parameters at hospital discharge (t2) out of psychometrical variables at hospital admission (t1).

3. RESULTS

3.1 Comparisons to healthy controls

Oneway analyses (table 1) yielded lower IgG levels for AN patients, lower CD26-positive cells for AN and BN patients and a higher DPP IV activity for AN patients, as compared to healthy controls.

Table 1. Immune parameters and mediators in 11 patients with anorexia nervosa (AN), 23 patients with bulimia nervosa (BN) at hospital admission, and 28 healthy controls (HC)

Parameter	AN		BN		HC		Oneway	
	M	SD	M	SD	M	SD	F	contrast
IgA (mg/dL)	192.2	40.2	165.3	57.0	206.0	70.2	ns	
IgG (mg/dL)	879.3	246.2	1025.7	227.0	1119.9	225.6	3.87*	AN<HC
IgM (mg/dL)	137.0	35.4	125.2	34.5	129.3	45.4	ns	
CD2+ %	75.9	11.6	75.2	12.1	80.8	4.9	ns	
CD3+ %	56.1	8.4	57.2	14.0	61.2	7.1	ns	
CD26+ %	40.1	6.8	40.1	7.3	47.4	7.4	5.53**	AN,BN<HC
DPP IV (mU/mL)	104.8	24.4	91.0	19.2	80.3	17.2	5.61**	AN>HC

3.2 Immunoglobulins

For all immunoglobulins a median-split in groups with high and low levels was performed. The group with lower IgG levels gave higher ratings of *commitment* at t1 ($t=2.25$, $p=.033$) and at t2 ($t=2.43$, $p=.021$) and the group with lower IgM levels reported a greater extent of *complaints* at t1 ($t=2.14$, $p=.041$). Yet, a Bonferroni correction of the alpha-error had to be performed, in order to reduce the risk of incorrectly positive results in the case of multiple tests. The reported differences did not prove robust against an alpha-error correction (critical $p=.008$).

Three separate regression analyses for the immunoglobulins as criteria, including the seven psychometric parameters as independent variables, were carried out (table 2). At t1 analyses yield an association between *anxious depression* and high IgA, *fatigue* and low IgA and between *commitment* and low IgG. At t2 an association between *listlessness* and IgG and *complaints pressure* and low IgG was found. All results (with the exception of the association between *fatigue* and low IgA) fulfil the Bonferroni-alpha corrected criterion of significance (p-value $< .017$).

None of the psychometrical variables at t1 could predict immunoglobulin parameters at t2. Neither could any of the psychometrical variables at t1 predict changes in immunoglobulin parameters over time.

An increase of *listlessness* correlated with an increase in all three immunoglobulins IgA (r=.52, p=.005), IgM (r=.50; p=.006), and IgM (r=.58, p=.001). Again, after the alpha-error adjustment only the latter association remains significant (critical p=.002).

3.3 T-cell surface markers

A median-split in groups with high and low levels of t-cell surface markers was performed. But merely a trend to higher ratings of *anger* at t2 (t=2.00, p=.055) was found for the group with lower levels of CD26.

Table 2. Overview of the significant regression equations of psychological variables on immune parameters and DPP IV activity at one time-point

Time-point	Criterion	Predictor	beta	p	Explained variance
t1	IgA	elevated mood	.21	ns	.26
		commitment	.04	ns	
		anger	-.19	ns	
		anxious depression	.63	.006	
		fatigue	-.44	.046	
		listlessness	-.22	ns	
		overall complaints	.23	ns	
t1	IgG	elevated mood	-.11	ns	.23
		commitment	-.48	.009	
		anger	.04	ns	
		anxious depression	-.09	ns	
		fatigue	-.02	ns	
		listlessness	-.03	ns	
		overall complaints	-.09	ns	
t2	IgG	elevated mood	-.09	ns	.42
		commitment	-.13	ns	
		anger	.22	ns	
		anxious depression	-.15	ns	
		fatigue	-.24	ns	
		listlessness	.70	.001	
		overall complaints	-.50	.005	
t2	CD2+ %	elevated mood	-.18	ns	.14
		commitment	.10	ns	
		anger	-.37	.050	
		anxious depression	-.01	ns	
		fatigue	-.02	ns	
		listlessness	.11	ns	
		overall complaints	.13	ns	
t2	CD26+ %	elevated mood	-.13	ns	.15
		commitment	.08	ns	
		anger	-.38	.042	
		anxious depression	.00	ns	

Time-point	Criterion	Predictor	beta	p	Explained variance
		fatigue	.26	ns	
		listlessness	.15	ns	
		overall complaints	.43	ns	
t2	DPP IV activity	elevated mood	-.23	ns	.17
		commitment	-.41	.022	
		anger	.07	ns	
		anxious depression	.09	ns	
		fatigue	-.24	ns	
		listlessness	-.07	ns	
		overall complaints	-.13	ns	

At t2 regression analyses yield an association between *anger* and CD2-positive cells (table 2) as well as between *anger* and CD26-positive cells. After a Bonferroni correction of the alpha-error for multiple tests (critical p-value=.017), both regressions do no longer fulfil the criterion of significance.

None of the psychometrical variables at t1 could predict t-cell surface markers at t2. Neither could any of the psychometrical variables at t1 predict alterations in the t-cell surface markers over time.

An increase of *commitment* correlated with an increase in CD26-positive cells (r=.40, p=.041). Yet, after a Bonferroni alpha-error correction (critical p=.002) this is no longer significant.

3.4 DPP IV

A median-split between high and low levels of DPP IV activity was performed. In the group with lower DPP IV activity T-tests yielded lower values of *anxious depression* at t1 (t=-2.14, p=.042) and higher values of *elevated mood* (t=2.27, p=.033) and *commitment* (t=2.59, p=.015) at t2. Yet, only the result for *commitment* proved robust against an alpha-error correction. Accordingly, at t1 *depression* and *elevated mood* did not prove predictive for DPP IV. At t2 regression analyses yielded a statistically significant association only between *commitment* and low DPP IV activity (table 2).

None of the psychometrical variables at t1 predicted DPP IV activity at t2. Neither could any of the psychometrical variables at t1 predict alterations in DPP IV activity over time.

Changes in the psychometrical variables between t1 and t2 were not correlated with changes in DPP IV activity.

4. CONCLUSION

Patients with eating disorders, especially those with AN, had lower levels of several immune parameters and higher levels of DPP IV activity. In addition to these alterations, associations with psychometrical variables were investigated.

In sum, higher levels of immunoglobulin appeared to be connected with an apathetic mood quality. No associations with moods or subjective complaints could be found for the t-cell surface markers. Surprisingly, a lower DPP IV activity, which seems to be untypical of patients with eating disorders, appeared to be connected with higher levels of commitment. Commitment can be considered a mood dimension that is fairly opposite to listlessness or depression. While the single t-test comparisons also yielded a trend to lower levels of DPP IV activity in lesser depressed and more elated patients, multiple correlational methods suggest that this effect was attributable to commitment. This result should not be simply compared with other findings of lower levels of DPP IV activity in patients with major or chronic depression[4,5], because there is no substantial knowledge of the underlying mechanisms of altered DPP IV activity in any of the groups so far.

The associations we did find were few in relation to the number of tests. Only 5 out of all 14 regression equations and only 7 out of 98 possible predictions proved significant and they did not follow patterns they followed seems not that were consistent with earlier findings. Finally, some of the effects did not prove robust to more rigorous statistical standards. Moreover, they could not be replicated at a second measurement on the same sample. On the whole, we therefore conclude that the data fail to support the assumption of a direct association between t-cell surface markers, DPP IV activity and subjective parameters of well-being, at least under the conditions of anorexia and bulimia nervosa.

REFERENCES

1. Hildebrandt, M., Rose, M., Mayr, C., Schuler, C., Reutter, W. et al. (1999). Alterations in expression and in serum activity of dipeptidyl peptidase IV (DPP IV, CD26) in patients with hyporectic eating disorders. Scand.J Immunol., 50, 536-541.
2. Hildebrandt, M., Reutter, W., Arck, P. Rose, M., and Klapp, B.F., 2000, A guardian angel: the involvement of dipeptidyl peptidase IV in psychoneuroendocrine function, nutrition and immune defence. *Clin Sci* **99**: 93-104.
3. Hildebrandt, M., Rose, M., Mönnikes, H., Reutter, W., Keller, W., and Klapp, B.F., 2001, Eating disorders: A role for dipeptidyl peptidase IV in nutritional control. *Nutrition* **17**: 451-454.

4. Maes, M., Bonaccorso, S., Marino, V., Puzella, A., Pasquini, M., Biondi, M., Artini, M., Almerighi, C., and Meltzer, H., 2001, Treatment with interferon-alpha (IFN alpha) of hepatitis C patients induces lower serum dipeptidyl peptidase IV activity, which is related to IFN alpha-induced depressive and anxiety symptoms and immune activation. *Mol Psychiatry* 6: 475-80.
5. Maes, M., De Meester, I., Vanhoof, G., Scharpe, S., Bosmans, E., Vandervorst, C., Verkerk, R., Minner, B., Suy, E., and Raus, J., 1991, Decreased serum dipeptidyl peptidase IV activity in major depression. *Biol Psychiatry* **30**: 577-86
6. Maes, M., De Meester, I., Scharpe, S., Desnyder, R., Ranjan, R., and Meltzer, H.Y., 1996, Alterations in plasma dipeptidyl peptidase IV enzyme activity in depression and schizophrenia: effects of antidepressants and antipsychotic drugs. *Acta Psychiatr Scand* **93**: 1-8.
7. Maes, M., Goossens, F., Lin, A., De Meester, I., Van Gastel, A., and Scharpe, S., 1998, Effects of psychological stress on serum prolyl endopeptidase and dipeptidyl peptidase IV activity in humans: higher serum prolyl endopeptidase activity is related to stress-induced anxiety. *Psychoneuroendocrinology* **23**: 485-95.
8. Rose, M., Hildebrandt, M., Fliege, H., Seibold, S., Mönnikes, H., and Klapp, B.F., 2002, T-cell immune parameters and depression in patients with Crohn's disease. *J Clin Gastroenterol* **34**: 40-8.
9. Hörhold, M., and Klapp, B.F., 1993, Testungen der Invarianz und der Hierarchie eines mehrdimensionalen Stimmungsmodells auf der Basis von Zweipunkterhebungen an Patienten- und Studentenstichproben. *Z. Med. Psychol* **1:** 27-35
10. 10. Prehler, M., Kupfer, J., Brähler, E., 1992, The Giessen Symptom Questionnaire for children and adolescents. *Psychother Psychosom Med Psychol.* **42(2):** 71-7.

DPP IV and Mental Depression in Crohn's Disease

MATTHIAS ROSE[*], OTTO B. WALTER[*], HERBERT FLIEGE[*], MARTIN HILDEBRANDT[*], HUBERT MÖNNIKES[#], and BURGHARD F. KLAPP[*]
Department of Internal Medicine, [*]Division of Psychosomatics and Psychotherapie, [#]Division of Gastroenterology, Charité, Humboldt University, Luisenstrasse 13a, Berlin, Germany

1. INTRODUCTION

The causes of the chronic inflammatory bowel diseases are still considered to be unknown. Nevertheless, it has become evident that an immune defect may be the dominant factor in the pathogenesis of inflammatory bowel disease. Two broad patterns of immune alterations leading to IBD-like immunological features and histopathological lesions have been proposed, although they appear contradictory: inflammatory reactions initiated by interleukin (IL-)12-driven processes resulting in the emergence of Th1 T-cells which produce a characteristic pattern of cytokines such as γ-interferon (γ-IFN) and tumor necrosis factor-alpha (TNF-α) is proposed on the one hand, and inflammatory processes initiated by an interleukin (IL-)4 driven, Th2 T-cell response on the other. A possible conclusion from the studies presented to date is that various, apparently contradictory immune alterations may ultimately contribute to a final common pathway of effector mechanisms, a concept proposed by Strober and colleagues that still retains the possibility of IBD as emerging from distinct entities of altered immune responses[1]. These entities are corroborated by murine models with targeted deletions of genes involved in immune functions. These, in turn, have resulted in histopathological patterns of mucosal inflammation resembling either transmural bowel inflammation known to occur in Crohn's disease or superficial inflammation resembling ulcerative colitis. Based on these findings, a Th1-inflammatory response involving excessive γ-IFN production and stimulation of macrophages appears to induce a transmural Crohn's-like

Dipeptidyl Aminopeptidases in Health and Disease, Edited by Hildebrandt et al.
Kluwer Academic/Plenum Publishers, New York, 2003

inflammatory reaction, while a Th2-cytokine profile preferably induces an ulcerative colitis-like inflammation with excessive production of interleukin (IL-)4.

Dipeptidyl peptidase IV (DPP IV, CD26) has been shown to modulate these immune responses. By virtue of its enzymatic activity, DPP IV is capable of expanding a T cell proliferative response *in vitro*. Crosslinking of CD26, i.e. membrane-bound DPP IV, with either CD2 or CD3 induces T cell activation and IL-2 production *in vitro*[2]. Potent cytokines such as RANTES, SDF-1 alpha, MCP-2 and TNF-alpha are among the substrates for DPP IV described to date, suggesting a complex immunomodulatory role for DPP IV activity in serum. The cleavage of chemokines such as eotaxin paves the way for an inhibition of Th2-like cytokine responses by DPP IV activity. This adds to the observation that high expression of DPP IV/CD26 defines a Th1/Th0 phenotype among T cells and correlates with enhanced production of Th1-like cytokines, namely γ-interferon. The net effect of DPP IV activity appears to be an enhanced Th1 cytokine response, in part attributable to a degradation of cytokines involved in Th2-like responses. Based on these experimental and clinical findings, we examined the DPP IV serum activity and the numbers of lymphocytes expressing CD26, i.e. membrane-bound DPP IV, additionally to the usual T-cell parameters[3].

It is well known that those parameters playing a role in the Th-1 immune response can, in particular, be affected by various psychosocial processes. Maes et al. were able to provide evidence that immunological changes, which could be understood as an activation of the Th-1 response, occur in the context of Major Depressive Disorders. Here, the number of activated T-cells (HLA-DR+, CD25+, i.e. the Interleukin-2 receptor α-chain) increases, as well as the number of CD4+ cells and the concentration of soluble Interleukin-2 receptors (sIL-2R) [4-6]. In addition, the DPP IV activity has been found to be reduced in depressive patients as well as in patients with IBD[4,7].

This could possibly be the decisive bridge for the proposed psycho-pathological explanation for the pathogenesis and exacerbation of IBD, which has existed for several decades and been based on individual obser-vations[8]. According to this psychotherapeutic approach, the illness occurs as a result of an attachment disorder[9,10] or is associated with a depressed, neurotic personality structure[11]. The illness would exacerbate in crisis situations in which the patient's particular vulnerability in social attachments is activated[12]. To date, the empirical evidence has hardly reflected this hypothesis. Very few of the newer, methodologically established, studies showed at least some evidence[13,14], most prospective studies do not support a relevant connection between psychosocial processes and the onset of disease[15-17] , nor do they show a positive effect resulting from specific psychotherapeutic interventions[18].

If the somatic outcome parameters are more limited, for example, to the extent of the mucosa defect in asymptomatic patients, as Susan Levenstein and her associates did[19], a psychopathological effect of general perceived Life Stress on morphological changes can be demonstrated. A relevant association between immune system parameters connected with morphological changes and particular psychological processes therefore continues to be likely.

For this reason, we examined (1) whether a connection can be found between depressive moods and immune parameters likely associated with the morphological changes and (2) whether or not a constellation of psycho-immunological parameters can be found which could predict a pending attack of inflammation.

2. METHODS

2.1 Sample

The study took place in the gastroenterological outpatient clinic of the Charité, Campus Virchow, Humboldt University Berlin in Germany. All patients gave written informed consent to the study. A control group for the immunological data was comprised of 28 healthy blood donors (71% women, 38±3 years old) and n=50 healthy students for the 'Short-Inflammatory Bowel Disease Questionnaire' (SIBDQ, 52% women, 26±2 years of age). Data gathered from two representative studies of the German population were also used as controls for the 'Center of Epidemiological Studies - Depression Scale' (CES-D, n=1298, 46% women, 33±15 year of age) and for the 'Giessentest' (GTS, n=1546, 54% women, 41±16 years).

2.2 The present state of illness

The present state of illness was determined by the attending physician, without knowledge of other data in the study, by means of the CDAI index as described by Best[20].

2.3 Immune parameters

Mononuclear cells were isolated from whole blood samples by density centrifugation using a Ficoll-Hypaque gradient (Seromed, Berlin, Germany).

Immunophenotyping was performed using a FACScan flow cytometer (Becton-Dickinson, Mountain View, CA, U.S.A.). After gating the lymphocyte population by size/complexity criteria, the results of the cytofluorometric analyses were recorded as the percent of cells showing positive for the respective antibody or, in double-staining techniques, for two different antibodies, as compared to an isotype control. Antibodies used for flow cytometry were obtained from Pharmingen (Heidelberg, Germany) and Coulter-Immunotech (Krefeld, Germany) and included the following: anti-CD2 (clone 39C1.5), anti-CD3 (clone UCHT1), anti-CD25 (clone B1.49.9), anti-CD26 (clone BA5).

Determination of DPP IV activity in serum was performed as described by Kreisel and colleagues[21]. In brief, 20 μl of serum were incubated with 10 μl of 2 mM Gly-Pro p-nitroanilide (Sigma Chemie GmbH, Deisenhofen, Germany) in 170 μl 0.1M TRIS-HCl pH 8.0 for 30 min. The reaction was stopped by addition of 800 μl sodium acetate buffer (1M, pH 4.5). The DPP IV activity was deduced from the increase of extinction at 405 nm due to the amount of chromogenic substrate metabolized by DPP IV.

2.4 Depression and Subjective Health Status

The assessment of "depression" was carried out with respect to various dimensions. A) The presence of a depressed mood over a medium length time period was assessed by means of the Center of Epidemiological Studies - Depression Scale (CES-D) a validated test translated into German with 15 items[22]. The questionnaire offers a cut-off point between clinically depressed and non-depressed patients ($>=17$ vs <17 units). B) Data suggesting the presence of a depressive personality was gathered by means of the Giessentest (GTS) a psychoanalytically oriented questionnaire with 40 items used to measure personality traits deemed to be connected to depression[23]. C) The FKV-lis (Freiburger Krankheitsverarbeitungs-Fragebogen) was used to determine depressive coping behavior (35 items)[24].

Additionally, the SIBDQ ('Short Form - Inflammatory Bowel Disease Questionnaire') was included to measure the disease-specific subjective health status or the health-related quality of life. The SIBDQ is the short form of the well-established IBDQ ('Inflammatory Bowel Disease Questionnaire') developed by Irvine[25]. We used the validated German short form[26].

3. RESULTS

71 patients with Crohn's Disease (62% women) between the ages of 20 and 73 years of age (mean 38±13 years old) were examined. On average, the patients had been suffering from their illness for 10.8 years (±8.1 years, min.: <1 year, max.: 38 years). 45 took aminosalicylacids, 23 corticosterioids, 4 immunosuppressives and 2 loperamid. 54.4% were employed at the time of the study, 14.9% in training, 15.9% retired. 76.7% were living with a partner. 60 of these patients could be re-examined 4.4±1.8 months later. Between both time points the treatment stayed the same for all patients in remission (CDAI<150) at the first examination.

3.1 Present state of illness

Separating patients at the initial data-gathering time period ("baseline") into patients with "active disease states" (CDAI > 150) and patients in "remission" (CDAI ≤ 150), we found an increased number of leukocytes in the peripheral blood of patients with active stages of illness, and an equivalent number of lymphocytes in both groups. The number of CD25+ and CD25+/CD26+ cells was considerably higher when compared to a control group, whereas the DPP IV activity was decreased (table 1).

The subjective feeling of health as assessed by the SIBDQ revealed the expected subjective feelings of being restricted in life by illness-specific symptoms. Patients with "active disease" states showed the poorest values in the overall health-related quality of life (HRQL), as well as in the dimensions "intestinal complaints" and "social problems" (table 1).

No differences could be found with respect to the various aspects of a depressive mood, depressive disposition, depressive illness-coping strategies, sociodemografic or treatment variables (table 1).

3.2 Depression

When we divided the patients based on the CES-D cut-off point of 17 units into "depressed" versus "not depressed" persons (22% vs. 78%) at baseline, no differences were found in the immune parameters, sociodemographic or treatment variables examined (table 2).

For the parameter "subjective health status" as well as for the other parameters assessing depressive experiences, we found that "depressed" Crohn patients showed depressed personality dispositions as well as depressive illness-coping strategies. They also reported considerably more physical complaints (table 2).

Table 1: Immunological and psychological parameters in 'Active Disease and Remission' at baseline

	patients at baseline				controls		variance analysis, t-test or Chi² test
	Active disease CDAI > 150 n=23		remission CDAI ≤ 150 n=48				
	M	sd	m	sd	m	sd	F/t (p)
Disease activity							
CDAI	199	35	88	37	-	-	-
C reactive protein mg/dl	***2.04***	***2.70***	***1.04***	***.94***	***.80***	***.23***	***4.29 (.017)***
Orosomucoide g/l	***1.40***	***.53***	***1.04***	***.30***	-	-	***3.23 (.002)***
Immune Parameters							
Leucocytes /nl	***9.50***	***3.69***	***7.83***	***2.70***	***6.65***	***1.83***	***5.80 (.004)***
Lymphocytes /nl	1.72	.81	1.63	.64	1.97	.54	ns
CD2+ %	74.59	11.01	75.99	13.06	81.27	6.76	ns
CD3+ %	64.86	17.30	65.54	13.88	76.18	8.16	ns
CD19+ %	13.45	8.38	14.68	11.91	11.40	7.42	ns
CD25+ %	***8.93***	***3.64***	***7.09***	***4.27***	***4.31***	***2.95***	***7.65 (.001)***
CD26+ %	39.65	15.79	42.93	17.51	41.36	12.02	ns
CD2+/CD26+ %	38.19	15.85	41.11	17.39	40.18	11.57	ns
CD25+/CD26+ %	***5.07***	***2.58***	***4.21***	***2.89***	***2.34***	***1.81***	***6.27 (.003)***
CD3+/CD28+ %	44.13	18.22	46.58	16.94	52.54	10.33	ns
CD28+/CD95+ %	23.16	9.11	21.88	7.76	22.40	5.23	ns
DPP IV mU/ml	***47.12***	***14.07***	***55.77***	***17.65***	***63.31***	***15.12***	***4.01 (.022)***
IgM mg/dl	107.07	44.14	140.55	103.42	-	-	ns
IgG mg/dl	995.77	231.59	1048.8	255.77	-	-	ns
IgA mg/dl	261.31	98.43	225.31	105.57	-	-	ns
Subjective Health Status							
Sum-Score (SIBDQ)	***5.40***	***1.11***	***5.73***	***.76***	***6.00***	***.55***	***2.63 (.011)***
Bowel function (SIBDQ)	***5.47***	***1.34***	***6.04***	***.82***	***6.57***	***.62***	***12.7 (.000)***
Emotional funct. (SIBDQ)	5.13	1.31	5.37	1.13	5.27	.85	ns
Social function (SIBDQ)	***5.97***	***1.52***	***6.44***	***1.07***	***6.99***	***.07***	***9.58 (.000)***
Systemic function (SIBDQ)	4.72	1.37	5.07	1.19	5.29	.94	ns
Mental Depression							
Depression – State (CES-D)	8.54	7.89	10.11	9.43	10.72	8.03	ns
Depression – Trait (GTS)	26.93	7.76	23.85	6.30	22.0	5.48.	ns
Depressive Coping (FKV)	1.88	.97	2.10	.97	-	-	ns

3.3 Predictors for the Course of Illness

For those patients who were in remission at baseline, we tried to identify a pattern based on the presented parameters that may predict an active inflammatory attack within the next four months. Various procedures of discriminance analyses were used. In cross-validations, none of the models examined could be seen to predict more than 70% of the cases correctly. We therefore decided not to present the details here. Patients in remission at baseline who developed active disease at follow up did not differ significantly in any of the investigated parameters at baseline from patients who remained in remission.

Table 2: Immunological and psychological parameters in depressed and non-depressed patients at baseline

	patients at baseline				t-test for independent samples or Chi² test
	depressive CES-D >= 17 n=14 [1]		not depressive CES-D < 17 n=52 [1]		
	m	sd	m	sd	t (p)
Disease activity					
CDAI	108	57	121	60	ns
C reactive protein mg/dl	1.07	1.14	1.63	2.21	ns
Orosomucoide g/l	.95	.30	1.23	.49	ns
Immune Parameters					
Leucocytes /nl	8.51	3.22	8.36	3.56	ns
Lymphocytes /nl	1.97	.71	1.62	.73	ns
CD2+ %	71.63	20.50	76.61	10.21	ns
CD3+ %	64.05	17.22	65.52	14.62	ns
CD19+ %	18.29	18.78	12.57	7.98	ns
CD25+ %	7.94	4.01	7.22	4.32	ns
CD26+ %	43.46	19.10	42.53	15.90	ns
CD2+/CD26+ %	41.29	19.19	40.82	15.58	ns
CD25+/CD26+ %	4.73	3.19	4.34	3.16	ns
CD3+/CD28+ %	47.46	19.31	48.19	17.27	ns
CD28+/CD95+ %	23.94	6.35	21.82	8.59	ns
DPP IV mU/ml	49.19	14.16	51.20	15.63	ns
IgM mg/dl	188.17	171.06	108.88	53.56	ns
IgG mg/dl	967.83	206.23	1013.8	265.99	ns
IgA mg/dl	182.17	31.44	220.28	91.76	ns
Subjective Health Status					
Sum-Score (SIBDQ)	***5.06***	***.71***	***5.87***	***.81***	***-2.88 (.006)***
Bowel function (SIBDQ)	5.60	.66	6.04	.88	ns
Emotional function (SIBDQ)	***4.47***	***1.38***	***5.64***	***.98***	***-3.08 (.003)***
Social function (SIBDQ)	6.15	1.18	6.46	1.11	ns
Systemic function (SIBDQ)	***4.05***	***1.07***	***5.34***	***1.20***	***-3.07 (.004)***
Depression					
Depression - State (CES-D)	***24.36***	***5.07***	***5.56***	***4.24***	-
Depression - Trait (GTS)	***30.82***	***6.01***	***22.92***	***6.08***	***3.81 (.000)***
Depressive Coping (FKV)	***2.73***	***.78***	***1.77***	***.87***	***3.27 (.002)***

[1] The CES-D offers a so-called lie criterion to identify persons with implausible answer patterns. 5 patients fulfilled this criterion and had to be excluded from this analysis.

4. DISCUSSION

Our results show decreased levels of serum DPP IV activity in patients with inflammatory bowel disease (IBD). Concomitantly, we could observe an increased number of CD26-positive lymphocytes coexpressing the activation marker CD25 (IL-2 receptor α chain).

Similarly, in experimental and clinical studies on rheumatoid arthritis[27,28] and systemic lupus erythematosus[29], changes in DPP IV activity like those in our study have been reported. The decreased activity in serum not only correlated inversely with the severity of the disease, but also contrasted with an enhanced DPP IV expression and activity at sites of inflammation[30] and with an increased number of CD26-positive lymphocytes[31], suggesting a functional compartmentalization of DPP IV expression[30]. As discussed recently, changes in DPP IV activity may impinge on the metabolism of peptide hormones known to be potent mediators of immune responses. Although merely hypothetical, a decrease in DPP IV activity in serum could negatively affect a Th1 inflammatory cytokine profile and, thus, be of crucial importance for a limitation of the inflammatory process to a local inflammatory reaction[2,3,7,32].

The fact that the same changes in the DPP IV and the Th-1 activity can be found in depressed persons[4-6], added to the long-standing clinical supposition that there is a connection between depression and the development or exacerbation of IBD, urged us to examine whether or not the immunological parameters investigated here could be connected with a depressed mood. In the sample under study here, we were not able to find a connection between the immunological parameters examined and parameters of a depressive mood, depressed personality or depressive illness coping strategies in a cross-sectional design.

Additionally we could not show an effect of any of the psychological parameters on the disease activity at follow up. Yet, we should point out here that the number of patients who changed from "remission" (n=48) at baseline to an "active disease" at follow up was very small (n=11). Consequently, in the longitudinal design the effect strength would have had to be very great so as to have reached a significant level.

However, this result corresponds to most other *empirical* studies, in which no connection was found between psychological parameters or life events and objective illness parameters of IBD[15-17]. As North et al.[17] and Levenstein et al.[19] noted, most of the studies assuming psychological causes for somatic symptoms are based on individual case study observations, are retrospective or methodologically insufficient work. We tend to agree with North et al. that evidence for a connection between psychological and somatic processes can only be produced in exceptional cases, at least when working with the established measurements of psychological constructs presently used.

The particular course of Crohn's Disease, characterized by phases or attacks, makes it understandable that physicians[33] as well as patients[34] look for situational influences as causes for a renewed attack, thereby making the illness appear to be more predictable. The popular concept of "stress" or the related construct of being psychologically overwhelmed by something, expressed in an individual's depressed mood, appear to be fitting explanatory models. Developments in psychoneuroimmunological research give cause to hope that evidence for such a connection may be found. However, the lack of such evidence at this point makes it also worthwhile to consider whether the clinical observations with respect to psychological causes of somatic symptoms in IBD may be overly biased by traditional theoretical models.

The negative findings presented here can be considered under methodological aspects, too. The patients included in our study were recruited in an outpatient facility, thereby excluding hospitalised patients with acute phases of inflammation and high CDAI scores. Any attempt to include such patients, however, must take into account that these inpatients bear some characteristics of known influence on immunological and psychological variables, which can hardly be controlled for, e.g. higher dosages of medications such as steroids, other antiinflammatory and analgetic drugs, antibiotics, drugs that influence the bowel motility, or artificial nutrition.

Another methodologically important issue requiring attention is the diagnosis of major depression as psychiatric disease in contrast to the assessment of a depressive mood or disposition performed here. A coincidence of a major depressive disorder and Crohn's disease is rather unlikely. Furthermore, the psychometric assessment of a disposition on the one hand and situational moods on the other may not be expected to lead to identical results, but can probably provide no more than an approximation.

REFERENCES

1. Strober, W., Fuss, I. J., Ehrhardt, R. O., Neurath, M., Boirivant, M., & Ludviksson, B. R., 1998. Mucosal immunoregulation and inflammatory bowel disease: new insights from murine models of inflammation. *Scand J Immunol.*, **48:** 453-458.
2. Tanaka, T., Duke-Cohan, J. S., Kameoka, J., Yaron, A., Lee, I., et al., 1994. Enhancement of antigen-induced T-cell proliferation by soluble CD26/dipeptidyl peptidase IV. *Proc.Natl.Acad.Sci.U.S.A,* **91:** 3082-3086.
3. Hildebrandt, M., Reutter, W., Arck, P., Rose, M., & Klapp, B. F.,2000. A guardian angel: the involvement of dipeptidyl peptidase IV in psychoneuroendocrine function, nutrition and immune defence. *Clin.Sci.*, **99:** 93-104.
4. Maes, M., De-Meester, I., Vanhoof, G., Scharpe, S., Bosmans, E., et al., 1991. Decreased serum dipeptidyl peptidase IV activity in major depression. *Biol.Psychiatry*, **30:** 577-586.
5. Maes, M., De Meester, I., Scharpe, S., Desnyder, R., Ranjan, R. et al., 1996. Alterations in plasma dipeptidyl peptidase IV enzyme activity in depression and schizophrenia: effects of antidepressants and antipsychotic drugs. *Acta Psychiatr.Scand.,* **93** : 1-8.

6. Maes, M., De-Meester, I., Verkerk, R., De-Medts, P., Wauters, A., et al., 1997. Lower serum dipeptidyl peptidase IV activity in treatment resistant major depression: relationships with immune-inflammatory markers. *Psychoneuroendocrinology*, **22:** 65-78.
7. Hildebrandt, M., Rose, M., Rueter, J., Salama, A., Reutter, W. et al., 2001. Dipeptidyl Peptidase IV (DPP IV, CD26) in Patients with Inflammatory Bowel Disease. *Scand.J Gastroenterol* **10:** 1067-72
8. Drossman, D. A., 1998. Presidential address: Gastrointestinal illness and the biopsychosocial model. *Psychosom Med*, **60:** 258-267.
9. Stewart, W. A., 1949. Psychosomatic aspects of regional ileitis. *N.Y.St.J.Med.*, **49:** 2820-4.
10. Wirsching, M., 1984. Familientherapeutische Aspekte bei Colitis ulcerosa und Morbus Crohn Family therapy of ulcerative colitis and Crohn's disease. *Zeitschrift fuer Psychosomatische Medizin und Psychoanalyse,* **30:** 238-246.
11. Helzer, J. E., Chammas, S., Norland, C.C., Stillings, W.A., & Alpers, D.H., 1984. A study of the association between Crohn's disease and psychiatric illness. *Gastroenterol*, **86:** 324-30.
12. Paulley, J. W., 1958. Crohn's disease. *Lancet*, **2:** 959-960.
13. Duffy, L. C., Zielezny, M. A., Marshall, J. R., Byers, T. E., Weiser, et al., 1991. Relevance of major stress events as an indicator of disease activity prevalence in inflammatory bowel disease. *Behav.Med*, **17:** 101-110.
14. Levenstein, S., Prantera, C., Varvo, V., Scribano, M. L., Andreoli, A., et al., 2000. Stress and exacerbation in ulcerative colitis: a prospective study of patients enrolled in remission. *Am.J Gastroenterol*, **95:** 1213-1220.
15. Riley, S. A., Mani, V., & Goodman, M. J. (1991). Why do patients with ulcerative colitis relapse? *Gut*, **32:** 832.
16. von Wietersheim, J., Overbeck, A., Kiel, K., Kohler, T., Jantschek, G. et al., 1994. The significance of recurrence-inducing events for patients with chronic inflammatory bowel diseases. Results of a prospective longitudinal study over three years. *Psychother Psychosom Med Psychol*, **44:** 58-64.
17. North, C. S., Alpers, D. H., Helzer, J. E., Spitznagel, E. L., & Clouse, R. E., 1991. Do life events or depression exacerbate inflammatory bowel disease? A prospective study. *Ann.Intern.Med*, **114:** 381-386.
18. Jantschek, G., Zeitz, M., Pritsch, M., Wirsching, M., Klor, et al., 1998. Effect of psychotherapy on the course of Crohn's disease. *Scand.J Gastroenterol*, **33:** 1289-1296.
19. Levenstein, S., Prantera, C., Varvo, V., Scribano, M. L., Berto, E., et al., 1994. Psychological stress and disease activity in ulcerative colitis: a multidimensional cross-sectional study. *Am.J Gastroenterol*, **89:** 1219-1225.
20. Best, W. R., Becktel, J. M., Singleton, J. W., & Kern, F., 1976. Development of a Crohn's disease activity index. Nat. Cooperative Crohn's Disease Study. *Gastroenterol*, **70:** 439-44.
21. Kreisel, W., Heussner, R., Volk, B., Büchsel, R., Reutter, W., et al., 1982. Identification of the 110000 Mr glycoprotein isolated from rat liver membrane as dipeptidylaminopeptidase IV. *FEBS Lett*, **147:** 85.
22. Hautzinger, M. & Bailer, M., 1993. ADS Allgemeine Depressionsskala. Weinheim: Beltz.
23. Beckmann, D., Braehler, E., & Richter, H. E., 1991. Der Giessen-Test (GT). Ein Test fuer Individual- und Gruppendiagnostik. Bern: Huber.

24. Muthny, F. A., 1989. Freiburger Fragebogen zur Krankheitsverarbeitung. Weinheim: Beltz.
25. Irvine, E. J., 1993. Quality of Life--rationale and methods for developing a disease-specific instrument for inflammatory bowel disease. *Scand.J Gastroenterol Suppl.*, **199:** 22-27.
26. Rose, M., Fliege, H., Hildebrandt, M., Korber, J., Arck, P., et al. (2000). Validation of the new German translation version of the "Short Inflammatory Bowel Disease Questionnaire" (SIBDQ). Z.Gastroenterol, 38, 277-286.
27. Gotoh, H., Hagihara, M., Nagatsu, T., Iwata, H., Miura, T., 1989. Activities of dipeptidyl peptidase II and dipeptidyl peptidase IV in synovial fluid from patients with rheumatoid arthritis and osteoarthritis. *Clin Chem*, **35(6):**1016-1018.
28. Muscat, C., Bertotto, A., Agea, E., Bistoni, O., Ercolani, R., Tognellini, R., et al., 1994. Expression and functional role of 1F7 (CD26) antigen on peripheral blood and synovial fluid T cells in rheumatoid arthritis patients. *Clin Exp Immunol*, **98(2):** 252-6.
29. Plana, M., Font, J., Vinas, O., Martorell, J., Ingelmo, M., Vives, J., 1994. Responsiveness of T lymphocytes from systemic lupus erythematosus to signals provided through CD26 antigen. *Clin Immunol Immunopathol.* **72(2):** 227-32.
30. Walsh, D.A., Mapp, P.I., Wharton, J., Polak, J.M., Blake, D.R.,1993. Neuropeptide degrading enzymes in normal and inflamed human synovium. *Am J Pathol*, **142(5):**1610-21.
31. Mizokami, A., Eguchi, K., Kawakami, A., Ida, H., Kawabe, Y., Tsukada, T., et al., 1996. Increased population of high fluorescence 1F7 (CD26) antigen on T cells in synovial fluid of patients with rheumatoid arthritis. *J Rheumatol*, **23(12):** 2022-6
32. Hildebrandt, M., Rose, M., Mayr, C., Schuler, C., Reutter, W. et al. (1999). Alterations in expression and in serum activity of dipeptidyl peptidase IV (DPP IV, CD26) in patients with hyporectic eating disorders. *Scand.J Immunol.*, **50:** 536-541.
33. Mitchell, C. M. & Drossman, D. A., 1987. Survey of the AGA membership relating to patients with functional gastrointestinal disorders. *Gastroenterology*, **92:** 1282-1284.
34. Robertson, D. A., Ray, J., Diamond, I., & Edwards, J. G., 1989. Personality profile and affective state of patients with inflammatory bowel disease. *Gut*, **30:** 623-626.

Microscopic Acid-Base Equilibra of Alanyl-boroAlanine

JACK H. LAI*, YUHONG ZHOU, JAMES L. SUDMEIER, WENGEN WU, DAVID G. SANFORD, MAW HLIANG, SARAH POPLAWSKI, and WILLIAM W. BACHOVCHIN
*Department of Biochemistry, Tufts University School of Medicine, 136 Harrision Avenue,. Boston, MA 02111, USA Email: jack.lai@tufts.edu

The *in vivo* introduction of DPP IV specific inhibitors has been shown to enhance the levels of intact endogeneous peptides, creating a new therapeutic paradigm in diabetes treatment.[1] Ala-boroAla (AbA) belongs to a class of very potent serine protease inhibitors known as "peptide boronic acids". Their high affinities for proteases are derived from close mimicry of boronyl-serine adducts to tetrahedral transition states in enzyme-catalyzed reactions[2]. Preliminary studies of AbA as a DPP IV inhibitor in our lab showed that the degree of inhibition was dependent upon the pH (either 2 or 8) and time duration (up to 24 hrs) of the pre-incubation, i.e. time prior to enzyme addition. This prompted us to use NMR to elucidate the various components, both active and inactive, of AbA at various pH values and their dissociation constants.

A study of the titration behavior of Ala-Ala showed the N-terminal methyl group to be the most reliable reporter of ionization in the dipeptide, because of its large intensity and large protonation shift (~0.30 ppm), affected only by its proximal functional group. That is, the methyl resonance on the C-terminal residue was somewhat affected by N-terminal ionization, and the alpha proton of the N-terminal residue was somewhat influenced by C-terminal ionization. In Ala-Ala the two acid-base dissociation steps were

Dipeptidyl Aminopeptidases in Health and Disease, Edited by Hildebrandt et al.
Kluwer Academic/Plenum Publishers, New York, 2003

well separated, exhibiting $pK_1 = 3.20$ and $pK_2 = 8.27$, making the above conclusion unambiguous (Fig. 1A).

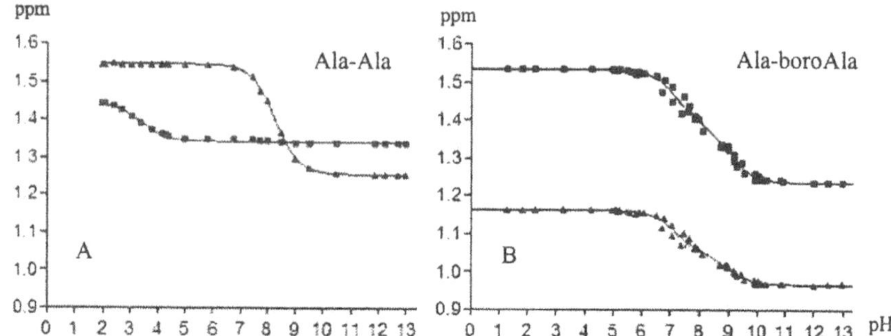

Figure 1. 1H NMR-pH titration curves of the methyl protons associated with N, C and B terminals with the fitted function for A. Ala-Ala (left) and B. Ala-boroAla (right). Triangles in the left are N methyl and squares are C methyl. Squares in the right are N methyl and triangles are B methyl.

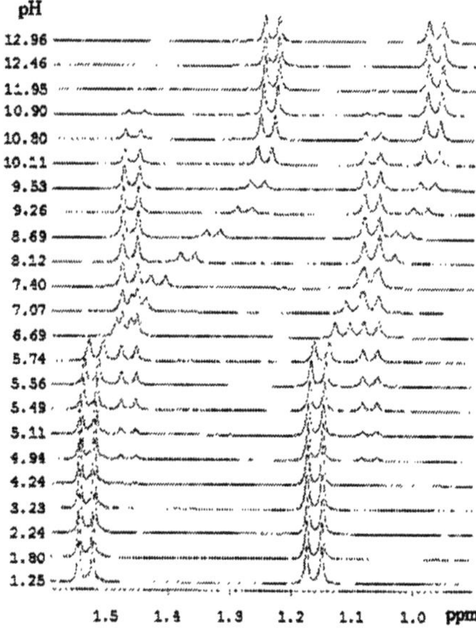

Figure 2. A typical 1H NMR titration spectrum of AbA with NaOH (in 10% D2O/H2O). The pH dependent and the independent resonances for alanyl methyl protons are well distinguished.

Unlike Ala-Ala, when AbA is titrated by proton NMR (Fig. 2), we see from a plot of the chemical shift of the two methyl doublets (Fig. 1B) two very close inflection points -- one at around 7 and one around 9. By least-squares fitting to a standard equation for diprotic acid titration curves (Eq. 1), we find values for the two breaks in the more downfield methyl resonance at values of $pk_1=7.21$ and $pk_2=9.01$, with 48% of the break occurring at the lower pH and 52% occurring at the higher pH. Thus the acidities of $-NH_2$ and $-B(OH)_3^-$ are very close, with the boronic acid winning by a nose.

$$\delta_{obs} = \frac{10^{-pH}10^{-pH}10^{pk_1}10^{pk_2}\,\delta[+NB] + 10^{-pH}10^{pk_2}\,\delta[NB] + \delta[NB-]}{10^{-pH}10^{-pH}10^{pk_1}10^{pk_2} + 10^{-pH}10^{pk_2} + 1} \quad (1)$$

Another feature of the NMR titration of AbA in Fig. 2 is the growing in of pH-independent resonances beginning around pH 5, reaching a maximum around 8 and then disappearing by pH around 11. These resonances clearly belong to the cyclic form of AbA, whose acid and base groups are tied up and no longer available for rapid exchange with protons. The relative abundances of the cyclic species of AbA was calculated from the integrals of the pH-independent methyl protons compared to the sum all methyl proton integrals. The fraction, $\alpha_{[NcycB]}$, of the cyclic species was fit by the non-linear least squares method to Eq (2).

Data collected at different incubation times were fitted to the alpha equation (Fig. 3). The results showed that the fraction of the cyclic species increases with time, reaching full equilibration and a maximum of ~90% after 18-24 hours of incubation. Accordingly, two microscopic equilibrium constants ($pk_{1C}=5.96$, $pk_{2C}=10.19$) and the cyclic fraction factor ($Fc=0.896$) were derived.

$$\alpha_{[NcycB]} = Fc\left[\frac{10^{-pk_{1c}}10^{-pH}}{10^{-pH}10^{-pH} + 10^{-pk_{1c}}10^{-pH} + 10^{-pk_{1c}}10^{-pk_{2c}}}\right] \quad (2)$$

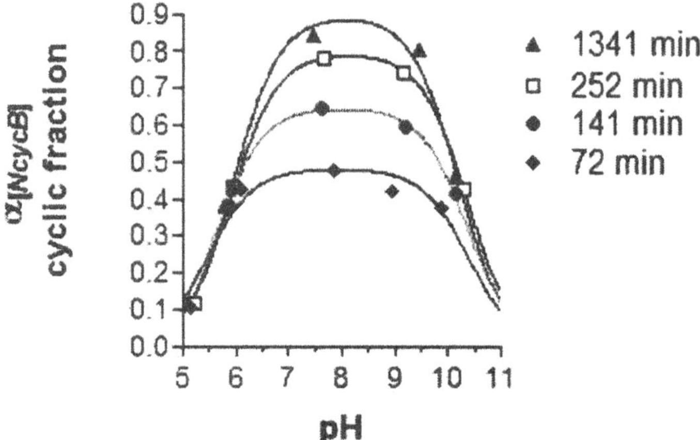

Figure 3. The relative abundance of cyclic form of AbA shown at various incubation times. The solid lines represent the non-linear square fit to Eq. 2 .(K1=5.96, K2=10.19 and Fc=0.896 for 1341 min).

The proton chemical shift of the N-terminal alanyl methyl group represents a weighted average of its chemical shift in the various forms which are in rapid exchange. Both the data for rapidly exchanging chemical shifts vs. pH (pk_{1N} and pk_{2N}) and that of the slowly exchanging integrals (pk_{1C} and pk_{2C}) produce microscopic equilibrium constants directly. How do we derive the overall macroscopic dissociation constants K_1 and K_2 from the microscopic constants? Boldface type is used to indicate macroscopic concentrations, which are summations of the microscopic species represented in Fig. 4:

$K_1 = [H^+]\,[\mathbf{NB}] / [\mathbf{+NB}]$ (3), and
$K_2 = [H^+]\,[\mathbf{NB-}] / [\mathbf{NB}]$ (4), where
$[\mathbf{+NB}] = [+N\ t\ B] + [+N\ c\ B]$ (5),
$[\mathbf{NB}] = [+N\ t\ B-] + [N\ cyc\ B] + [N\ t\ B]$ (6), and
$[\mathbf{NB-}] = [N\ t\ B-] + [N\ c\ B-]$ (7).

K_1 is the sum of the three microscopic constants, and the reciprocal of K_2 is obtained by summation of the reciprocals of the microscopic constants as follows:

$K_1 = k_{1N} + k_{1B} + k_{1C}$ (8), and
$1/K_2 = 1/k_{2N} + 1/k_{2B} + 1/k_{2C}$ (9).
Thus $K_1 = 10^{-7.21} + 10^{-7.18} + 10^{-5.96} = 10^{-5.91}$ (10), and
$1 / K_2 = 1 / 10^{-9.01} + 1 / 10^{-8.98} + 1 / 10^{-10.19} = 1 / 10^{-10.24}$ (11).

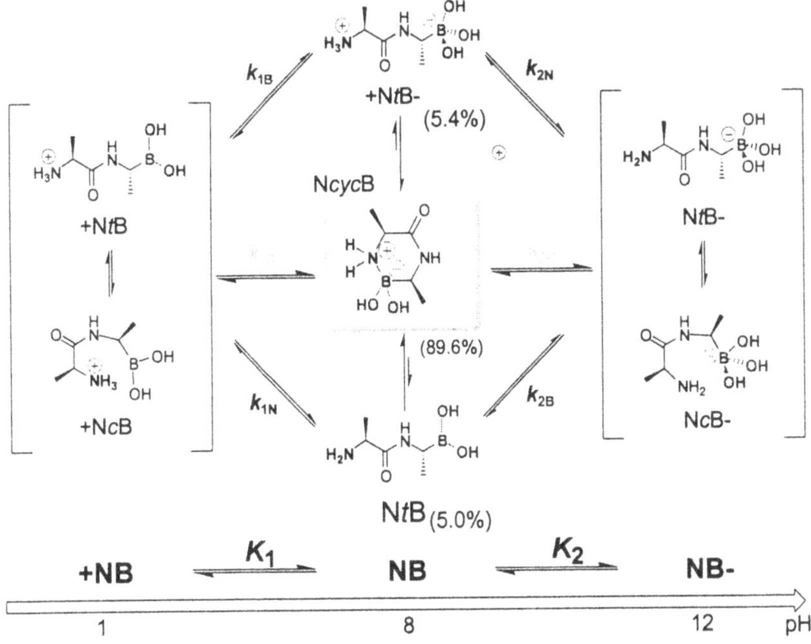

Figure 4. Macroscopic and microscopic acid-base equilibrium of AbA. k1B, k1C and k1N are the microscopic acid-base constants associated with macroscopic K1 of the diprotic acid. k2B, k2C and k2N are the microscopic constants associated with macroscopic K2. The use of "NB" symbols gives a pictorial representation of the N-terminal and B-terminal functional groups, their ionization state, and thus their ionic charges. The italic symbols "t", "c", and "cyc", denote trans and cis with respect to the peptide bond, and the cyclic species.

Therefore, the calculated overall pK_1 = 5.91 and pK_2 = 10.24. The relationship between all the equilibrium constants and their corresponding ionization species are summarized in Fig. 4. and Table 1. The much lower microscopic dissociation constant for the cyclic compared to the linear species in the first dissociation step shows the acid-strengthening effect of cyclization.

Table 1. Summary of microscopic and macroscopic dissociation constants of AbA.

	Microscopic Constants						Macroscopic	
Related site	**N**		**B**		*cyc*		**N**	**B**
Constants	**Pk_{1N}**	**pk_{2N}**	**pk_{1B}**	**pk_{2B}**	**pk_{1C}**	**pk_{2C}**	**pK_1**	**pK_2**
Obs.	7.21	9.01	7.18	8.98	5.96	10.19	5.91	10.24

REFERENCES

1. Holst, J. J., Deacon, C.F., 1998, Inhibition of the activity of dipeptidyl-peptidase IV as a treatment for type 2 diabetes. *Diabetes*. 47(11):1663-70.
2. Tsilikounas, E., Rao, T., Gutheil, W. G., Bachovchin, W. W., 1996, ^{15}N and ^{1}H NMR spectroscopy of the catalytic histidine in chloromethyl ketone-inhibited complexes of serine proteases. *Biochemistry*. 35(7): 2437-2444.

Acylated Hydroxamates as Selective and Highly Potent Inhibitors of Dipeptidyl Peptidase I

ANDRÉ J. NIESTROJ, DAGMAR SCHLENZIG, ULRICH HEISER, KERSTIN KÜHN-WACHE, BLAS CIGIC‡, MICHAEL WERMAN, TORSTEN HOFFMANN, BERND GERHARTZ, and HANS-ULRICH DEMUTH

probiodrug AG, Weinbergweg 22, 06120 Halle, Germany and ‡Institute Jozef Stefan, Departement of Biochemistry and Molecular Biology, Jamova 39, 1000 Ljubljana, Slovenia

1. INTRODUCTION

Dipeptidyl peptidase I (DP I), also known as cathepsin C (EC 3.4.14.1), was discovered in 1948 by Gutman & Fruton[1]. DP I removes dipeptides sequentially from unsubstituted N-termini of polypeptide substrates with a broad substrate specificity[2,3]. DP I is inhibited only weakly by unspecific reversible and irreversible cysteine protease-inhibitors such as leupeptin and E-64[4]. Stronger reversible inhibitors are stefin A, chicken cystatin, and other inhibitors of the cystatin super-family[5]. Specific inhibition has been achieved with the *a priori* reactive affinity labels of the diazomethyl ketone and sulphonylmethyl ketone type[6-8]. DP I is known to release active granulocyte serine proteases of lymphatic cells from their pro-forms[9,10]. The inhibition of DP I leads to a reduced concentration of granzyme A and decreases the lymphocyte-mediated cytotoxicity[9,10]. DP I represents a lysosomal cysteine peptidase belonging to the C1 family of the papain-like cysteine peptidases. In mammals, DP I is present in a variety of tissues with the highest concentrations in lung, spleen, liver, placenta, and kidney[11-14]. Being involved in intracellular protein breakdown in lysosomes, DP I fulfils regulatory functions by limited proteolysis. The role of DP I may be of potential therapeutic interest as a central co-ordinator for the activation of

Dipeptidyl Aminopeptidases in Health and Disease, Edited by Hildebrandt et al.
Kluwer Academic/Plenum Publishers, New York, 2003

many serine proteases in cytotoxic T-lymphocytes (CTL), by activating granzymes triggering the apoptotic "death"-cascade of the caspases during a CTL-target cell interaction. Here, the synthesis and kinetic characterisation of a set of active site directed inhibitors containing an acylated hydroxamate moiety with different residues of the structure TFA*R-Gly-Phe-NHO-CO-R′ (*Figure* I) is reported.

2. RESULTS AND DISCUSSION

Cysteine protease inhibitors such as peptide aldehydes or peptide diazomethanes allow only variations of the peptide moiety to modulate their specificity. Compared to these inhibitors, acylated hydroxamates permit the variation of the "war head" moiety R′ (*Figure* I) as well. Kinetic characterisation of the compounds **10-14** indicates that they are selective and potent inhibitors of DP I. Additionally, variation of the residue R′ results in compounds with varied activities. Surprisingly, in contrast to the literature, compound **14** with a methylated *N*-terminus is recognised by the protease.

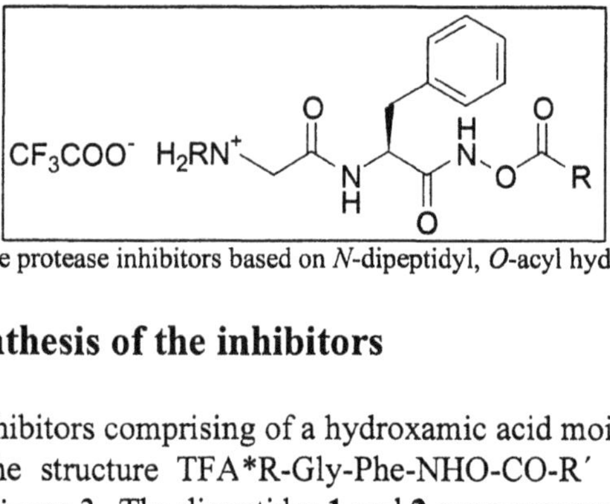

Figure 1. Cysteine protease inhibitors based on *N*-dipeptidyl, *O*-acyl hydroxylamines

2.1 Synthesis of the inhibitors

A set of inhibitors comprising of a hydroxamic acid moiety with different residues of the structure TFA*R-Gly-Phe-NHO-CO-R′ was prepared as described in *Figure 2*. The dipeptides **1** and **2** were prepared by the coupling reaction starting from HCl*H-L-Phe-OMe and Boc-Gly-OH or Boc-Sar-OH with CAIBE and NMM. The dipeptides were converted into the peptidylhydroxylamines **3** and **4** by treatment with freshly prepared hydroxylamine. Acylation with various carbonic acid chlorides in the presence of triethylamine gave the corresponding acetyl derivatives **5-9**. Treatment of the compounds **5-9** with trifluoroacetic acid generated the inhibitors **10-14**.

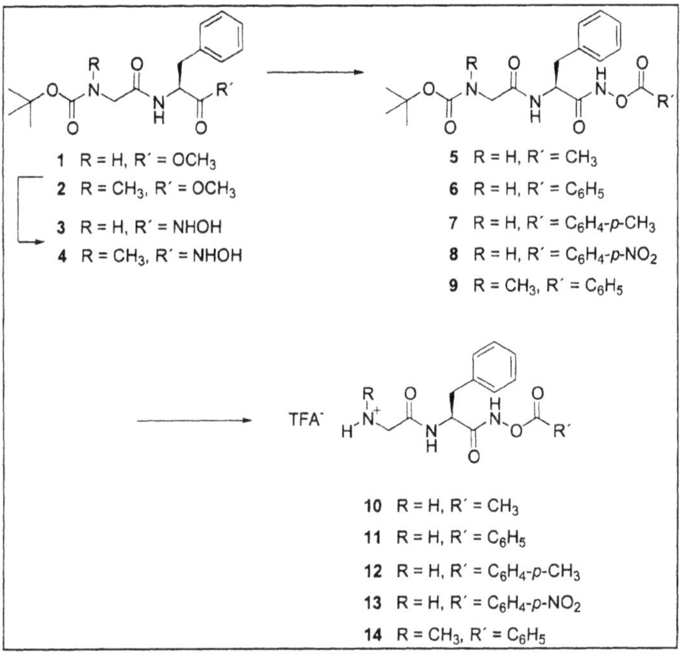

1 R = H, R′ = OCH_3
2 R = CH_3, R′ = OCH_3
3 R = H, R′ = NHOH
4 R = CH_3, R′ = NHOH

5 R = H, R′ = CH_3
6 R = H, R′ = C_6H_5
7 R = H, R′ = C_6H_4-*p*-CH_3
8 R = H, R′ = C_6H_4-*p*-NO_2
9 R = CH_3, R′ = C_6H_5

TFA⁻

10 R = H, R′ = CH_3
11 R = H, R′ = C_6H_5
12 R = H, R′ = C_6H_4-*p*-CH_3
13 R = H, R′ = C_6H_4-*p*-NO_2
14 R = CH_3, R′ = C_6H_5

Figure 2: Preparation of the inhibitors 10-14

2.2 Kinetic characterisation

The compounds **10-14** were tested as inhibitors of DP I and checked for their cross-reactivity against other proteases namely cathepsin B, H, L and DP IV. No inhibition of cathepsin B, H, L and DP IV was observed at concentrations below 0.5 μM (Table 1). The kinetic constants are summarised in Table 2. The compounds inhibit the target in an irreversible manner.

Table 1. Protease inhibition by selective *N*-dipeptidyl, *O*-acylhydroxylamines

compound	R	R′	IC 50 [μM]				
			DP I	Cath. B	Cath. H	Cath. L	DP IV
10	H	CH_3	0.258	117	n.i.	156	n.i.
11	H	C_6H_5	0.034	37.7	n.i.	36.9	n.i.
12	H	C_6H_5-p-CH_3	0.203	34.05	154	33.05	n.i.
13	H	C_6H_5-p-NO_2	0.0026	5.31	34.7	7.32	n.i.
14	CH_3	C_6H_5	0.157	41.8	205	4.61	n.i.

n.i. = no inhibtion

Table 2. Kinetic constants of the inactivation of DP I by *N*-dipeptidyl, *O*-acylhydroxylamines

compound	K_i [μM]	k_{inact} [min]	k_{inact}/K_i [$mM^{-1}s^{-1}$]
10	0.518	0.21	6.74
11	0.0043	0.34	132.0
12		not detectable*	
13	2.6	0.10	0.648
14	5.25	0.72	2.28

* unstable during the time of the characterisation

3. CONCLUSION

The compounds **10-14** which are *N*-dipeptidyl derivatives of *O*-acyl hydroxamates proved to be potent, selective and irreversible inhibitors of DP I.

REFERENCES

1. Gutman, H.R. & Fruton, J.S. 1948, On the proteolytic enzymes of animal tissues VIII. An intracellular enzyme related to chymotrypsin. *J. Biol. Chem.*, **174**: 851-858.
2. McDonald, J.K., Callahan, P.X., Ellis, S., and Smith, R.E., 1971, Polypeptide degradation by dipeptidyl aminopeptidase I (cathepsin c) and related peptidases. In *Tissue Proteinases* (Barrett, A.J. & Dingle, J.T., eds.), North Holland Publishing, Amsterdam, pp. 69-107.
3. McDonald, J.K. & Schwabe, C., 1977, Intracellular exopeptidases (Barrett, A.J., ed.), North Holland Publishing, Amsterdam, pp. 311-391.
4. Nikawa, T., Towatari, T., and Katunuma, N., 1992, Purification and characterization of cathepsin J from rat liver. *Eur. J. Biochem.* **204**: 381-393.
5. Nicklin, M.J.H. & Barrett, A.J., 1984, Inhibition of cysteine proteinases and dipeptidyl peptidase I by egg-white cystatin. *Biochem. J.* **223**: 245-253.
6. Angliker, H., Wikstrom, P., Kirschke, H., and Shaw, E., 1989, The inactivation of the cysteinyl exopeptidases cathepsin H and C by affinity-labelling reagents. *Biochem. J.* **262**: 63-68.
7. Green, G.D.J. & Shaw, E., 1981, Peptidyl diazomethyl ketones are specific inactivators of thiol proteinases. *J. Biol. Chem.* **256**: 1923-1928.
8. Hanzlik, R.P. & Xing, R., 1998, Azapeptides as inhibitors and active site titrants for cysteine proteinases. *J. Med. Chem.* **41**: 1344-1351.
9. McGuire, M.J.; Lipsky, P.E.; Thiele, D.L., 1993, Generation of active myeloid and lymphoid granule serine proteases requires processing by the granule thiol protease dipeptidyl peptidase I. *J. Biol. Chem.* **268**: 2458-2467.
10. Mabee, C.L.; McGuire, M.J.; Thiele, D.L., 1998, Dipeptidyl peptidase I and granzyme A are coordinately expressed during CD8+ T cell development and differentiation. *J. Immunol.* **160**: 5880-5885.

11. Ishidoh, K.; Muno, D.; Sato, N.; Kominami, E., 1991, Molecular cloning of cDNA for rat cathepsin C. Cathepsin C, a cysteine proteinase with an extremely long propeptide. *J. Biol. Chem.* **266**: 16312-16317.
12. Rao, N.V.; Rao, G.V.; Hoidal, J.R., 1997, Human dipeptidyl-peptidase I. Gene characterization, localization, and expression. *J. Biol. Chem.* **272**: 10260-10265.
13. McGuire, M.J.; Lipsky, P.E.; Thiele, D.L., 1997, Cloning and characterization of the cDNA encoding mouse dipeptidyl peptidase I (cathepsin C). *Biochim. Biophys. Acta* **1351**: 267-273.
14. Pham, C.T.; Armstrong, R.J.; Zimonjic, D.B.; Popescu, N.C.; Payan, D.G.; Ley, T.J., 1997, Molecular cloning, chromosomal localization, and expression of murine dipeptidyl peptidase I. *J. Biol. Chem.* **272**: 10695-10703.

CD26-/DPP IV-Positive Lymphocytes in Murine Acute Experimental Colitis

SABIHA FATIMA, JENS RÜTER, JAN H. NIESS, BURGHARD F. KLAPP, PETRA C. ARCK, and MARTIN HILDEBRANDT
Department of Internal Medicine, Division of Psychosomatics and Psychotherapy, Charité, Humboldt University, Luisenstrasse 13a, Berlin, Germany

1. INTRODUCTION

Stress has long been postulated to influence the progression of inflammatory bowel disease (IBD) by mechanisms such as the interaction of neuroendocrine and immune system, abnormalities of epithelial ions secretion and passage of macromolecules into the lamina propria[1,2]. There is some evidence of a dysregulation of the balance of pro- and anti-inflammatory cytokines and T helper (T_H) lymphocyte subsets in chronic intestinal inflammation[3-8]. Studies in humans and in animal models indicate that stress skews the $T_{H1}\backslash T_{H2}$ balance to T_{H1}-inflammatory response[9].

CD26 is a cell surface ecto-enzyme with dipeptidyl peptidase IV (DPP IV; EC 3.4.14.5) activity expressed in different tissues, including a subset of human resting T cells[10]. DPPIV is a serine protease with unique enzyme activity that can cleave off terminal dipeptides from polypeptides and proteins having either L-proline or L-arginine at the penultimate position[11]. CD26 is involved in T cell migration through endothelial cell monolayers[12] and also serves as membrane binding protein for ecto adenosine diaminase (ADA) [13-15]. It has earlier been demonstrated that high expression of CD26 defines a $T_{H1}\backslash T_{H0}$ type phenotype with enhanced production of T_{H1} like cytokines, namely INF-γ[16].

Furthermore, overexpression of some neuropeptides and their receptors, in particular substance P (SP) and its receptor, the neurokinin-1 (NK-1), may play a significant role in initiating and modulating the inflammtion associated with ulcerative colitis[17]. SP preferentially binds to neurokinin receptor with the highest affinity for NK-1 receptor[18] and the systematic application of SP NK-1 receptor antagonist in rats reduced the severity of inflammatory process in the intestine[19,20].

Based on these clinical and experimental findings we examined the number of $CD26^+$, $CD4^+$ and $CD8^+$ cells Stress and SP induced colitis in mice model. The effect of NK1-RA was also assessed in animals exposed to stress post induction of colitis.

2. MATERIAL AND METHODS

Female Balb\C mice were purchased from BGVV, Berlin Germany. The animals were housed in community cages with free acess to mouse chow ad libitum. The mice were divided into 8 groups. Group 1) used as a control, Group 2) treated with 50% EtOH intrarectally (i.r), group 3) 50% EtOH (i.r) and exposed to sound stress, groups 4) dinitrobenzenesulphonic acid (DNBS) (Sigma, Deisenhofen, Germany) i.r 36 μg\g body weight dissolved in 0.1 ml PBS and 0.1 ml EtOH 5) DNBS and highly specific NK1 receptor antagonist 200 μg in 200μl PBS (Rhone-Poulenc, Antony, France) injected intraperitonially (i.p) every second day, group 6) DNBS (i.r) and exposed to stress, 7) DNBS (i.r) and injected once with recombinanat substance P (SP), 100nmol\ml (i.p) (Sigma, Deisenhofen, Germany) in non-stressed mice as a single dose on day 3 after induction of colitis. All experimental stressed mice were approved by the local animal care committee (LaGetSi, Berlin, Germany).

The mice were exposed to stress for 24 hrs or SP was injected 3 days after induction of colitis. Mice were exposed to sound stress emitted by a rodent repellent device (conrad Electronic, Berlin, Germany) at frequency of 300 HZ in intervals of 15 seconds.

After 9 days of DNBS injection mice were sacrificed and colon were harvested and carefully covered with embedding medium. The cryostat sections (~6μm) were fixed in acetone for 10 mins at -20^0C and stored at -80^0C until use.

Immunohistochemistry. Anti-mouse CD26 at concentration 1:50, anti-mouce CD4 1:100, ant-mouse CD8 1:100, anti-mouse major histocompatibility complex (MHC) II 1:100 were used in this experiment.

Immunohistochemical staining was performed following a standard protocol, using a biotinylated secondary goat-anti-rat antibody. The site of antigen antibody complex was visualized by the use of avidin-biotin-alkaline phosphatase complex (ABC) method (ABC kit, Vector laboratries, germany) using substrate Neo-fuchsin substrate The sections were then counterstained with hymalaun blue, dehydrated, and mounted.

All sections were analysed using a zeiss Axioscope light microscope (Zeiss, Jena, Germany); photo documentation was performed using Zeiss KS400. Immunorecative cells were counted by using a scaled eye piece.

Statistical analysis. All the calculations were performed using SPSS 9.0 computer software. Mean of all counts±SEM were calculated. Differences were juged as significant if the p values were $p \leq 0.05$ or $p \leq 0.01$, as determined by the Mann Whitney U Test. Non-parametric data were analysed by Kruskal-Wallis Test.

3. RESULTS

Effect of stress on $CD26^+$, $CD4^+$, $CD8^+$ and MHC II^+ cells in the intestine. Animals exposed to sound stress showed significant increase in the number of $CD26^+$ (61±12 vs 12±8 cells mm^2 $p \leq 0.05$) cells in the distal colon 9 days post induction of colitis when compared with group treated with DNBS alone (Fig 1,2,3). The number of $CD4^+$ and $CD8^+$ cells when assessed in animals exposed to stress after induction of colitis showed significant change. The effect of stress after DNBS treatment led to increase in the number of $CD4^+$ (64±15 vs 29±4 cells mm^2 $p \leq 0.05$) and $CD8^+$ (37±10 vs 12±2 cells mm^2 $p \leq 0.05$) cells 9 days post induction of colitis compared to non-stressed DNBS treated mice. No differences were detectable in the number of $CD26^+$, $CD4^+$, $CD8^+$ cells between the EtOH and DNBS treated groups after the induction of colitis.

Effect of Substance P on $CD26^+$, $CD4^+$, $CD8^+$ and MHC II^+ cells in the intestine

Figures 1-3 show changes in $CD26^+$ (79±19 vs 12±7 cells mm^2 $p \leq 0.05$), $CD4^+$ (86 ±12 vs 29±4 cells mm^2 $p \leq 0.01$) and $CD8^+$ (32±8 vs 12±2 cells mm^2 $p \leq 0.01$) cells in intestine of the animals injected with SP post induction of colitis compared to non-stressed DNBS treated rats.

Effect of Substance P neurokinin 1 (NK1-RA) receptor antagonist on stress-induced alterations of CD26⁺, CD4⁺, CD8⁺ and MHC II⁺ cells. Administration of NK1-RA in stressed DNBS animals showed reductions in the numbers of $CD26^+$ (22±1 vs 61±12 cells mm^2 $p \leq 0.01$), $CD4^+$ (26±2 vs 64±15 cells mm^2 $p \leq 0.01$) and $CD8^+$ (11±8 vs 37±10 cells mm^2 $p \leq 0.05$) cells in the colon 9 days post induction of colitis compared to the stressed DNBS treated mice.

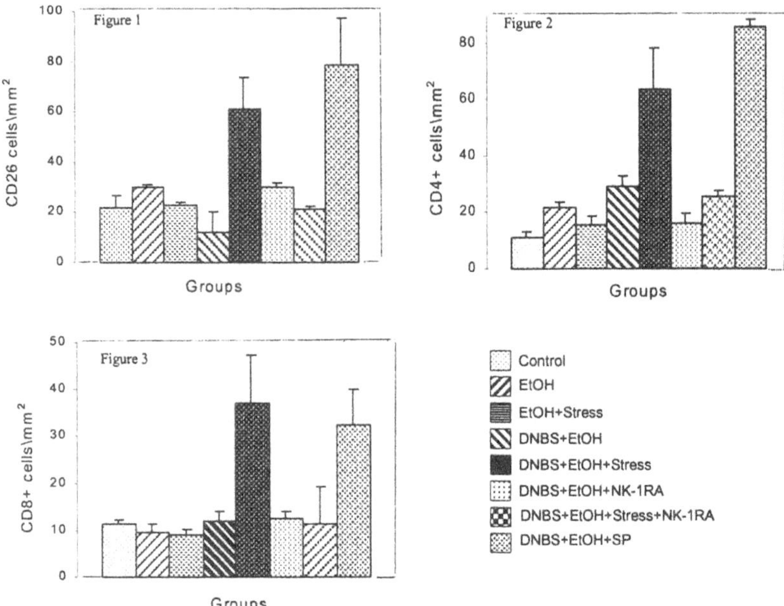

Figures 1-3. The effect of Stress on the number of $CD26^+$ (fig-1), $CD4^+$ (fig-2) and $CD8^+$ (fig-3) cells\mm^2 in the diatal colon 9 days post indiction of colitis (mean±SE). The SP injection mimicked the observed effect of stress on $CD26^+$, $CD4^+$ and $CD8^+$ cells. The injection of NK-1RA abrogated the effect of stress.

4. DISCUSSION

We report that activation of immune parameters are the pathways by which stress and SP effect colonic inflammation. Increased number of CD26 lymphocytes results in increased secretion of cytokines. Substance P, a potent mediator Stress induced colitis[17] is also the substrate for DPPIV[21] ,and further studies will be required to elucidate the interaction between SP and CD26 in inflammatory bowel disease.

REFERENCES

1. Saunders, P.R., Kosecka, U., McKay, D.M.,and Perdue, M.H., 1994, Acute stressor stimulate ion secretion and increase epithelial permeability in rat intestine. *Am J Physiol* 267:G794-G799.
2. Kiliaan, A.J., Saunders, P.R., Bijlsma, P.B., Berin, M.C., Taminiau, J.A., Groot, J.A., and Perdue, M.H., 1998, Stress stimulates transepithelial macromolecular uptake in rat jejunum. *Am J Physiol* 275:G1037-G1044.
3. MacDermott, R.P., 1884, Alterations in mucosal immune system in ulcerative colitis and cohn's disease. *Med Clin North Am* 78:1207-1231.
4. Sartor, R.B., 1994, Cytokines in intestinal inflammation: Pathophysiological and clinical considerations. *Gastroenterology* 106:533-539.
5. Sartor, R.B., 1995, Current concept of the etiology and pathogenesis of ulcerative colitis and crohn's disease. *Gastroenterol Clin North Am* 24:475-507.
6. Sartor, R.B., 1996, Cytokine regulation of experimental intestinal inflammation in genetically engineered and T-lymphocyte reconstituted rodents. *Aliment Pharmacol Ther* 2:36-42.
7. Jewell, D.P., 1995, Immunology of inflammatory bowel disease: an update. *J Gastroenterol* 8:78-82.
8. Powrie, F., 1995, T cells in inflammatory bowel disease: Protective and pathogenic roles. *Immunity* 3:171-174.
9. Arck, P.C., Merali, F.S., Chaouat, G., and Clark, D.A., 1996, Inhibition of immunoprotective CD8+ T cells as a basis for Stress-triggered substance P mediated abortions in mice.Cell. *Immunol* 171:226-230.
10. Fleischer, B., 1994, CD26: a surface protease involved in T cell activation. *Immunol Today* 15:180-184.
11. De Meester, I., Korom, S., Van Damme, J., and Scharpe, S., 1999, CD26, Let it cut or cut it down. *Immunol Today* 20:367-375.
12. Dang, N.H., Torimoto, Y., Schlossman, S.F., and Morimoto, C., 1990, Human CD4 helper T cell activation;functional involvement of two distinct collagen receptors, IF7 and VLA integrin family. *J Exp Med* 172:649-652.
13. Kameoka, J., Tanaka, T., Nojima, Y., Schlossman, S.F., Morimoto, C., 1993, Direct association of adenosine diaminase with a T cell activation agent , CD26. *Science* 261:466-469.
14. Blanco, J., Marie, I., Callebaut, C., Jacotot, E., Krust, B.,and Hovanessian, A.G., 1996, Specific binding of adenosine deaminases but not HIV-1 transactivator protein Tat to human CD26. *Exp Cell Res* 225:102-111.
15. Franco, R., Valenzuela, A., Lluis, C., and Blanco, J., 1998, Enzymatic and extraenzymatic role of ecto-adenosine deaminase in lymphocytes. *Immunol Rev* 161:27-42.

16. Willheim, M., Ebner, C., Baier, K., Kern, W., Schrattbauer, K., Thien, R., Kraft, D., Breiteneder, H., Reinisch, W., and Scheiner, O., 1997, Cell surface characterization of T lymphocytes and allergen-specific T cell clones: correlation of CD26 expression with T(H1) subset. *J Allergy Clin Immunol* 100:348-355.
17. Stucchi, A.F., Shofer, S., Leeman, S., Materne, O., Beer, E., McClung, J., Shebani, K., Moore, F., O'Brien, M., and Becker, J.M., 2000, NK-1 antagonist reduces colonic inflammation and oxidation stress in dextran sulphate-induced colitis in rats. *Gastrointest Liv Physiol* 279:G1298-G1306.
18. Holzer, P., 1998, Implications of tachykinins and calcitonin gene-gene related peptide in inflammatory bowel disease. *Digestion* 59:269-283.
19. Kataeva, G., Agro, and A., Stanisz, A.M., 1994, Substance P mediated intestinal inflammation : inhibitory effect of CP 96, 345 and SMS 201-995. *Neuroimmunomodulation* 1:350-356.
20. Pothoulakis, C., Castagliuolo, I., LaMont, J.T., Jaffer, A., O'keane, J.C., Snider, R.M., and Leeman, S.E., 1994, CP-96345, a substance P antagonist, inhibits rat intestinal responses to *clostridium difficile* toxin A but not cholera toxin. *Proc Natl Acad Sci* 91:947-951
21. Nausch, I., Mentlein, R., and Heymann, E., 1990, The degradation of bioactive peptides and proteins by dipeptidyl peptidayse IV from human placenta. *Biol Chem Hoppe Seyler* 371:1113-1118.

Neuroprotective Effects of Inhibitors of Dipeptidyl Peptidase-IV In Vitro and In Vivo

YONG-QIAN WU, DAVID C. LIMBURG, DOUGLAS E. WILKINSON, PAUL JACKSON, JOSEPH P. STEINER, GREGORY S. HAMILTON, and SERGEI A. BELYAKOV
Guilford Pharmaceuticals, Inc., Tributary St., Baltimore, MD, USA

1. INTRODUCTION

Recent findings of potent DPP-IV inhibitors, which have emerged as a novel concept in the treatment of NIDDM (type II diabetes), spurred the quest for other therapeutic perspectives of DPP IV inhibition. The potential of DPP-IV inhibitors as immunosuppressants for the treatment of autoimmune diseases, deterrence of transplant rejection, prevention of HIV infection, cancer and rheumatoid arthritis treatments, etc., was recently reported[1]. The role of DPP-IV in the CNS and PNS has not been extensively explored. It was claimed that DPP-IV is capable of metabolizing neuropeptides, particularly, peptide YY and substance P[2], and that inhibition of DPP-IV induces a three- to four-fold increased secretion of latent TGF-β[3]. Since enhancement of TGF-β signaling is associated with neuroprotective effects, we investigated the possibility of potent DPP-IV inhibitors to serve as novel neuroprotective agents.

2. CHEMISTRY

Typical structure of potent DPP-IV inhibitor has to include an amino acid moiety in the S2 area, with peptide bond evolving into pyrrolidine (or thioazolidine) moiety in the S1 area. Most crucial from both selectivity and

Dipeptidyl Aminopeptidases in Health and Disease, Edited by Hildebrandt et al.
Kluwer Academic/Plenum Publishers, New York, 2003

activity standpoint is a type of $S1^I$ substituents. Numerous studies reported elsewhere showed that phosphonates, boronic acids, and cyanopyrrolidines were among most active DPP-IV inhibitors.

Table 1. SAR in the series of synthesized DPP-IV inhibitors

Cmpd. No	X	R	R^1	R^2	IC_{50}, uM
1	CH_2	CN	H	Adamantyl	0.032
2	S	CN	H	Adamantyl	0.032
3	CH_2	COOH	H	Adamantyl	>100
4	S	H	H	Adamantyl	4.820
5	CH_2	CN	CH_3	Adamantyl	>100
6	CH_2	CN	C_6H_{12}	H	0.005
7	CH_2	CH_2OH	C_6H_{12}	H	10
8	CH_2	CN	H	H	1.030
9	CH_2	CN	CH_3	CH_3	1,000
10	CH_2	CH_2CN	C_6H_{12}	H	1,010
11	CH_2	CN	H	$(CH_2)_2NH$-Py(4-CN)	0.039

Activity of DPP IV inhibitors is granted when terminal amino group is displaced with bulky substituents (such as adamantane), and substituent in R position is a cyano group (*cf. Table*, compounds **1, 2**), or at least a proton (**4**). Surprisingly, when R=COOH (**3**) or $CONH_2$ (not shown), no inhibition was observed. Same negative effect was found when compound **1** did contain an additional methyl substituent in α-position (compound **5**). We also prepared and tested glycine derivatives, variously substituted at R and R^1 positions. Bulky cyclohexyl substituent at the α-position enhances activity when pared with cyano group in $S1^I$ area (compound **6**), while replacement of the latter with hydroxymethyl group significantly diminishes activity (compound **7**). Unsubstituted pyrrolidine derivative (**8**) has acceptable activity, while substitution at both α-carbon and nitrogen makes the compound completely inactive (**9**). Interestingly, if cyano group is separated from pyrrolidine nucleus by a methylene link, no activity was observed as well (compound **10**). For comparison purposes, we also prepared and tested compound **11**, recently disclosed potent DPP-IV inhibitor NVP-DPP728[4].

3. BIOLOGY

The inhibition of DPP IV protease activity was characterized in an *in vitro* screening assay, using rat plasma as the source of DPP-IV and the peptide, and H-Gly-Pro-AMC as peptide substrate[4]. Inhibitory IC_{50} values for these compounds are presented in *Table*.

From the DPP-IV inhibitors we synthesized and evaluated as selective protease inhibitors (*Table*), we chose compounds **2**, **6**, and **11** for further additional characterization of neuroprotective actions. We evaluated these inhibitors as neuroprotective agents against excitotoxic lesions in organotypic spinal cord cultures[5]. Cultured spinal cord slices of nine-day old SD rats were treated with glutamate-reuptake inhibitor, threohydroxyaspartate (THA), and either increasing concentrations of drug or vehicle. Treatment of the cultures with THA resulted in 35-40% cell death, while treatment of the cultures with THA and the DPP-IV inhibitors **2** and **6** (concentrations of 0.5 uM and 1 uM) spared greater than 50% of the vulnerable neurons. This result demonstrated that DPP-IV inhibitors do protect motor neurons against excitotoxic cell death.

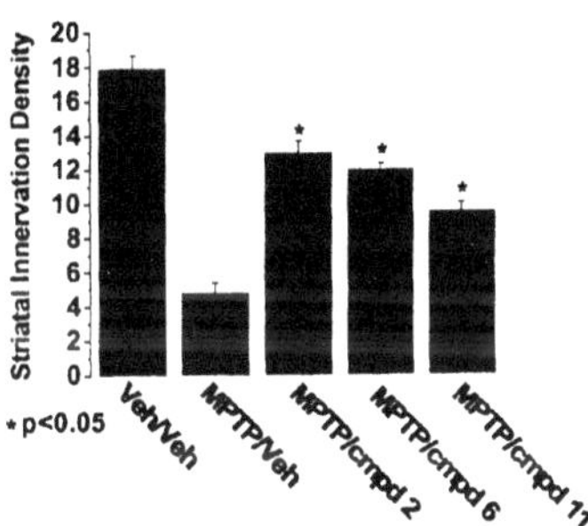

Figure 1. Neuroprotective action of DPP-IV inhibitors

We have further studied the neuroprotective potential of selected DPP-IV inhibitors in the concurrent drug dosing MPTP paradigm[6]. Mice were dosed with MPTP (30 mg/kg, i.p.) and compounds **2**, **6**, and **11** daily (10 mg/kg in Intralipid, s.c.) on days 1 to 5. On day five animals were perfused, and tissue was sectioned/stained with anti-tyrosine hydroxylase Ig. When mice were treated with MPTP and vehicle, striatal tyrosine hydroinnervation density dropped almost 70% compared to control, while concurrent administration of DPP-IV inhibitors resulted in substantial (50-70%) protection for all 3 tested compounds (**2**, **6**, and **11**), compared to MPTP/vehicle (*Fig.)*. Thus, DPP IV inhibitors are systemically active and protect striatal innervation of dopaminergic neurons in MPTP protective model.

We also established a significant neuroregenerative effect of DPP-IV inhibitors in post-MPTP recovery model of Parkinson's disease[6]. Four-week old male CD1 white mice were dosed with MPTP (30 mg/kg, i.p.) once daily on days 1 to 5, followed by administration of the DPP IV inhibitors **2** and **6**, subsequently, once daily on days 8-12 (10 mg/kg in Intralipid, s.c.). On day 18, animals were perfused and treated as mentioned above. The level of neurodegeneration of the dopaminergic terminal innervation density was about 60%, compared to TH innervation density in the vehicle/vehicle treated mice. Each of the DPP IV inhibitors promoted 40-45% recovery of the striatal TH innervation, when compared to MPTP/vehicle.

4. CONCLUSION

For the first time, we demonstrated both neuroprotective and neuroregeneratrive effects of common DPP-IV inhibitors *in vitro* and *in vivo*. DPP IV inhibitors protect motor neurons from excitotoxic cell death. They are systemically active and protect striatal innervation of dopaminergic neurons, when administered concurrently with MPTP. Furthermore, DPP-IV inhibitors promote recovery of striatal innervation density when given in a therapeutic manner following MPTP treatment. These data suggest that DPP IV inhibitors may provide protective effects on neurons and promote their use as therapies for treatment of neurodegenerative disorders.

REFERENCES

1. Hildebrandt, M., Reutter, W., Arck, P., Rose, M., and Klapp, B.F. A Guardian Angel: The Involvement of Dipeptidyl Peptidase IV In Psychoneuroendocrine Function, Nutrition, and Immune Defence. 2000, *Clinical Sci,* **99**: 93-104.
2. Medeiros, M.S., and Turner, A.J. Processing And Metabolism of Peptide YY. 1993, *Biochem. Soc. Trans.* **21**: 248S; Kato, T., Nagatsu, T., Fukazawa, K., Harada, M., Nagatsu, I., and Sakakibura, S. Successive Cleavage of N-Terminal Arg-Pro and Lys-Pro From Substance P But No Release of Arg-Pro From Bradykinin, By X-Pro Dipeptidyl-Aminopeptidase. 1978, *Biochim. Biophys. Acta* **525**: 417-422.
3. Reinhold, D., Bank, U., Buhling, F., Lendeckel, U., Faust, J., Neubert, K., and Ansorge, S. Inhibitors of Dipetidyl Peptidase Induce Secretion of Transforming Growth Factor-β_1 in PWM-stimulated PBMC and T-cells. 1997, *Immunology* **91**: 354-360
4. Hughes, T.E., Mone, M.D., Russell, M.E., Weldon, S.C., and Villhauer, E.B. (1-[[[2-[(5-Cyanopyrrolidin-2-yl)amino]ethyl]amino]acetyl]-2-cyano-(*S*)-pyrrolidine, a Slow-Binding Inhibitor of Dipeptidyl Peptidase IV. 1999, *Biochemistry* **38**: 11597-11603.
5. Rothstein, J.D., Jin, L., Dykes-Hoberg, M. and Kuncl, R. W. Chronic inhibition of glutamate uptake produces a model of slow neurotoxicity. 1993, *Proc. Natl. Acad. Sci. USA* **90**: 6591-6595.

6. Steiner J.P., Hamilton G.S., Ross D.T., Valentine H.L., Guo H., Connolly M.A., Liang S., Ramsey C., Li J.H., Huang W., Howorth P., Soni R., Fuller M., Sauer H., Nowotnik A.C., and Suzdak P.D. Neurotrophic immunophilin ligands stimulate structural and functional recovery in neurodegenerative animal models. 1997, *Proc Natl. Acad. Sci. USA* **94**: 2019-2024.

Index

GPSR Compliance
The European Union's (EU) General Product Safety Regulation (GPSR) is a set of rules that requires consumer products to be safe and our obligations to ensure this.

If you have any concerns about our products, you can contact us on

ProductSafety@springernature.com

In case Publisher is established outside the EU, the EU authorized representative is:

Springer Nature Customer Service Center GmbH
Europaplatz 3
69115 Heidelberg, Germany

www.ingramcontent.com/pod-product-compliance
Ingram Content Group UK Ltd.
Pitfield, Milton Keynes, MK11 3LW, UK
UKHW020321200726
13857UKWH00001B/241
* 9 7 8 1 4 7 5 7 8 7 2 9 0 *